SPRINGER SERIES
IN PERCEPTION ENGINEERING

Series Editor: Ramesh C. Jain

Springer
New York
Berlin
Heidelberg
Barcelona
Budapest
Hong Kong
London
Milan
Paris
Santa Clara
Singapore
Tokyo

Springer Series
in Perception Engineering

Besl: *Surfaces in Range Image Understanding*

Fan: *Describing and Recognizing 3-D Objects Using Surface Properties*

Gauch: *Multiresolution Image Shape Description*

Jain and Jain (eds.): *Analysis and Interpretation of Range Images*

Krotkov: *Active Computer Vision by Cooperative Focus and Stereo*

Landy, Maloney, and Pavel (eds.): *Exploratory Vision: The Active Eye*

Masaki (ed.): *Vision-based Vehicle Guidance*

Rao: *A Taxonomy for Texture Description and Identification*

Sanz (ed.): *Advances in Machine Vision*

Strat: *Natural Object Recognition*

Skifstad: *High-Speed Range Estimation Based on Intensity Gradient Analysis*

Vogt: *Automatic Generation of Morphological Set Recognition Algorithms*

Michael S. Landy
Laurence T. Maloney
Misha Pavel
Editors

Exploratory Vision

The Active Eye

With 133 Illustrations

Springer

Editors
Michael S. Landy
Laurence T. Maloney
Department of Psychology and
Center for Neural Science
New York University
6 Washington Place, Room 961
New York, NY 10003
USA

Misha Pavel
Department of Electrical Engineering
and Applied Physics and
Department of Computer Science
and Engineering
Oregon Graduate Institute
Portland, OR 97291-1000
USA

Series Editor
Ramesh C. Jain
Electrical Engineering and
Computer Science Department
University of Michigan
Ann Arbor, MI 48109
USA

Library of Congress-in-Publication Data
Landy, Michael S.
Exploratory vision : the active eye / Michael S. Landy, Laurence T. Maloney, Misha Pavel.
p. cm. – (Springer series in perception engineering)
Includes bibliographical references and index.

ISBN-13: 978-1-4612-8460-4 e-ISBN-13: 978-1-4612-3984-0
DOI: 10.1007/978-1-4612-3984-0

1. Computer vision. 2. Vision – Computer simulation. 3. Eye – Movements – Computer simulation. I. Maloney, Laurence T. II. Pavel, Misha. III. Title. IV. Series.
TA1634.L36 1995
006.3′7 – dc20 95-30508

Printed on acid-free paper.

Softcover reprint of the hardcover 1st edition 1996

Production managed by Frank Ganz; manufacturing supervised by Joe Quatela.
Photocomposed pages prepared from the author's LaTeX files.

9 8 7 6 5 4 3 2 1

Preface

Advances in sensing, signal processing, and computer technology during the past half century have stimulated numerous attempts to design general-purpose machines that see. These attempts have met with at best modest success and more typically outright failure. The difficulties encountered in building working computer vision systems based on state-of-the-art techniques came as a surprise. Perhaps the most frustrating aspect of the problem is that machine vision systems cannot deal with numerous visual tasks that humans perform rapidly and effortlessly.

In reaction to this perceived discrepancy in performance, various researchers (notably Marr, 1982) suggested that the design of machine-vision systems should be based on principles drawn from the study of biological systems. This "neuromorphic" or "anthropomorphic" approach has proven fruitful: the use of pyramid (multiresolution) image representation methods in image compression is one example of a successful application based on principles primarily derived from the study of biological vision systems. It is still the case, however, that the performance of computer vision systems falls far short of that of the natural systems they are intended to mimic, suggesting that it is time to look even more closely at the remaining differences between artificial and biological vision systems.

One striking difference between the two is that most early work in computer vision concentrated on passive processing of single images, or short sequences of images taken by a stationary camera. In contrast, people and many species of animals can move about in their environment while independently moving their eyes and heads. This self-motion with added eye or head motion places an enormous demand on the visual system: The reconstruction process must cope with images that are contaminated, and possibly impaired, by image motion and uncertainties of position and timing. In partial recompense, the motion of the camera/eye can potentially make it easier to extract certain kinds of information about the scene: The reconstruction process can be aided by analysis of input images taken from multiple, spatially proximal viewpoints (*active vision*). In addition, further simplifications in the reconstruction process result if the vision system can control the location of the camera/eye and use it to actively explore the scene (*exploratory or purposive vision*).

The thesis underlying this volume is that neuromorphic and anthropomorphic engineering is most effectively employed in solving the kinds of problems that biological vision systems solve: using moving, active cameras/eyes to explore the environment. The book is divided into four parts, each addressing a different aspect of exploratory or active vision in biological and machine vision systems. We describe the contents of each of these parts next.

Part I: Active Human Vision. In normal human vision we see little effect of head and eye movements. This approximate constancy suggests that we compensate well, but the question remains, how well? The first part of this book summarizes what we know about the interplay of eye and head movements with normal human vision and cognition. The chapter by Steinman (*Chapter 1: Moveo ergo video: Natural retinal image motion and its effect on vision*) provides a historical introduction to the study of the effects of eye and head motion on visual acuity and on other measures of visual performance. It describes recent work based on the first accurate measures of eye and head movement when the head is allowed to move freely. The chapter by Kowler (*Chapter 2: Cogito ergo moveo: Cognitive control of eye movement*) examines human capabilities and strategies in control of the direction of gaze. Human observers appear to explore their environment using a well-balanced combination of cognitive and visual control, operating with low-level oculomotor constraints.

Part II: Solving Visual Problems. Given that many biological visual systems have developed sophisticated ways of moving their eyes, it is likely that there are benefits associated with eye movements. Some of these benefits are obvious, such as stabilization of moving images or distance estimation from motion parallax. The second part of the book is concerned with methods and algorithms that make use of the motion of an eye or camera, or other active exploration, to improve vision qualitatively and quantitatively in subtle ways.

A recurring problem for most biological and artificial sensors is the maintenance of their spatial correspondence and calibration. The calibration problem is particularly important under the sampling limitation imposed by the visual sensor array (i.e., the retinal mosaic). Maloney (*Chapter 5: Exploratory vision: Some implications for retinal sampling and reconstruction*) develops the mathematics necessary for understanding visual coding and reconstruction given the sampling limitations imposed by the sensor, and describes how these sampling limits may be ameliorated by eye movements and by the assumption of invariance of the image across small eye movements. That eye movements may have a role in development and repair of the visual system following disease or accident was suspected by Helmholtz (1866). Ahumada and Turano (*Chapter 6: Calibration of a visual system with receptor drop-out*) suggest that the calibration algorithms of Maloney and Ahumada might be used to recalibrate the visual system as receptors are lost (e.g., in retinitis pigmentosa).

Pavel and Cunningham (*Chapter 4: Eye movements and the complexity of visual processing*) discuss the general topic of the computational complexity of visual tasks. One question they address is whether computational complexity can predict the difficulty encountered by humans performing these tasks. They show how eye movements could mediate a tradeoff between computational complexity of parallel architectures and speed by use of sequential algorithms. The human visual system achieves the serialization of visual tasks by scanning of the visual scene using eye movements or covert shifts of visually guided selective attention. Thus, the structure and organization of the visual system may be organized around the demands of exploration.

Another important problem associated with a moving sensor is the possible increase in overall uncertainty due to uncertainty of sensor location and motion, and due to image noise. The chapter by Schunck (*Chapter 3: Robust computational vision*) discusses the general problem of making vision algorithms robust to outlier observations caused by sensor noise or mistaken assumptions. The techniques that he develops are applicable to a variety of areas, and examples of their use are given for surface reconstruction, image-flow calculation, and dynamic stereo.

A problem related to sensor calibration is that of determination of sensor motion relative to the world. The determination of observer motion (called heading) from optical flow in the presence of sensor movements is an especially difficult problem. Thomas, Simoncelli and Bajcsy (*Chapter 7: Peripheral visual field, fixation and direction of heading*) show how a machine vision system can determine qualitative aspects of heading relative to an actively fixated object simply, using the retinal flow in the periphery. The chapter by Langer and Zucker (*Chapter 8: Local qualitative shape from active shading*) includes a discussion of qualitative aspects of surface shape that can be computed easily by use of multiple light sources controlled by the observer — a technique that Langer and Zucker call active shading.

Part III: Robots that Explore. Part III complements the first by outlining the problems whose solution is prerequisite to the design of robots that see as they move and explore. Many of the requirements of such active vision systems echo human capabilities. Fermüller and Aloimonos (*Chapter 9: The synthesis of vision and action*) recommend that the field of computer vision move beyond the initial "active vision" paradigm originally introduced by Aloimonos (1988), Bajcsy (1988) and Ballard (1991). They assert that the goal of modeling and estimating all the metric aspects of the world is too difficult; instead, they argue for qualitative visual representations of object shape, layout, and motion that are tied to the purposes and activities of the observer — an approach that they call "purposive vision". Katkere and Jain (*Chapter 10: A framework for information assimilation*) argues along similar lines, and further suggest that visual information (e.g., in a mobile robot) be gathered at a number of levels by multiple sensors. These multiple sources of information then can be assimilated into a common set of representations to model the environment dynamically. Ikeuchi and Hebert (*Chapter 11: Task-oriented vision*) continue this discussion and add weight to the argument for purposive vision. They recommend a task analysis for the design of robotic vision systems, and outline how they conducted such analyses for two example vision applications.

Part IV: Human and Machine: Telepresence and Virtual Reality. The design of robots that explore is an exciting challenge that is likely to lead to and benefit from a deeper understanding of how humans process visual information during eye and head movements. Another challenge to our understanding of human active vision is the design of telepresence and virtual reality systems. Part IV describes two applications that attempt to match the active, exploratory behavior of the human eye to virtual reality environments. Madden and Farid (*Chapter 12: Active *vision and virtual reality*) show how a telepresence system may benefit from the use of multiple active sensors (with the ability to pan, zoom and focus). The focus mechanism allows for a reasonably quick computation of range to generate a three-dimensional model of the environment. The pan and zoom capabilities allow the cameras to get information on aspects of the environment as the information is needed. By combining the image and depth information from multiple camera viewpoints, a telepresence system can interpolate views of the environment from viewpoints other than those provided by the cameras. Finally, Darrell, Maes, Blumberg and Pentland (*Chapter 13: A novel environment for situated vision and behavior*) describe a virtual reality system that uses active, purposive vision techniques to place observers in a virtual world where those observers can interact with simulated agents that respond to their gestures and other behaviors.

In 1988, the Cognitive Science and Machine Intelligence Laboratory at the University of Michigan in Ann Arbor hosted a week long workshop titled *Exploratory Vision: the Active Eye*. The participants in the original workshop were drawn from several different disciplines: eye movement research, psychophysics, physiology, computational modeling of vision and robotics. They shared an interest in studying the consequences of eye or camera motion for a visual system.

This book is not a record of that workshop. Many of the authors of the chapters in this book did not participate in the original workshop, and a topic area not considered in the original conference (*Telepresence and Virtual Reality*) has been added. However, this volume reinforces the belief that active and exploratory vision in humans and machines have much in common. Our goal has been to provide a much-needed source that summarizes the consequences of eye and head movement for human vision, and that presents what is known about algorithms that can compensate for, or take advantage of, eye or camera movement.

We thank Gary Olson, Terry Weymouth, Brian Schunck and Ramesh Jain for help in organizing the original workshop (*Exploratory Vision: The Active Eye*, Ann Arbor, Michigan, June 1988). We are grateful to the Cognitive Science and Machine Intelligence Laboratory of the University of Michigan, Ann Arbor, for providing funds. Editing and preparation of this work were supported in part by grant EY08266 from the National Eye Institute to Michael S. Landy, by grant F49620-92-J-0187 from the Air Force Office of Scientific Research to Laurence T. Maloney, and by the National Aeronautics and Space Agency Grant NAG 2-931 to Misha Pavel.

Michael S. Landy
New York

Laurence T. Maloney
Paris

Misha Pavel
Portland, Oregon

References

Aloimonos, Y., Weiss, I. & Bandopadhay, A. (1988). Active vision. *International Journal of Computer Vision*, *2*, 333–356.

Bajcsy, R. (1988). Active perception. *Proceedings of the IEEE*, *76*, 996–1005.

Ballard, D. H. (1991). Animate vision. *Artificial Intelligence*, *48*, 57–86.

von Helmholtz, H. L. F. (1866). *Handbuch der Physiologischen Optik*, Volume 2. Hamburg-Leipzig: Voss.

Marr, D. (1982). *Vision*. San Francisco: W. H. Freeman.

Contents

List of Contributors

Albert J. Ahumada Jr.: NASA-Ames Research Center, MS 262-2, Moffett Field, CA 94035–1000

Yiannis Aloimonos: Computer Vision Laboratory, Center for Automation Research, Department of Computer Science, and Institute for Advanced Computer Studies, University of Maryland, College Park, MD 20742-3275 and Institute of Computer Science, FORTH, P.O.Box 1385, Heraklio, Crete, GR-711-10, Greece

Ruzena Bajcsy: General Robotics and Active Sensory Perception (GRASP) Laboratory, Department of Computer and Information Science, University of Pennsylvania, Philadelphia, PA 19104

Bruce Blumberg: Massachusetts Institute of Technology, Media Lab, 20 Ames Street, Cambridge, MA 02139

Helen A. Cunningham: Apple Computer, Inc., 1 Infinity Loop, Cupertino, CA 95014

Trevor Darrell: Massachusetts Institute of Technology, Media Lab, 20 Ames Street, Cambridge, MA 02139

Hany Farid: General Robotics and Active Sensory Perception (GRASP) Laboratory, Department of Computer and Information Science, University of Pennsylvania, Philadelphia, PA 19104

Cornelia Fermüller: Computer Vision Laboratory, Center for Automation Research, University of Maryland, College Park, MD 20742

Martial Hebert: The Robotics Institute, School of Computer Science, Carnegie Mellon University, Pittsburgh, PA 15213

Katsushi Ikeuchi: Computer Science Department, School of Computer Science, Carnegie Mellon University, Pittsburgh, PA 15213

Ramesh Jain: Visual Computing Laboratory, Electrical and Computer Engineering, University of California at San Diego, La Jolla, CA 92093-0407

Arun Katkere: Visual Computing Laboratory, Electrical and Computer Engineering, University of California at San Diego, La Jolla, CA 92093-0407

Eileen Kowler: Department of Psychology, Rutgers University, New Brunswick, NJ 08903

Michael S. Langer: Center for Intelligent Machines, McConnell Engineering, McGill University, 3480 University St., Montréal, H3A2A7, Canada

Brian C. Madden: General Robotics and Active Sensory Perception (GRASP) Laboratory, Department of Computer and Information Science, University of Pennsylvania, Philadelphia, PA 19104

Pattie Maes: Massachusetts Institute of Technology, Media Lab, 20 Ames Street, Cambridge, MA 02139

Laurence T. Maloney: Department of Psychology and Center for Neural Science, New York University, New York, NY 10003

Misha Pavel: Department of Electrical Engineering and Applied Physics and Department of Computer Science and Engineering, Oregon Graduate Institute, Portland OR 97291-1000

Alex P. Pentland: Massachusetts Institute of Technology, Media Lab, 20 Ames Street, Cambridge, MA 02139

Brian G. Schunck: 1026 West Liberty, Ann Arbor, MI 48103

Eero Simoncelli: General Robotics and Active Sensory Perception (GRASP) Laboratory, Department of Computer and Information Science, University of Pennsylvania, Philadelphia, PA 19104

Robert M. Steinman: Department of Psychology, University of Maryland, College Park, MD 20742-4411

Inigo Thomas: General Robotics and Active Sensory Perception (GRASP) Laboratory, Department of Computer and Information Science, University of Pennsylvania, Philadelphia, PA 19104

Kathleen Turano: Wilmer Institute, Johns Hopkins University, Baltimore, MD 21205

Steven W. Zucker: Center for Intelligent Machines, McConnell Engineering, McGill University, 3480 University St., Montréal, H3A2A7, Canada

Part I

Active Human Vision

Under normal viewing conditions we rarely confuse our own eye and head movements with movements of objects in the scene before us. The two chapters in this section summarize what is known about the effects of eye and head movements on human vision, and how we plan and guide eye movements within a scene.

1

Moveo Ergo Video: Natural Retinal Image Motion and its Effect on Vision

Robert M. Steinman[1]

ABSTRACT
This paper describes highlights in the nearly century and a half long history of continuous research on the role of retinal image motion in limiting or influencing visual acuity. This review is followed by a summary of recent work in which it has been shown that oculomotor compensatory actions leave appreciable retinal image motion that, unexpectedly, has only minor detrimental effects on the capacity to resolve fine details in the visual scene. This paper is based on a talk delivered at an interdisciplinary conference under somewhat unusual conditions — conditions that encouraged adoption of a novel organizing theme to allow description of such a lengthy and voluminous body of work. The coherence of the material in this published version of the talk required preserving and explaining the organizing theme adopted. Participants in the collaborative research described in this paper may have quite different memories of why and how they became involved with the author in the various projects described. The reader is warned that the "motivations" described are, in fact, not fictitious, but they may be entirely true only when seen from the author's viewpoint.

1.1 Prologue

This chapter is based on a talk which was the first time I had been asked to be an after dinner speaker. I only discovered this the night before I spoke. After dinner speaking is usually assigned to someone good at telling jokes. Well, I know only two, neither really good. Surely, I was not asked to speak after dinner for this reason. The organizers of the workshop knew me and my two jokes too well to make this mistake. I also learned the night before my talk that dinner would be accompanied by wine and preceded by a "hosted" reception. Could it be that the organizers gave me the after-dinner slot because they thought that my topic would keep sated listeners ("bombed" in the vernacular) alert despite the hour and their condition? Probably not. Consider, . . .

[1]Department of Psychology, University of Maryland

I had been asked to talk about the way in which eye movements influence how we see the world. The modern history of this problem goes back 136 years. I was also asked to be sure to include a description of research I, and a number of collaborators (most notably, Han Collewijn, Jack Levinson and Alex Skavenski) had done during the last decade on eye movements and vision when a subject supports or moves his head in a relatively natural manner. Offhand, it isn't obvious that this topic, in itself, guarantees rapt attention. It seems more suitable for a 14 week graduate seminar, or, if not for a seminar, at least for a morning session in front of an alert or, at least, sober audience. In the present circumstances, what was I to do?

For a start, I made a great personal sacrifice, eschewing all alcohol and substituting 6 cups of coffee. In short, I arranged to keep myself awake well past my preferred bedtime. This solved only half of the problem. I was still faced with the inescapable fact that the audience would fall asleep if I didn't do something startling. Sex and violence, the traditional crowd stimulators, were rather tangential to my topic. How could I discuss physiological nystagmus and liven up an audience suffering from physiological nystagmos? I decided to try something unusual in discourse on serious scientific matters. I decided to tell the truth. Here, I describe how our research on natural retinal image motion actually came to be done, what actually motivated our work. We did not self-actualize. We had no theories, no profound thoughts, no important historical precedents, no serendipitous observations, not even sudden insights. We got phone calls — periodic, unlikely, phone calls. I will describe the source and nature of these electronic revelations as we proceed.

1.2 Introduction

Let me begin the substantive portion of this chapter by pointing out that interest in the role of eye movements in vision goes back a long way. Part of the reason for this persistent interest — the reason that most of us working in this area think about today — is rooted in the heterogeneity of the human retina. By way of quick reminder, each of your eyes has about a 100° monocular visual field on the horizontal meridian, and about a 90° visual field on the vertical meridian. Your foveal floor, containing most of the 5 to 6 million cone receptors found in a single human eye, is a densely packed region that only has a diameter of about 90 minutes of visual angle — only a degree and a half — this works out to a "region of best detail and color vision" that occupies something like 1/40th of 1% of the area of the retinal surface. Why did Charles Darwin make our visual detector in this fashion and how do we manage to survive within its limitations?

This question will not be answered in this paper. It will only be mentioned briefly once again when we consider the role of very small eye movements in enhancing contrast. Interest in my topic actually began with a different, and much more fundamental problem, namely, how do we, as perceivers, come to know that the visual world is extended in the 3-dimensional space around us. Modern discussion of this problem starts in 1852 with Lotze, a philosopher-psychologist, who approached it from the position of the Associationist Philosophers, who continued a tradition begun in the 17th C. by John Locke. This tradition in perception emphasizes the important role of learning in constructing a perceptually organized world. Lotze's idea was that you are born with no knowledge of visual extent. You have the capacity to move your eyes and you use your eye movements to construct what he called "local signs" — these are markers that can tell you where things are located in the visual world relative to yourself.

Now, what was the basis for Lotze's model? It was not linear systems or AI symbols; not even neural networks. There were only 2 minor variations of a single model in 1852. You adopted either the St. James version or the Roman version. The central issue, often called the nature-nurture problem, is how much knowledge about His world are you given by the Creator and how much must you learn? Not much is assumed to be given innately in the Locke-Lotze tradition; only sensations of qualities (such mental states as red or sour) and sensations of intensity (such mental states as degree of brightness or sourness). Everything else must be learned. Lotze started his treatment of visual extent by postulating that eye movements were used to learn retinal "local signs" ("place tags" in Koenderink's contemporary usage). Their positions relative to one another were not known a priori. Eye movements established this relationship through a learning process. By the turn of the present century Hering (1899), who preferred a nativistic approach in much of his theorizing about perceptual processing, postulated that retinal "local signs" contained a priori knowledge of the absolute and relative positions of objects in 3-D space that were represented in the visual array (a 2-D representation of the visual field on the retinal surface, i.e., the image plane). Hering used these built-in local signs in seminal ways, as will be shown later, but at the turn of the present century eye movements were still believed by some to play an important role in the perception of relative size. For example, Wundt (1910), the founding father of Structural Psychology, explained the Müller-Lyer illusion by reference to differences in the size of eye movements used to examine each of the figures. His claim was discredited by researchers who measured eye movements while the illusion was viewed. A role for eye movements in perceiving size will not pop up again for more than 60 years when Festinger proposed his "efferent readiness theory", which I will describe after a brief treatment of Hebb's motor theory of form perception.

Hebb (1949) revived the empirical tradition in perception after the Gestalt revolution "died of success" (Boring, 1942). Hebb's model fell out of favor, following the work of Hubel and Wiesel in the 1960s. Hebb required that newly-sighted kittens (as well as human infants) would only perceive what he called "primitive unity", i.e., a shapeless smudge bearing only a crude figure-ground relationship with its surround. It had no contour or shape. Everything beyond this "primitive unity" had to be learned by fixating and making eye movements. First lines and corners or angles were learned. Hebb called these learned features "cell assemblies". Once these features were learned the young animal started to learn eye movement patterns, called "phase sequences", by repeating over and over again the pattern of saccades required to scan from feature to feature in a particular shape. "Engrams" (hypothetical brain correlates of a memory), representing these learned oculomotor patterns, or "phase sequences" provided the neural substrates for perceiving and discriminating shapes. Hubel and Wiesel (1962, 1963) found that the primary cortical monocular receptive field organization of newly-sighted kittens did not differ in fundamental ways from the monocular organization of their parents. Kittens had everything except functional binocular input. This is a fatal problem for Hebb's theory but, interestingly, many contemporaries (including Hubel and Wiesel, 1963) confused Hebb's emphasis on line and corner features with the basis upon which he said that these features were formed. Specifically, for Hebb it was the organism's oculomotor behavior, not pre-existing brain circuitry, that provided the basis on which the perception of shape was learned. Hubel and Wiesel's work with newly-sighted kittens discredited Hebb's claim.

Festinger (1971) revived motor theory of shape and size perception. He explicitly avoided considering the problem of the ontogeny of these capacities and proposed that the oculomotor program that was made ready by the visual representation of a given shape provided the basis of the percept of its shape. How the eye would move to explore a triangle, for example, was the proposed mechanism. This program would be different from the program that would be loaded to explore a round or rectangular form. The question of whether the Creator or the individual infant wrote these programs was ignored. Festinger's approach seemed timely because terminology like "loading programs" had a nice modern ring in the late 1960s and early 70s. Such terms were at least as compelling as "model" or "representation" or "module" are today. None of these terms is quite up to inspiring the glazed looks in both speaker and listener that could be aroused by slowly incanting "massively distributed parallel processing" a few years ago. But talking about "loading programs" did get Festinger's oculomotor theory of shape and size more attention then it deserved on the basis of the long, and clearly fruitless, history of similar, earlier attempts. Festinger's "theory" had the advantage of being almost incapable of falsification because it does not require that any eye movements be made; it was sufficient merely to load the appropriate programs to perceive or discriminate shapes. The one test possible, namely, an examination of spontaneous eye movements that were made in the presence of different forms, did not support the theory (Murphy, Haddad & Steinman, 1974) and it dropped out of sight during the 80s.

At present, motor theories of shape and size are not prominent in current research on human perception. At least for shape and size perception, the phoenix hatched by Locke has no active support in the contemporary oculomotor community. Robots, however, are beginning to moves their eyes and may, therefore, be learning to discriminate directions, sizes and shapes. At this point I will put aside further discussion of the role of eye movements in higher perceptual processes such as direction, size and shape perception and turn to the role of eye movements at a more fundamental level of visual science, namely, their role in the discrimination of contrast or, using an older, and somewhat broader term, their role in visual acuity.

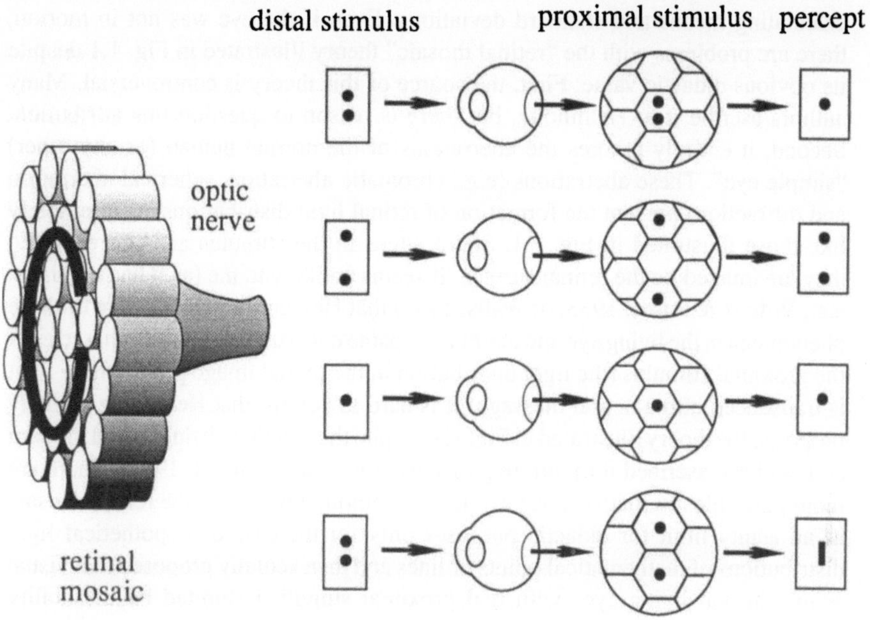

FIGURE 1.1. The retinal mosaic theory of visual acuity. Distinguishing a C from an O requires that one receptor be unstimulated. Distinguishing a point from two points requires at least one unstimulated receptor (redrawn from Hochberg, 1964).

1.3 Relation Between Eye Movement and Visual Acuity Circa 1900

Ideas about this relationship were well-established as the 20th Century began. Here, as well as in many other areas, Helmholtz and Hering adopted alternative views. Helmholtz ignored the role of eye movements entirely and Hering gave them an important role in spatial vision. Helmholtz (1866) is often said to be the author of the "retinal mosaic" theory of visual acuity — an approach that holds that the factor limiting the ability to discriminate spatial details is imposed by the fineness of the receptor grain in the retina. The main idea is illustrated in Fig. 1.1, which illustrates how this idea is presented in introductory treatments of perception.

Fig. 1.1 is largely self-explanatory. It shows, on the left, that the gap in a Landolt C will only be discriminated if the size of the gap allows at least a single receptor in the mosaic to remain unstimulated. On the right it shows that a gap in the retinal mosaic between 2 point stimuli will be necessary for the observer to make out that there are 2, rather than 1, stimuli falling on his receptor surface. Clearly, if the eye were to be in motion, if there were eye movements, the story would become much less straightforward. The receptors near the region of the gap would receive, on average, less light than their neighbors and the students in the introductory perception class would require a prerequisite course in elementary statistics before they would be prepared to deal with how the brain might handle visual acuity by calculating means and standard deviations. Even if the eye was not in motion, there are problems with the "retinal mosaic" theory illustrated in Fig. 1.1 despite its obvious didactic value. First, the source of this theory is controversial. Many authors ascribe it to Helmholtz, but there is reason to question this attribution. Second, it entirely ignores the aberrations of the normal human (or any other) "simple eye". These aberrations (e.g., chromatic aberration, spherical aberration and diffraction) prevent the formation of retinal light distributions even remotely like those illustrated in Fig. 1.1. Sharp edges in the stimulus are blurred when they are imaged on the retinal surface. It seems unlikely to me (as it has to others, e.g., Wilcox & Purdy, 1933, or Walls, 1943) that Helmholtz was unaware of these phenomena in the living eye and their inescapable consequences for the character of the proximal stimulus (the light distribution in the retinal image plane where light is transduced into a neural message).It is hard to believe that Helmholtz actually proposed the theory illustrated in Fig. 1.1 despite the fact that distinguished modern authors have ascribed it to him (e.g., Riggs, 1965, or Le Grand, 1967). There are more plausible alternatives, namely, that Helmholtz implicated the retinal mosaic as an acuity limit for didactic purposes only for the case of hypothetical light distributions of mathematical points or lines and then actually proposed that visual acuity, in real living eyes with real proximal stimuli, is limited by the ability to discriminate differences in light intensity falling on adjacent receptors rather than by the presence of unstimulated retinal elements (see refs. cited just above or Steinman & Levinson, 1990, for a discussion of the controversy surrounding Helmholtz's use of the mosaic concept in his treatment of visual acuity). The idea that visual acuity is limited by the ability to discriminate differences in the intensity of various regions in the retinal light distribution was developed by subsequent investigators who, like Helmholtz, also ignored the potential importance of eye movements. Hartridge (1922) and Hecht (1927, 1928; Hecht & Mintz, 1939) were the most prominent proponents.

Hering's treatment of visual acuity was different. He did not ignore eye movements and introduced the approach that would lay the foundation of what will come to be called "dynamic", as contrasted with "static", theories of visual acuity (Falk, 1956). Hering (1920) distinguished two kinds of spatial vision — "resolving power" as studied in traditional acuity tests (the kind of tasks illustrated in Fig. 1.1) and the "spatial sense" — the remarkably keen capacity to detect minute offsets in vernier and stereo acuity targets, where resolution of offsets was possible of elements differing laterally or in depth by only a few seconds of visual angle, that is, by amounts very much smaller than the grain of the receptor surface (Westheimer, 1981, recently renewed interest in such tasks, calling these capacities "hyperacuity"). It was while considering problems of the spatial sense that Hering introduced his treatment of local signs (mentioned above) that was subsequently picked up in the 1920s by Weymouth and developed into a dynamic theory of visual acuity. Hering's use of local signs to explain the straightness of an edge in a living, moving eye is illustrated in Fig. 1.2.

Hering was concerned with how it was possible to perceive a straight edge when its retinal light distribution would fall on an irregular spaced mosaic of retinal

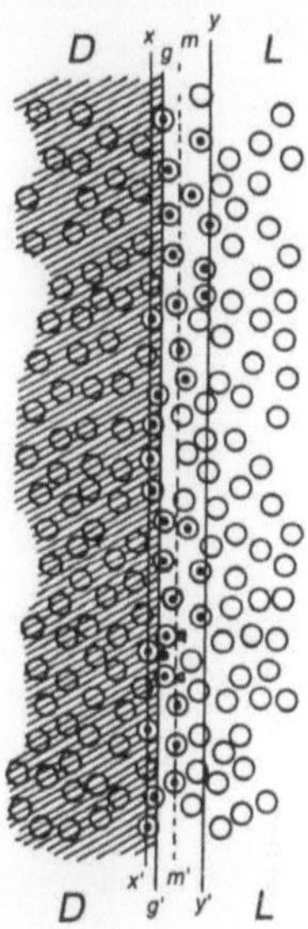

FIGURE 1.2. Illustrates Weymouth's theory of vernier and stereo-acuity. The diagram shows retinal conditions at the margin of a stimulated area. *D–D* is in darkness, *L–L* is illuminated. The geometrical margin of the image is *g–g′*. The cones are shown as circles. Cones *a*, *b* and *c* (near the bottom of region *g–g′*) have local signs whose "center of gravity" is amidst them, tending to pull b to the right. This action among all of the cones cut by *g–g′*, smoothes the percept of the contour despite the raggedness of the line of cones concerned. Furthermore, normal nystagmus shifts *g–g′* back and forth between the between the extreme positions *x–x′* and *y–y′*, so that *m–m′* represents the center of gravity of all the points stimulated, and is the "local sign" of the percept. The localization of this percept is independent of such factors as the size of one cone (from Walls, 1943).

receptors while the edge moved back and forth through an appreciable distance (the receptor mosaic at the center of best vision in the fovea was believed to be somewhat irregular until quite recently; it is now known that these irregularities were caused primarily by histological artifacts). The edge extending from *x–x′* in Fig. 1.2 moves over to the right to position *y–y′* and back again. The line, *g–g′*, in this figure is the physical edge before it moves and *m–m′* is the average of the positions of the edge, oscillating across the jagged receptors shown as circles. (Fig. 1.2 is taken from Walls' 1943 illustration of how Weymouth's dynamic theory worked. Weymouth, in turn, credited Hering, 1899, with the basic ideas illustrated in this figure.) Hering suggested the idea of averaging local signs to improve the apparent straightness of an edge but did not provide experimental support for the basic idea or for its extension to vernier acuity, omissions Averill and Weymouth (1925) proceeded to correct. The way they did this is illustrated in Figs. 1.3 and 1.4.

The basic idea of their experiments was rather modern. They did a simulation of what should be happening on the retina during a test of vernier acuity. They then had an observer (they called him a "reagent") detect the offset of an edge, which could be stationary or moving in the way they thought the eye would move during maintained fixation. They also varied exposure duration, the length of the edge, and whether the edge was seen with one or with both eyes. These manipulations, like oscillations of the edge, should improve the estimate of the mean positions of features of the edge and thereby facilitate detecting any vernier offset that the experimenters may have introduced. Their apparatus is shown schematically in Fig. 1.3(A). A motor-driven cam, C, carries an edge with a variable offset, V, that could be oscillated in front of a replica of the fovea, R, containing irregularly spaced

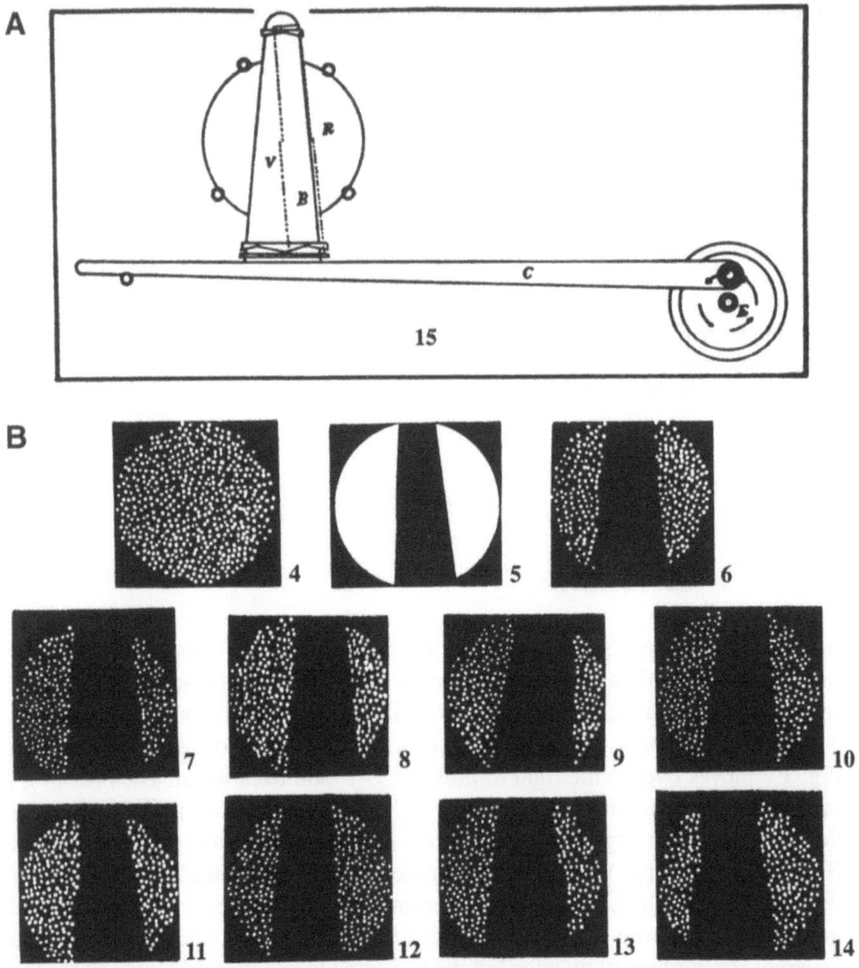

FIGURE 1.3. (A) Diagram of apparatus. R, replica of fovea (Fritsch, 1908) on an aluminum disc, cones being represented by minute perforations (see just below for details). V, inverted V-shaped shield used to produce the image-shadow. B, brass rod with offset (dotted) held in such a position that its broken edge projects just beyond the margin of the shield. C, wooden cross-bar to which the shield is attached. The cross-bar and shield move in an elliptical path whose horizontal diameter is 8 mm (5 cone diameters) and vertical diameter about one-third as great. E, motor-driven eccentric which produces oscillation of the cross-bar and shield in an elliptical path (from Averill & Weymouth, 1925). (B) Stimuli mounted on the apparatus shown in Fig. 1.2. (4) Diagram of retinal mosaic after Fritsch. Note the irregular arrangement of the cones and the great variation in inter-cone distances. This diagram is a replica of the perforated aluminum disc. (5) Appearance of the image shadow with displacement along the left margin and with the wider portion above the offset. (6 to 14) Representations of the retinal field as observed by the subject (reagent) who was required to judge the presence of an offset and its location when an offset was present. For example, in (14) there was a relatively large offset on the left which was wider in the lower part of the retinal field (from Averill & Weymouth, 1925).

perforations that represented the receptors. Fig. 1.3(B) shows the replica of the receptor surface with the receptors shown as minute trans-illuminated holes drilled in a thin sheet of aluminum, 4, and the shadow produced by the edge, 5. Examples of complete test stimuli seen by the reagent are illustrated in Fig. 1.3(B,6-14). The reagent was asked to indicate the position of the offset, and give a confidence judgment, under the conditions of stimulation described above. In Fig. 1.3(B,14), it is easy to see the large offset on the shadow's left edge. The offsets in most of the other test stimuli illustrated in Fig. 1.3(B) are harder to make out.

Averill and Weymouth reported that oscillating targets had lower thresholds than stationary targets, and that longer exposures, longer lines and using two eyes were also better. The first of these findings supports the idea that eye movements favor visual processing, the other results support the general averaging idea but are capable of other interpretations (e.g., probability summation for the binocular case).

Weymouth's dynamic theory remained a curiosity until the Second World War. It was mentioned in textbooks primarily as a minor problem for Hecht's dominant static theory of visual acuity that was built on the retinal intensity discrimination tradition, extending back through Hartridge to Helmholtz (see Steinman & Levinson, 1990, for a more complete review of Hecht's theory). Dynamic theory was presaged by two very influential papers on the electrophysiology of the frog retina published by Hartline (1938, 1940). In these papers, Hartline reported that the most common output of the ganglion cells of this simple eye was "phasic", that is, most neural activity signalled changes in the stimulus. Fifty percent of his units signalled at stimulus onset and offset. Thirty percent signalled only at stimulus offset. The remaining 20% of the units were like those of the ommatidium of the compound eye of the horseshoe crab. They were tonic, that is, they signalled the presence of a stimulus, beginning with a burst when the stimulus came on and continuing to respond as long as it remained, firing all the while at a somewhat reduced rate. Hartline's observations made stimulus transients particularly significant for generating visual neural messages. What better way to produce them then by allowing the eye to move? By 1941 eye movements were beginning to be taken very seriously.

1.4 The Marshall-Talbot Dynamic Theory of Visual Acuity

This theory was based primarily on two observations. First, Adler and Fliegelman's (1934) measurements of the miniature eye movements made during carefully maintained fixation, which had shown a high frequency (up to 100 Hz) oscillation of the eye ("physiological nystagmus"), whose average amplitude they reported to be about 2 minutes of visual angle. Second, the report of an anatomical "cortical magnification factor". This factor is based on the fact that 2 minutes of arc of the rhesus monkey's fovea (about 9 micrometers) was found to project to about 1 linear

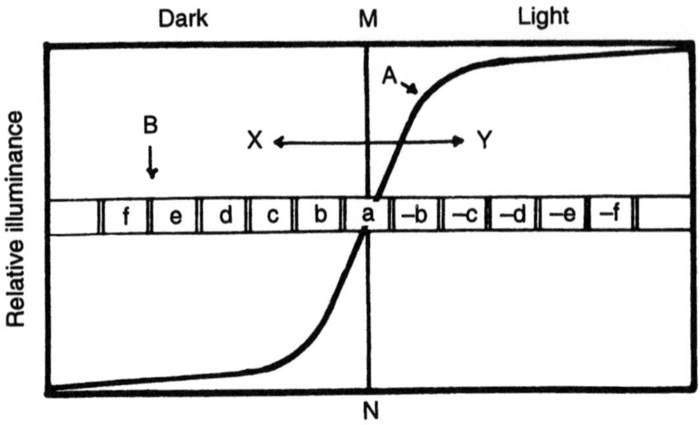

FIGURE 1.4. The distribution of illuminance on the retina across the geometrical boundary (MN) separating Light and Dark halves of a field (from Falk, 1956). See the text for details.

millimeter of its primary visual cortex. This has the consequence of making the effective ratio of cortical visual cells to foveal cone receptors at least 600:1. These observations, taken along with the importance of phasic stimulation demonstrated in frog retina by Hartline, encouraged Talbot and Marshall (1941; Marshall & Talbot, 1942) to develop a theory that made eye movement a necessary condition for acuity. Their basic scheme is illustrated in Fig. 1.4.

This figure shows the proximal stimulus of an edge made by a light and a dark region. The proximal stimulus, curve A, is the intensity distribution on the retina after the edge has been "smeared" by the aberrations of a normal eye (the physical or "distal" stimulus would be an intensity step). The row labeled B is the receptor surface containing foveal cones, having a center to center separation of about 19 seconds of arc (about 1.5 micrometers). These values of cone separation at the center of best vision in the fovea were based on Polyak's (1941) influential studies of primate retina (a somewhat larger value, perhaps about 30 seconds, would probably be preferred today). The arrow labeled "$X \longleftrightarrow Y$" represents the average 2 min arc amplitude of physiological nystagmus suggested by Adler and Fliegelmann. High frequency eye movements of this size would stimulate a row of about 6 cones. The maximum rate of change of stimulation, and hence the maximal firing rate, would be found at cone a, where the slope of the intensity function is maximal. Firing rate would fall off progressively on adjacent cones. The relationship of the firing rate in the maximally stimulated cone, a, to its neighbors, coupled with anatomical "magnification" in the ascending pathway and assumptions about the duration of the neural recovery cycle allowed recovery of the sharpness of the intensity step that was degraded in the proximal stimulus. The size of the cones relative to the size of the high frequency oscillations is critical to this theory, which has sometimes been called a statistical theory of visual acuity, because the physiological process correlating with the critical detail is the average or peak of the distribution of outputs of the neural elements firing in the cortex. (See Steinman & Levinson, 1990, for a detailed critique of this theory.)

1.5 Empirical Tests of the Marshall-Talbot Theory

There were two lines of attack. The first made careful measurements of the miniature eye movements characteristic of steady fixation. The question here was: Does physiological nystagmus have the properties required by the theory? This led to the development of the contact lens-optical lever eye movement recording technique, which was capable of resolution well under one minute of visual angle and was free from translational artifacts. This new method of recording also provided a means for stabilizing the retinal image of a test target — a development that permitted test of the Marshall-Talbot claim that eye movements were necessary for good acuity. Both lines of attack led to clear refutations of the theory.

For example, Ratliff and Riggs (1950) found that physiological nystagmus actually had an average amplitude somewhat less than 20 seconds of visual angle, which meant that it could not stimulate a population of cones. Its frequency was high enough (about 30–80 Hz) but its excursion was so small that the maximum of the oscillating light intensity distribution would be confined to a single receptor. At least part of the problem with Adler and Fliegelman's (1934) eye movement measures, which were used by Marshall-Talbot to devise their theory, was, as Ratliff and Riggs pointed out, an apparent scaling error. Adler and Fliegelman apparently did not realize that an optical lever has an inherent amplification factor of two, that is, a 1° rotation of the eye causes a 2° angular shift of the beam reflected from the mirror mounted on the contact lens.

The work with stabilized images also made trouble for the Marshall-Talbot theory. Riggs, Ratliff, Cornsweet and Cornsweet (1953) found that stabilized and normal viewing both permitted good vision of fine details. It was only when the display remained stabilized for long periods of time and began to fade that natural viewing became better than stabilized viewing. Eye movements were necessary to prevent fading but the discrimination of details was the same with or without eye movements until fading began. These authors also reported that image motion twice "normal" was better for maintaining visibility over prolonged periods than normal image motion. This became a common, and mysterious, finding in subsequent work with stabilized images (see Kowler & Steinman, 1980, for the likely explanation of this mystery).

As I see it, this line of evidence from stabilized image experiments against the Marshall-Talbot theory is less compelling than the measurements of physiological nystagmus (described above) because it has been quite clear since Barlow (1963) examined the quality of the best stabilization obtainable (with the contact lens and Yarbus sucker methods) that excursions of the eye as small as physiological nystagmus had never been stabilized on the retina. I, following Barlow's (1963) lead, am inclined to believe that high contrast targets, confined entirely to the central fovea, probably only lose their sharp edges, but never disappear completely (see Steinman & Levinson, 1990, for a discussion of the vast, and more often than not, controversial literature on stabilized images). But, regardless of the particular reason one prefers for rejecting the Marshall-Talbot theory of visual acuity, it was rather generally agreed about 30 years ago that physiological nystagmus was not a functionally significant eye movement — sufficiently long ago to guarantee that the basic idea will crop up with increasing frequency as the people who know this literature first forget its details and then die off. (The basic findings of Riggs and his coworkers have been replicated many times. See, for example, Ditchburn, 1973, for a review of work he began independently in England in 1953 and Yarbus, also for independent work, done in Russia (Yarbus, 1957a,b, 1967). Krauskopf (1957, 1962, 1963) in this country and Gerrits and Vendrik (1970, 1972, 1974) in the Netherlands added a great deal to our understanding of these phenomena during the heyday of research with images stabilized by means of "invasive" methods. Recent work with noninvasive methods will be described in some detail later.

1.6 The Phone Rang

And I had the pleasure of talking, for the first time, with Fran Volkmann, who was well-known for her work on threshold elevations associated with planning and making saccadic eye movements — "saccadic suppression" in trade jargon (see Volkmann, 1986, for a recent review of this topic and Sperling, 1990, for an alternative point of view). I was flattered to receive this phone call when I heard that Prof. Volkmann wanted me to participate in a workshop to be held at Princeton in April, 1974 that was being organized under the auspices of the prestigious Committee on Vision of the National Research Council of the National Academy.

The proposed plan was to have a number of people sit around and engage in a panel discussion about how we can see clearly as we look and move about in the real world. The panel was to include Ethel Matin, Ulker Tulunay-Keesey, Lorrin Riggs and myself. I was concerned about this topic because, as I pointed out to Fran, all of us worked with contact lens-optical levers, which required that the head be fixed on a biteboard. This might make it difficult, perhaps even dangerous, to extrapolate from this kind of research to the real world — actually I think I said something like "nobody knows anything about this, regardless of what we like to tell people at cocktail parties or claim in grant proposals." My recollection after almost 15 years was that there was agreement, or at least acquiescence, on the other end of the line. I think that Fran said something like: "Yes, but, it would be interesting and valuable to discuss what we do know or at least consider the problems we are facing in answering such a question." Who could disagree with this and I, cheerfully (at least my intent was to be cheerful), agreed to participate in the panel. This phone call probably came sometime in February during an exceptionally busy Spring. There were a number of research projects to get ready for the ARVO meeting in Sarasota at the beginning of May, lots of teaching, and the preparation of a review paper on oculomotor effects on vision I had agreed to deliver at a symposium in Stockholm during June (Steinman, 1975). Planning material for an informal panel discussion was the least of my concerns until the phone rang again sometime late in March. It was Fran Volkmann again. She began by saying: "Hello Bob, I was talking with Lorrin and we agreed that this idea of an informal panel discussion wasn't likely to work so well. We think that it would be better if we each talk about our specialized interests. I'll do a general review of saccadic suppression; Lorrin will talk about some exciting new experiments showing saccadic suppression with electrical phosphenes, rather than light as input; Ulker will talk about acuity with stabilized targets and Ethel will explain how saccadic suppression helps us perceive the direction of objects. We would like you to cover the more general issue of the role of eye movements in maintaining a phenomenally clear and stable world."

In short, I was expected to give a lecture on a topic I believed to be a complete mystery. What was I to do in these circumstances? I did what everyone I know does in such circumstances. I agreed to do the talk, knowing full well that I could begin with a disclaimer about actually being able to answer the big question and then move quickly to talk about what I actually was doing and could say something about. This lecture was not going to end there. It was to be published along with any discussion it engendered. This fact made it imperative that it included some new material. I was compulsive about this when I was young professionally (I know better now) probably because my doctoral mentor and role model (Jack Nachmias) had not been enthusiastic when confronted with rehashes of old stuff at meetings. What was I to do? Less than a month was available for generating some new and, at least superficially, relevant data. After discussion with my colleagues we decided that the best that could be done in the circumstances would be to find out the scope of the problem facing the oculomotor system. Specifically, when a human being sits still with the head free from artificial supports, the head was sure to move. These irreducible head movements would have to be compensated by the oculomotor system if the person is to be able to maintain gaze steadily on some stationary feature in the visual environment. Put differently, how much additional work did the oculomotor system have to do when the head was not supported by a biteboard? Fig. 1.5 shows how we tried to find this out.

Fig. 1.5(A) shows the late Brian Murphy (Steinman, 1976, p. 136) sitting in David Robinson's magnetic field eye movement recording apparatus at Johns Hopkins in Baltimore. He looks a little scrunched-in because this set-up is usually used to measure the eye movements of young rhesus monkeys, whose heads are held near the center of the wooden framework by means of a metal ring and bolts screwed into their skulls. Brian is more or less centered within 2 pairs of Helmholtz magnetic field coils, one pair above his head and below his elbows and the other pair to his left and right. These field coils were driven by sinusoidal A.C. signals in quadrature mode (orthogonal in space and time). In this type of instrument, the amplitude (voltage) of the signal induced in a sensor coil located within this magnetic field is proportional to the sine of the angle of the sensor coil with respect to the direction of the magnetic field. The induced signal is zero when the sensor coil's windings are parallel and maximal when the windings are perpendicular to the direction of the magnetic field. The horizontal and vertical components of the magnetic field are 90° out of phase and independent measurement of the sensor coil's orientation along each meridian can be measured by using a phase-lock amplifier tuned to the orthogonal phases of the induced signal. In Fig. 1.5(A), the sensor coil can be seen just in front of Brian's mouth where it was held by attaching it to a biteboard

A 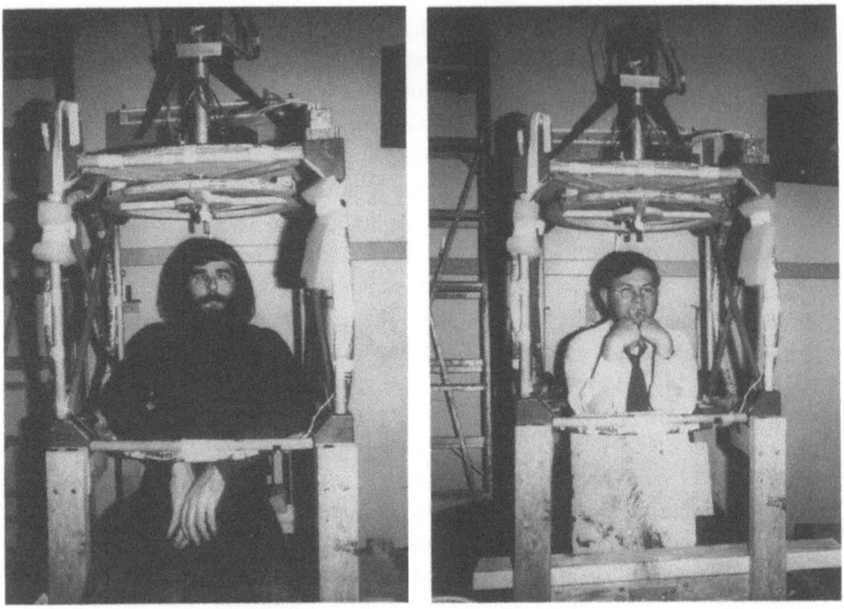 B

FIGURE 1.5. (A) Brian Murphy sitting in Robinson's magnetic field instrument having movements of his head measured by means of a sensor coil mounted on a biteboard (from Steinman, 1976). (B) David Robinson sitting in his magnetic field instrument having movements of his head measured by means of a sensor coil mounted on a biteboard (from Steinman, 1976).

(a silicone dental impression made on a sheet of plastic) clenched between the teeth. The lead from the sensor coil can be seen passing near the left side of his head to a connector located just above his hair. Brian's task was to sit as still as possible, breathing normally and also while holding his breath, using only natural supports to help him remain still. In this posture, Brian had only neck muscles to steady his head on his torso. His torso could be stabilized by resting his arms on the framework of the apparatus — a posture much like the one used while sitting still in an armchair. Fig. 1.5(B) shows another subject and another natural posture studied. The subject in Fig. 1.5(B) is David A. Robinson, the bioengineer, who developed the magnetic field-sensor coil technique that has proven to be so valuable in contemporary oculomotor research. He is shown using another commonly employed natural posture for supporting the head while sitting still and examining objects in the real world.

We found that the head rotated quite a bit even when the subject tried to sit as still as possible. This "fact" was published in the book covering the proceedings of the workshop (Steinman, 1976). I won't say anything further about these measurements because they were not what we thought them to be at the time, namely, they were not actually head rotations. We got this wrong because Robinson did not appreciate, or more likely had forgotten, that his, and similar instruments, were not suited for making such measurements. This became apparent only when I decided

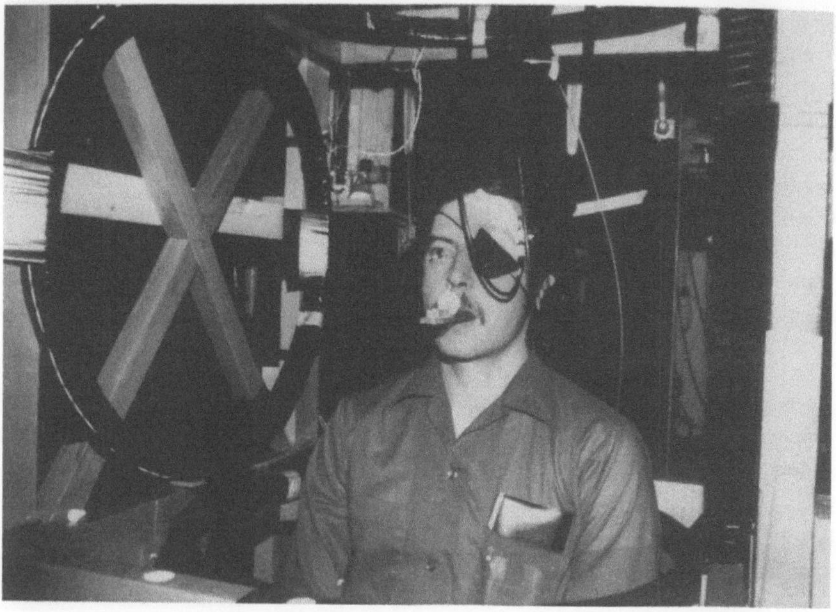

FIGURE 1.6. Alex Skavenski sitting in his magnetic field instrument having movements of his head and right eye measured by means of the amplitude-detecting technique (from Steinman, 1975).

to go one step further and show some free-headed compensatory eye movements in Stockholm. This decision began a period of jet-set science between College Park, Boston and Rotterdam. The first series of flights were to Alex Skavenski's lab in Boston where attempts to measure free-headed eye movements began in earnest. Fig. 1.6 shows Alex sitting in his version of the Robinson amplitude-detecting magnetic field-sensor coil apparatus. His set-up was a bit larger than Robinson's because Alex routinely records eye movements in human, as well as in monkey, subjects. The Helmholtz coils have an outer wrapping of black vinyl electrical tape that covers the aluminum foil Faraday shields that are visible in Fig. 1.5.

Skavenski is wearing a tight-fitting scleral contact lens that is held in place on his right eye by means of suction. Suction is established with a syringe filled with contact lens fluid and applied to the contact lens by means of the thin plastic tube visible as it passes across his nose near the patch over his left eye. A sensor coil is embedded in the surface of the contact lens. Its lead (a very thin twisted pair) can be seen near the right side of his nose as it passes directly upwards to a connector overhead. The white round object just below his nose is the sensor coil used to measure head rotations. It is mounted on a biteboard — the same placement used in Baltimore. Fig. 1.7(A) shows the author in the same apparatus. I am wearing a motorcycle helmet with a very tight chin-strap. Its purpose can be seen in Fig. 1.7(B). This figure shows that a loudspeaker was mounted at the back of the helmet with a mass of bolts stuck to its voice-coil. This rig was used to oscillate the head passively at frequencies up to 30 Hz with very small amplitudes (up to about 5 minutes of arc). These arrangements were used to study oculomotor compensation for head rotations beyond the range of head movements that could be induced naturally.

I reported the initial observations made with this instrumentation in Stockholm (Steinman, 1975) but was very careful to limit comment to saying that it was obvious that the oculomotor system had a lot to do and that a lot was going on. More was not said because it was becoming very clear when we started analyzing our records quantitatively that something did not make sense. The nature of the problem, as well as its solution is illustrated in Fig. 1.8.

Fig. 1.8(A) shows Skavenski sitting in his new, large Helmholtz field coils. His head was free as he maintained fixation on a small point of light located at optical infinity in an otherwise dark room. One of two sighting tubes (at left center) that were used to position Alex's experimental eye in a precalibrated portion of the magnetic field, and a wooden framework around, but not touching his head, can also be seen in this photo. Fig. 1.8(B) shows a close-up of Alex rigged and ready for recording. The wooden framework around his head permitted 3-D movements of about 1 cm in any direction. He was careful to avoid touching any part of the wooden framework with his head while recordings were made. A white surgical stocking cap was worn to allow some room between his curls and the wooden framework (men wore hair long in the 1970s, see the author in Fig. 1.7(A)). The head sensor coil was still supported by means of a biteboard, but the round, white sensor coil was now attached to a plastic extension that located the head coil at the bridge of the nose where it was near the experimental eye. This arrangement placed the eye and head coils about the same distance from the head's center of rotation and also placed them near each other in the magnetic field. The scleral contact lens is very prominent in this photo because it was necessary to increase the number of turns of wire in its sensor coil in order to maintain a reasonable S/N ratio with the relatively weak magnetic field that could be generated with the large field coil arrangement and the field-driving amplifiers available. Alex covered the new multi-turn sensor coil with a white, dental plastic tooth-filling compound, which made the appearance of his eye rather dramatic.

Why all these changes in instrumentation (Steinman, 1986a)? Simply because the Helmholtz coil arrangement only provides a minuscule region near its center where the magnetic flux is homogeneous, that is, the magnetic vectors are parallel within each of the orthogonal directions. If the head is free in a small Helmholtz field coil arrangement of the kind used in all primate research prior to Skavenski's new set-up, translations, as well as rotations, of the head will cause changes in the amplitude of signals induced in the sensor coil. If the translations are large relative to the rotations, it's a real mess. It was the use of small Helmholtz field coils that made interpretation of our initial free-headed eye movements impossible (Steinman, 1975) and the initial report of head "rotations" artifactual (Steinman, 1976). The large field coils used by Skavenski, Hansen, Steinman and Winterson

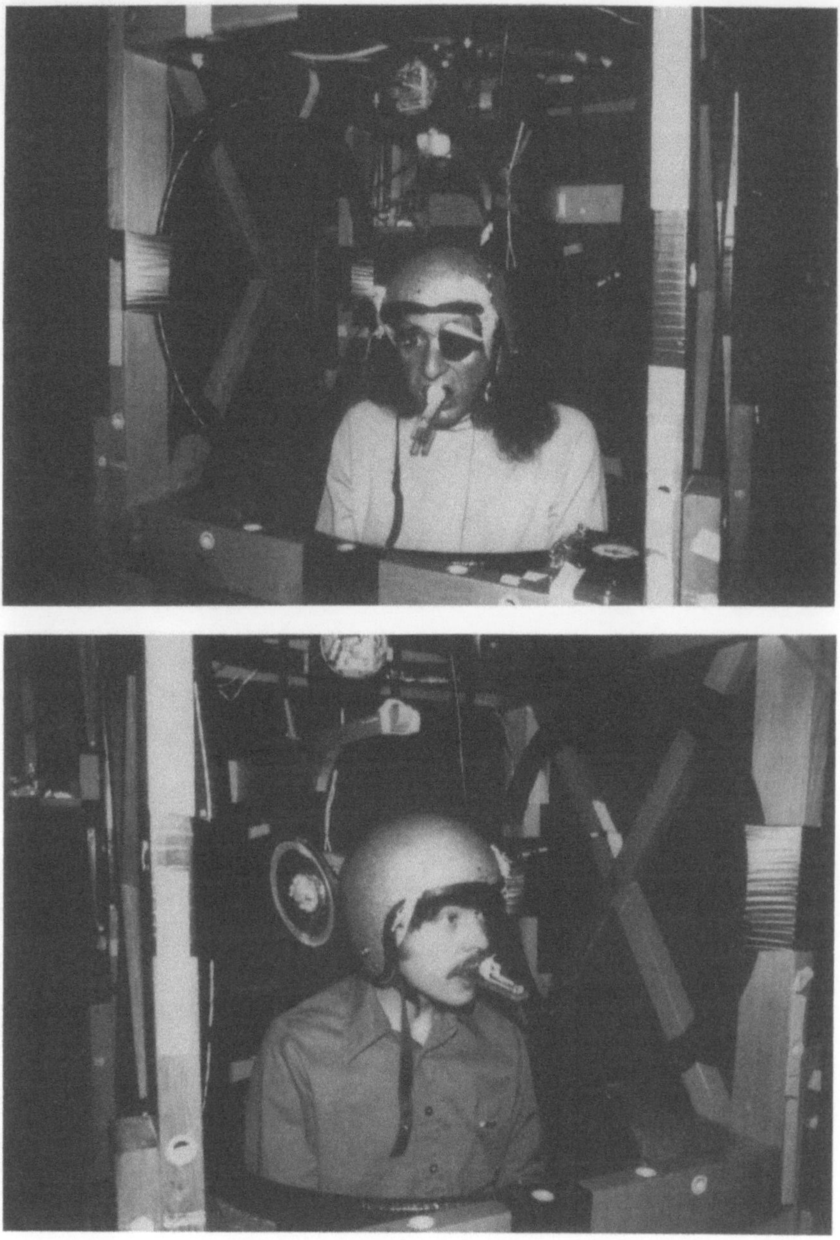

A

B

FIGURE 1.7. (A) Steinman sitting in Skavenski's apparatus having his head and right eye movements recorded. See the text for an explanation of the purpose of the helmet. (B) Skavenski sitting in his apparatus having his head and right eye movements recorded. See the text for an explanation of the purpose of the helmet.

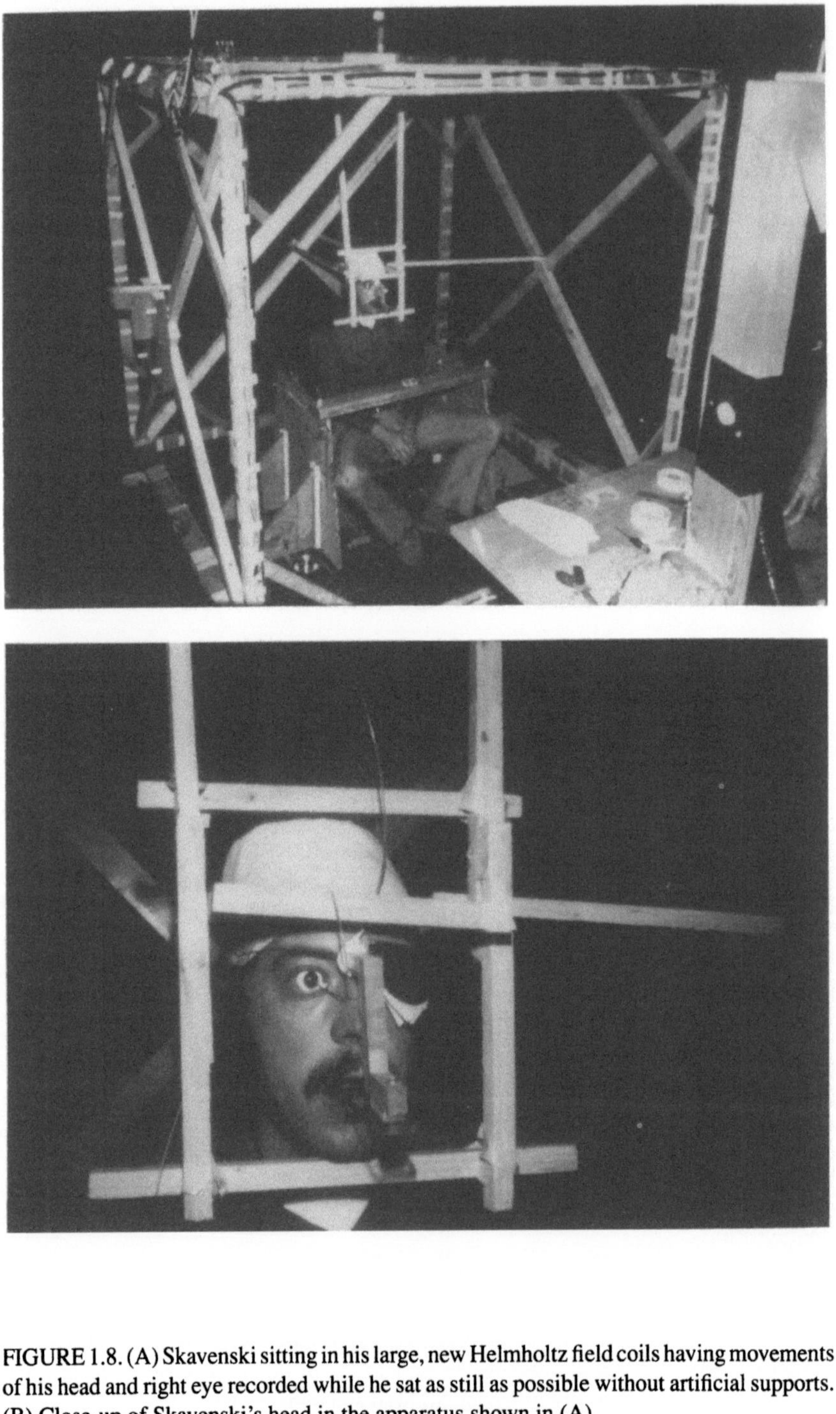

A

B

FIGURE 1.8. (A) Skavenski sitting in his large, new Helmholtz field coils having movements of his head and right eye recorded while he sat as still as possible without artificial supports. (B) Close-up of Skavenski's head in the apparatus shown in (A).

(1979) only provided a small, but practicable, homogeneous region in the sense that translations of the head did not produce artifacts equivalent to rotations of the eye larger than 1 minute of arc. (N.B., the standard deviation of a fixating eye, when the head is supported on a biteboard, is only about 2–3 minutes of arc; see Steinman, Haddad, Skavenski & Wyman, 1973). This means that very good instrumentation is required to study human fixation with accuracy and precision better than the oculomotor system itself. Pretty good bandwidth, as well as accuracy and precision, is needed because saccades (the eye movements used to shift gaze rapidly), which achieve maximal average peak speeds of about 525°/sec in human beings, can cover distances approaching 80° of visual angle in a single step with pretty fair accuracy (see Collewijn, Erkelens & Steinman, 1988a,b, for measurements of binocular saccades over their entire range of operation on both vertical and horizontal meridians with bandwidth = 244 Hz; accuracy and precision = 1'; and linearity > 0.01% and Erkelens, van der Steen, Steinman & Collewijn, 1989a and Erkelens, Steinman & Collewijn, 1989b, for equally accurate measurements of "vergence" eye movements).

Fig. 1.9 illustrates the results obtained with Skavenski's elaboration of Robinson's method (Skavenski et al., 1979). These records show that the unsupported head during sitting, as well as during standing, provided an unsteady platform (our head spectrum measurements showed considerable power in a range extending from D.C. to 7 Hz). These head movements were only partially compensated by eye movements, meaning that there was considerable retinal image motion of the fixation target when the head depended entirely on natural supports. Specifically, the retinal image velocity of the fixation target increased by a factor of 2–4 over velocities observed with the head on a biteboard and the standard deviation of fixation on a single meridian increased from about 2–3' to about 30'. Perfect oculomotor compensation during steady fixation would produce horizontal straight lines in the traces reproduced in Fig. 1.9 because these records show the angular orientations of the eyes and head with respect to an earth-fixed coordinate system. If the line of sight of the eye stayed exactly on-target, the trace shown in the record would not move. It would be a horizontal straight line even when the head trace indicated that the head was oscillating. All subsequent eye and head movement records will use this earth-fixed coordinate system so all of the traces have the same significance, that is, perfect oculomotor compensation for oscillations of the head will produce a horizontal straight line in the eye traces in all records shown.

Furthermore, when movement is apparent in the eye trace when the head moves, the degree and source of such departures from perfect oculomotor compensation can be inferred from the amount of motion of the eye and its direction relative to the head. Namely, if the eye moves in the same direction as the head in the record, the eye, trying to maintain fixation on a distant, stationary object, is under-compensating for the head movement (gain is too low). This allows the fixation target to move on the retina. When the eye moves opposite to the direction of the head in the record, the eye is over-compensating (gain is too high), and the fixation target image will also move on the retina. Keeping these features in mind will help the reader interpret the recordings reproduced from here on.

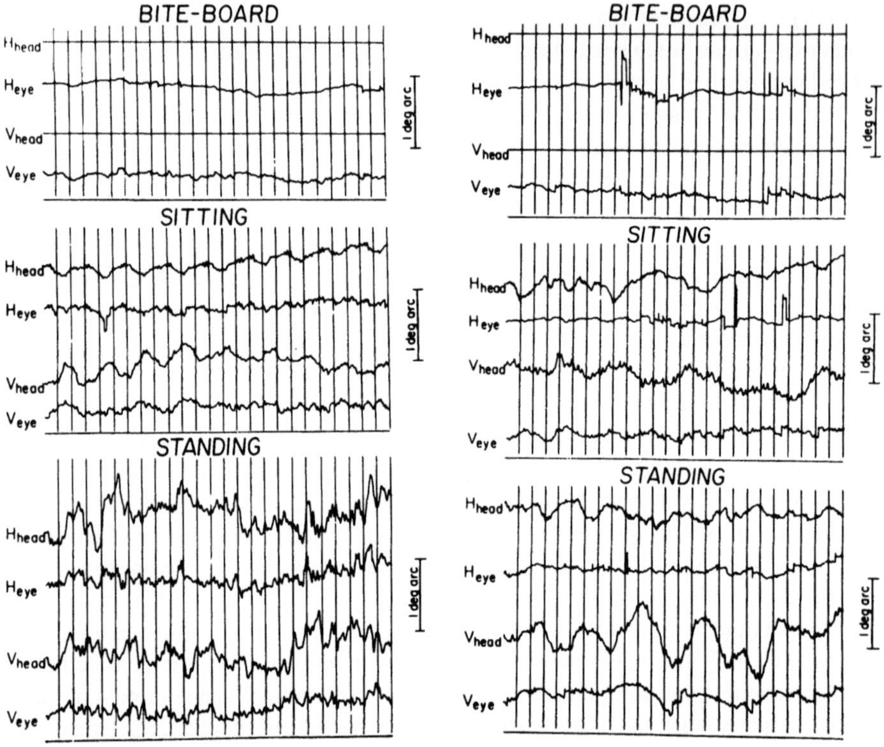

FIGURE 1.9. Representative horizontal (H) and vertical (V) head position and gaze (retinal image position) of subject Skavenski (left) and Steinman (right) while they fixated a target at optical infinity with their heads supported on a bite-board, or while they were sitting or standing as still as possible. Records begin on the left. The vertical time-lines show 1 sec intervals and the vertical scales on the right side of each record show 1 degree of visual angle. Upward changes in these position traces signify rightward or upward rotations (from Skavenski et al., 1979).

We also found that a larger proportion of rotational head movement was compensated when head oscillations increased in vigor over those observed when the subject sat or stood as still as possible. This result encouraged us to conclude that the goal of oculomotor compensation is not the reduction of retinal image motion to some minimum, near zero, value but, rather, the goal was to keep retinal image motion at some higher level that was optimal for visual processing because this goal would prevent fading whenever the observer wished to see well while sitting still. This idea, which will eventually have good direct support, will be described later (Collewijn, Martins & Steinman, 1981, 1983). Other important developments intervened and the next step in this story took place in Rotterdam when Han Collewijn (1977) measured the eye and head movements of freely-moving rabbits — an undertaking that required major improvements in existing recording methodology. I will first describe the controversy that led to the beginning of what has become a long and stimulating collaboration and then say a few words about Collewijn's new methods. These methods have made it possible to record eye and head movements under increasingly natural conditions and to examine the role of natural retinal image motion in visual processing.

Collewijn (1977) reported that the eye movements of his rabbits compensated, virtually perfectly, for movements of their heads as they moved about. We, on the other hand, reported that human beings failed to compensate anywhere near perfectly once their heads were free to move (Skavenski et al., 1979). Were humans less competent than rabbits in oculomotor as well as in reproductive skills? This seemed unlikely. Perhaps humans would do better if they were allowed to move their heads actively, rather than to sit quietly or have their heads oscillated passively through small amplitudes: the way we had done our work. There was also the fact that Collewijn had observed the eye movements of his rabbits on a rather coarse scale because the dynamic range of his recording instrument was limited and the animals tended to move their heads (and/or bodies) through large angles. It's not easy to explain to a rabbit that it should always face north and only oscillate its head through amplitudes rarely exceeding 20° peak to peak. Human beings can be instructed to do this quite easily. Han was as curious about our disagreement as I and we arranged for me to visit his lab in Rotterdam where his new instrument would be used with ourselves and his colleagues as subjects in a replication of his experiment with rabbits. The visit was planned for early June of 1978, just after I had returned from the annual ARVO meeting in Sarasota. As I was packing my bags for my trip to the Netherlands, . . .

1.7 The Phone Rang Again

It was Jarus Quinn, the Executive Secretary of the Optical Society, who called to thank me for agreeing to participate in a Symposium on Binocular Vision at the annual meeting of the Society in San Francisco in October. He assured me that I wouldn't have to pay to register at the meeting because I was an invited speaker, but that there were no funds for my expenses. This came as somewhat of a surprise. Not the business about expenses, I knew that physicists were tight with money. I was surprised because this was the first I had heard of the symposium, much less agreed to participate. I told Jarus that it must be some kind of mistake, but he insisted that Prof. Gerald Westheimer had made these arrangements and that I was expected to participate. Jarus apologized for the mixup and explained by saying that Prof. Westheimer probably had expected him to contact me and that he had thought that Prof. Westheimer would contact me. He suggested that I should phone Prof. Westheimer and discuss my participation with him. I stopped packing my bag and phoned Dr. Gerald (he calls me Dr. Bob) and pointed out that I was somewhat at a loss because I really didn't know much about binocular eye movements, had never measured them, and, in fact, didn't even have an instrument that could be used for the purpose. Dr. Gerald was unperturbed and suggested that I do something appropriate (or some such thing). Then I knew! Quinn had been chosen to make one of those unlikely phone calls. Suddenly, it came to me. I realized that I could make binocular recordings easily in the Netherlands with Collewijn's technique and said goodbye to Dr. Gerald, assuring him that I would come up with something by October. That's how Collewijn and I got into the human binocular eye movement business. We were assigned the task — no questions asked. How we began to do this is shown in Fig. 1.10.

Fig. 1.10(A) shows the author (tan after a week in Sarasota) looking off into the distance with something taped to his forehead and a suggestion that he may be wearing something on his eyes. Connectors overhead and parts of a couple of magnetic field coils can also be seen. Fig. 1.10(B) shows a close-up of the eyes, which can now be seen to be wearing silicone annulus sensor coils. Collewijn, van der Mark and Jansen (1975) developed these ingenious devices as substitutes for custom-fitted scleral contact lenses in oculomotor research. One size fits all (annuli for research are available from SKALAR in Delft). Tight-fitting scleral research contact lenses, even when held to the eye with as much as 100 mm of mercury suction, move around when large (> 10°) eye movements are made. Collewijn's silicone annulus, on the other hand, has been shown to adhere to the eye during even the largest eye and head movements (see, for example, Collewijn et al., 1981; de Bie, 1985; in addition to the original description of the new device by Collewijn et al., 1975). Each annulus contains a 9-turn coil of very fine copper magnet wire (approx. 47 AWG). The coils and their very delicate twisted-pair leads can be seen in this photo. The easiest lead to see in this photo is coming from the subject's right eye near the inner canthus and goes in front of the eyebrow on its way to one of the connectors overhead. The 9-turn sensor coil can be seen best in the other eye on the right side of the photo. A similar annulus was taped to the forehead above the eyes (see Fig. 1.10(A)). The eyes are anesthetized with drops before the annulus is inserted. Annuli can be worn for periods up to 40 minutes, individual subjects experiencing very different degrees of discomfort. Their effects on vision are minor in most subjects, who have been tested in concurrent visual psychophysical experiments, but wearing these annuli is not likely to catch on as a popular pastime, all things considered. They do, however, make it possible to study eye movements over their entire range of rotation and place no constraints on the vigor or speed of oculomotor responses. Ferocious blinks or clumsy experimenters can break leads, but the leads are relatively robust and it is not unusual to be able to sterilize an annulus and use it for 10, and often more, recording sessions. They can even be used to measure eye movements accurately during blinks and also when the eyes move under eyelids that are kept closed (Collewijn, van der Steen & Steinman, 1985). In short, the development of these annuli was a very large advance in human oculomotor research. In the author's opinion, it was larger than the introduction of the tight-fitting scleral contact lens, which got the modern eye movement recording era going 40 years ago. These annuli fit almost all adult human eyes. They adhere to the eye during large eye movements and, when used with the phase-detecting method introduced by Collewijn (1977), are capable of absolute calibration — a very convenient feature in the basic science oculomotor laboratory,

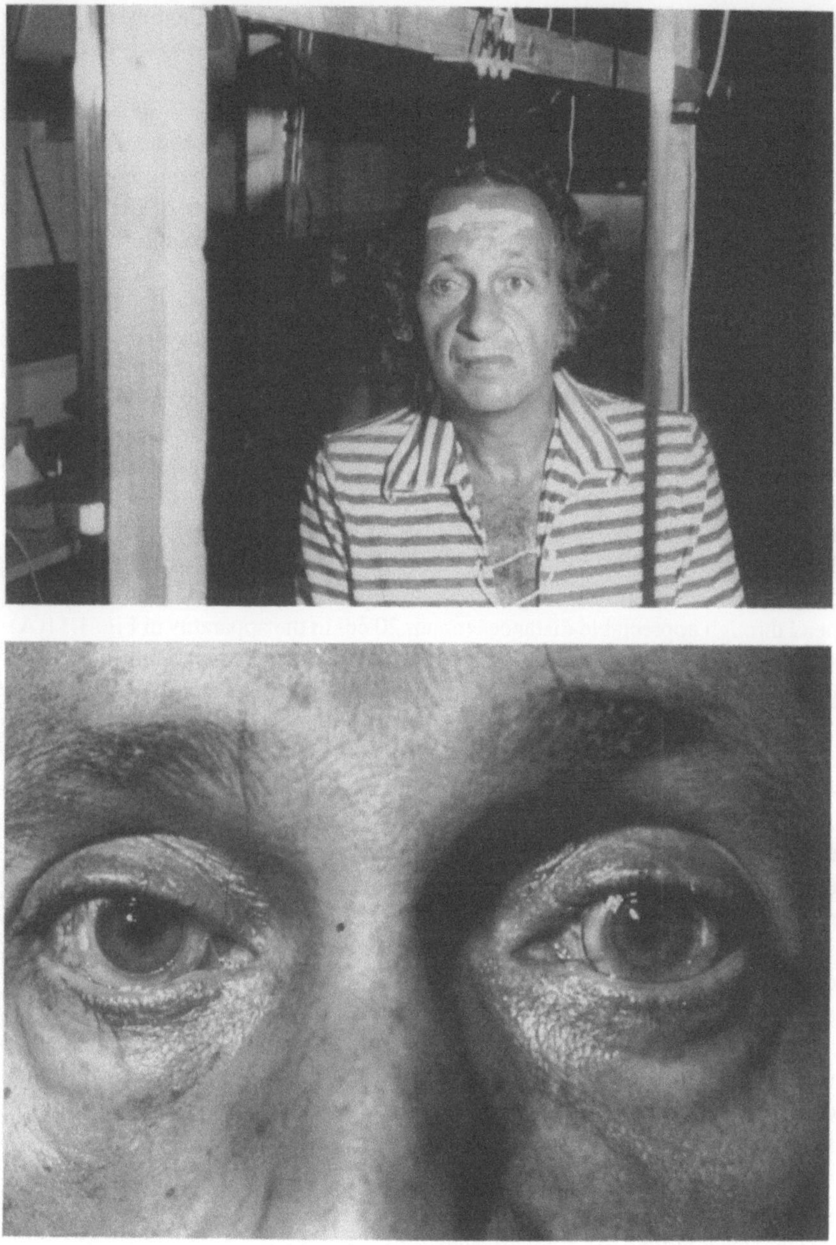

A

B

FIGURE 1.10. (A) Steinman in Collewijn's cube-surface field coil, phase-detecting rabbit apparatus in Rotterdam having movements of his head and both eyes recorded. (B) Close-up of the silicone annuli worn on both eyes.

and a necessity in the clinic where one cannot count on a patient understanding instructions or being able functionally of following instructions to shift gaze or to track.

The silicone annulus sensor coil was only one of three major improvements in the technology for measuring eye movements that were introduced during the late 1970s by Collewijn's group in Rotterdam. Another is illustrated in Fig. 1.11(A), which shows the author sitting with his head near the center of the apparatus Collewijn had used to measure the eye and head movements of unrestrained rabbits. It looks like I am sitting in jail, surrounded by coils of wire. Actually, I was at the center of a pair of cube-surface field coils. The cube-surface field coil configuration was developed during World War II as a practical way to degauss ships to protect them from magnetic mines. The optimal coil-arrangement (i.e., a good approximation to a true solenoid) was described theoretically just after the war by Rubens (1945) and introduced into eye movement research by Collewijn in his rabbit paper (1977). The cube-surface field coil arrangement makes it possible to achieve a relatively large homogeneous central region of magnetic flux by using 5 coils on a meridian, rather than the Helmholtz pair of coils traditionally used (see Rubens, 1945, or Collewijn, 1977, for the required ratios of windings in each of the 5 coils). The cube-surface field coil arrangement permits the subject to move his head through appreciable distances (about 20 cm in the apparatus in Fig. 1.11(A)) without introducing disconcerting translational artifacts (the area of good, i.e., 1 part in 1000, homogeneity depends on the overall size of the cube). Collewijn also developed the technique of phase- rather than amplitude-detecting the signal produced in a sensor coil by a rotating magnetic field (an idea suggested by Hartmann & Klinke, 1976). A homogeneous rotating magnetic field can be made by having 2 cube-surface coil arrangements oriented 90° to each other (in spatial quadrature) and driving them out of phase with each other (in phase quadrature). This arrangement will generate a magnetic vector that rotates through 360° during every period of the frequency of the a.c. currents that are driving each cube-surface field coil. The phase-detecting technique has a number of advantages over amplitude detection; namely, it gives linear indications of the orientation of the sensor coil throughout 360°, it is insensitive to fluctuations in the strength of the revolving magnetic field and it is capable of absolute calibration. These properties, coupled with the solution of the translational artifact by means of the cube-surface field coil arrangement just described, opened the way for beginning to study how eye movements allow us to maintain a phenomenally clear and stable world as we move about — the question Prof. Volkmann raised 3 years earlier in her fateful phone call.

The visual stimuli we used are shown in Fig. 1.11(B). This is the view to the North-West from the window just outside the rabbit phase-detecting, cube-surface field coil apparatus on the of 15th floor of the Medical Faculty in Rotterdam. Our fixation stimulus was the control tower at the Rotterdam airport (indicated by an arrow). It was 5000 m from where the subject was seated. On particularly clear days we used a building near the horizon in the suburbs of Den Haag (35000 m away). The subject's task was to fixate either of these far targets and to oscillate

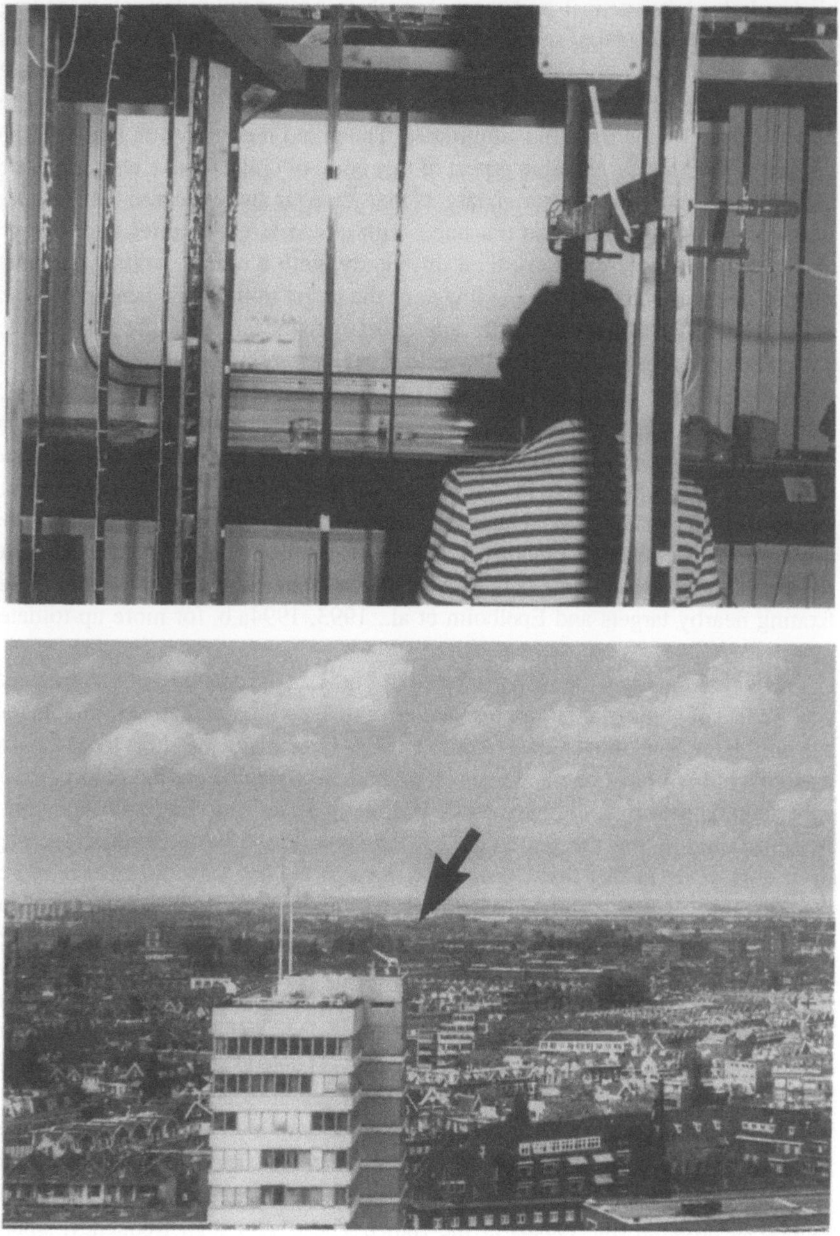

A

B

FIGURE 1.11. (A) Back view of Steinman sitting in the rabbit apparatus while looking out of the window at a distant target. (B) The view from the window shown in (A). The arrow points to the control tower at the Rotterdam airport.

his head about its vertical axis through peak to peak amplitudes not exceeding 20°, increasing frequency and simultaneously reducing amplitude until the limits of active head oscillations had been reached. The subject was also asked to note the appearance of the visual scene as he moved his head throughout this entire range of head frequencies and amplitudes. The photo reproduced in Fig. 1.11(B) illustrates the most remarkable aspect of this body of collaborative research done in Rotterdam. We needed fixation targets that were far away because maintaining steady fixation when the head translates with a near target requires that the eye makes compensatory rotations. Said differently, with a nearby target gaze must shift in space in order to keep the image of the target in the same position on the retina. This comes about because the angles between the target and the eyes change a lot when the head translates relative to a nearby target. This does not happen when the target is far away. It is impossible to assess the steadiness of fixation, that is, the amount of motion of the retinal image of the target, if the target is near and the amount of head and eye translation is not known. We had no way to measure this at the time with the accuracy required to interpret eye movement records (10 years of additional instrumentation development were required before this could be done; see Collewijn et al., 1990; Kowler et al., 1990, for the first reports of accurate measurements of eye and head translations and rotations while fixating nearby targets and Epelboim et al., 1993, 1994a,b, for more up-to-date reports).

The view from the window reproduced in Fig. 1.11(B) was the only convenient way we had of getting far targets because an optical collimator that was suitable for use with large head movements required a very large diameter, short focal length lens. We did not have one and even if we had had one, the configuration of the experimental set-up would have made it difficult to set it up properly. The view from the window was the way to go, but who goes to the Netherlands, expecting to be able to see farther than 10 or 20 m through the fog, rain and/or haze more than an hour or two, once or twice each month. June of 1978 was unusual in Rotterdam. The weather remained as clear as is shown in Fig. 1.11(B) day after day after day. Readers familiar with the climate of the Netherlands will appreciate that this was a supernatural event. Based on the author's prior and subsequent visits to Rotterdam, such weather was almost as miraculous and every bit as helpful as Moses' parting of the Red Sea in biblical times. The head and eye movements of 4 subjects recorded under these conditions are illustrated in Fig. 1.12.

Each recording began with low frequency, large amplitude head movements (the panel on the left) and continued through intermediate frequencies and reduced amplitudes (the middle panels) and ended with the subject oscillating his head as fast as possible (the panels at the right). These very high frequency, small amplitude oscillations were accomplished by clenching the teeth and tensing the platysmas muscles. The separation of what were, in fact, continuous recording runs in Fig. 1.12 was done for convenience during plotting. The head trace at the top of each subject's record shows head position divided by 10, the right eye trace is below the head trace, the left eye below it and the "vergence", that is, the difference between the gaze directions of each eye is shown in the lowest trace.

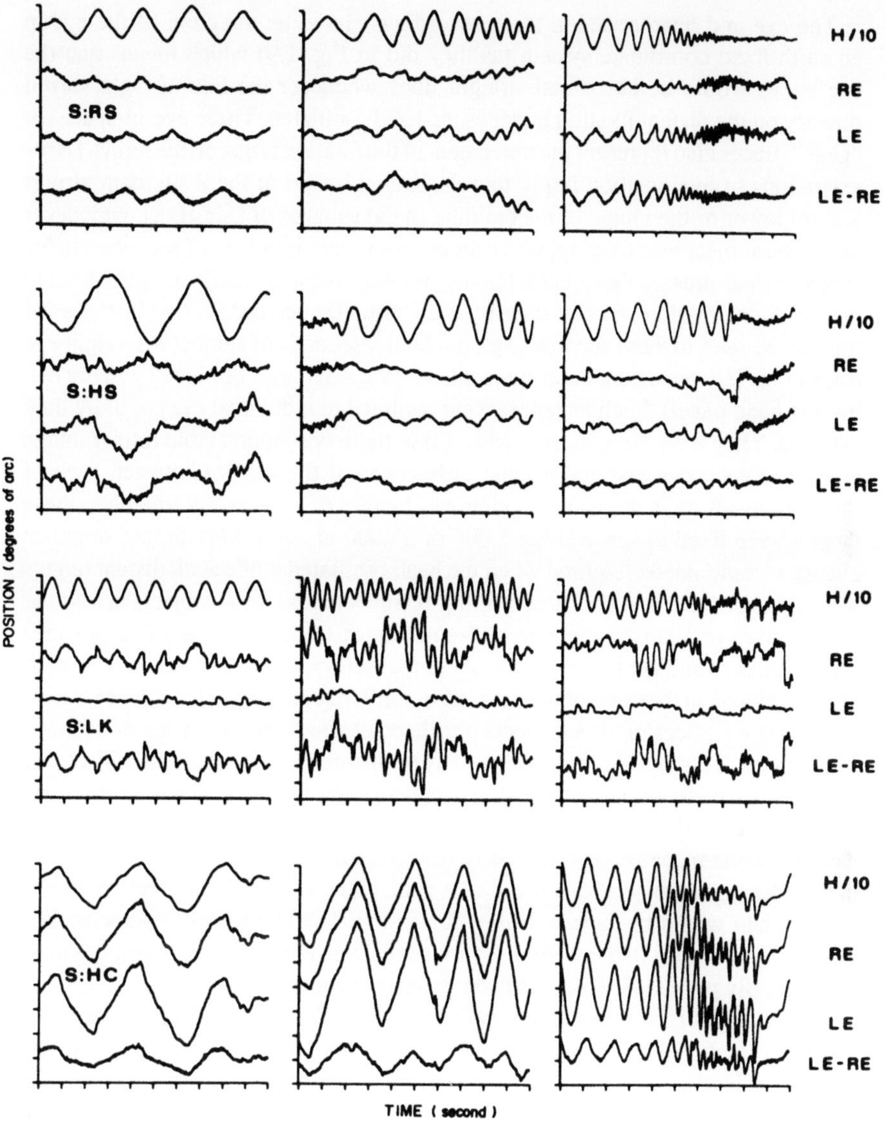

FIGURE 1.12. Representative recordings of horizontal head (H) and binocular gaze (retinal image position) of 4 subjects while they fixated a distant target and actively oscillated their heads about its vertical axis. Each of the 12 second records begins on the left and tick-marks on the abscissa indicate 1 sec intervals. Tick-marks on the ordinate indicate 1 degree of visual angle. The movements of the head have been scaled to 1/10 of their actual value and the trace LE-RE signifies vergence eye movements with convergence shown by upward changes in position. Upward changes of the right eye (RE), left eye (LE) and scaled head traces (H/10) signify movements to the right (from Steinman & Collewijn, 1980).

The eye and head traces in these recordings show the direction of the eye in an earth-fixed coordinate system (as they did in Fig. 1.9) which means that the eye traces would be horizontal straight lines whenever the line of sight stayed directly on the distant fixation target as the head oscillated. These eye-in-space (or "gaze") traces also represent the movement of the fixation target on the retina. Large excursions of eye position imply that the retinal image of the Rotterdam airport control tower or the image of the building in the suburbs of Den Haag, whichever target the subject was fixating, was dancing over large numbers of receptors in his foveal retinal mosaic. Very occasionally, retinal image motion was quite limited in extent, one might even say, stability was "virtually perfect". A few instances of this can be seen in these records (e.g., the final 4 seconds of subject RS's right eye trace in his left-hand panel and subject LK's first 4 seconds of his left eye trace in his left-hand panel). Such instances were confined to individual eyes of individual subjects. They were rare, indeed. Most often there was appreciable retinal image motion in each eye and appreciable differences of the motion between each of the eyes, resulting in large and continuous changes of vergence. Remember, these targets were fixed in space either 5,000 or 35,000 m away. Measurable vergence changes would not be required when the head oscillated while such distant targets were fixated. Note that the most striking features across all 4 subjects across the entire range of head frequencies and amplitudes were the large idiosyncracies shown by each subject in the degree of compensation for head movement by each of his eyes and, at the same time, the relative similarity among their vergence traces. The vergence traces of all 4 subjects oscillated at about the frequency of his head oscillations. This was accomplished in very different ways by each subject, for example, by having relatively modest compensation in each eye but, at the same time, a modest difference in compensation between them (subject HC), or by having excellent compensation in one eye and relatively incomplete compensation in the other eye (subjects RS and LK). Overall, there was much more retinal image motion than would be expected from conventional wisdom about compensatory eye movements during this period. Conventional wisdom had it that compensation was virtually perfect (see, e.g., Wilson & Melvill Jones, 1979, p.267).

Note that all of the subjects reported that they continued to see a phenomenally clear and stable world up until the very highest frequency oscillations, where the world remained clear but began to be seen as oscillating slightly. The reader should try this experiment himself by oscillating his head while fixating a distant target. The world will continue to look good under all conditions, providing only that the head oscillations remain in the range of frequencies and amplitudes illustrated in Fig. 1.12. We now know that perfection of oculomotor compensation is not the basis upon which the world looks fused and clear under such conditions. These results were quite unexpected by most, if not all, "experts" a decade ago. These days most, if not all, "experts" claim that they knew that it would come out the way that it did. Despite these claims, the fact remains that we are periodically asked to prove that our "generally expected" results do not reflect some artifact in our recording method (see Ferman, Collewijn, Jansen & van den Berg, 1987, for the most recent experiments designed to calm these fears).

The next direction our research would take was revealed in a phone call from New York. Bernie Cohen, from Mount Sinai (the medical school, not the mountain), was arranging the program for an International Meeting of the Barany Society to be held under the auspices of the New York Academy of Science in the Fall of 1980. The meeting would deal with oculomotor and vestibular physiology and we were invited to contribute something suitable. I am not sure after so many years whether I got the phone call or whether Han got it and then called me. In any case, the vestibular system was the way to go next. What should we do?

1.8 Retinal Image Slip Following Adaptation of the VOR

. . . was a perfect experiment. It was well within the theme of the meeting and was, at the same time, quite important for progress towards understanding natural retinal image motion. Our idea for the meeting followed, in part, from the work in Boston where we had proposed that the goal of oculomotor compensation might be some nonzero retinal image motion sufficient to prevent fading when the subject sat, as still as possible, without artificial supports for his head. It also followed on the work just completed in Rotterdam where the head had moved actively under relatively uncontrolled conditions while fixation was maintained on a distant target. Here I will describe what we did by quoting directly from our report (Collewijn et al., 1981): "First, we extended our results on natural retinal image motion to a larger sample of subjects, making these observations with passive, as well as active, rotations. Second, we measured binocular retinal image motion with sinusoidal stimulation of fixed frequencies and amplitude, allowing estimates of retinal image speed at each of several frequencies. Third, we determined the degree with which the VOR [the vestibulo-ocular response] (in the dark and when supplemented by vision) could be modified by changing the correlation between the amount of retinal image motion and a given degree of eye rotation. These adaptation experiments were undertaken to determine whether the observed departures from virtually perfect compensation arose from limitations inherent in the compensatory subsystems or from the desire of the compensatory subsystems to maintain retinal image motion at some nonzero value that might be optimal for vision" (p. 312). Fig. 1.13(A) illustrates the pre-adaptation natural eye and head movement records of a subject who had not served in the original experiments either in Boston or Rotterdam. His performance was similar to what had been seen earlier in other subjects in all important respects. Fig. 1.13(B) shows the results of experiments in which magnifying and minifying spectacles were used to require the VOR to adapt to novel arrangements. (See Collewijn et al., 1983; Steinman, Cushman & Martins, 1982, for more complete analyses then available in the original paper.)

The results illustrated in these figures were obtained at the University of Maryland at College Park where a new cube-surface, phase-detecting, eye and head monitoring instrument was under development. This instrument used primarily

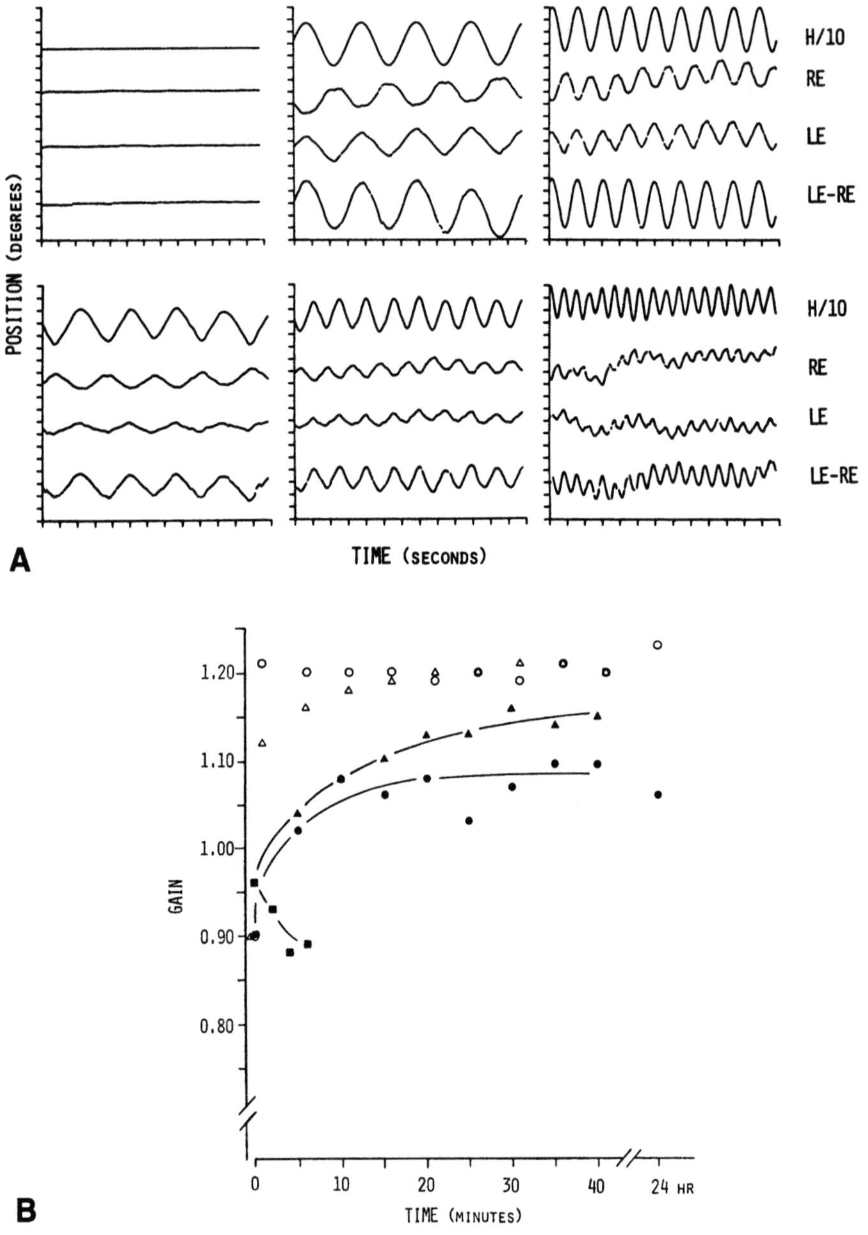

FIGURE 1.13. (A) Six representative records of horizontal head position and gaze (retinal image position) of Albert Martins, fixating a target while his head was passively or actively oscillated (from Collewijn et al., 1981). (B) VOR gain in the light (open symbols) and in the dark (closed symbols) as a function of adaptation time. Collewijn's performance is shown by circles, Martins' by triangles and Steinman's by squares.

digital circuitry and, a decade ago when this work was done, its noise level and "dead band" were less than 1 minute of arc, its bandwidth was 178 Hz (−3 dB) and its linearity was 1%. It only operated along the horizontal meridian, but its dynamic range on the horizontal was 360° with the characteristics just described. In short, this new instrument at an early stage of development was good enough to provide the best measurements of these kinds of phenomena available at the time. There were two findings that went beyond confirmation and extension of the general features of natural retinal image motion reported by Skavenski and colleagues (1979) and Steinman and Collewijn (1978, 1980). First, there was our discovery that the adaptation of the VOR (in the dark as well as in the light) was virtually complete in minutes rather than incomplete after days, providing the requirements for adaptation were kept modest, that is, less than 40%, rather than the +100% or −50% (i.e., 2 times or 1/2 magnification) requirement used by prior investigators. Second, we found that each subject adapted to his idiosyncratic level of natural retinal image motion (2–8% of head speed). When he did this, the experimental conditions had forced the amount of retinal image motion to pass through a value approximating zero retinal image motion during the course of adaptation. All 3 subjects tested in this experiment continued to adapt, ending at their preferred, clearly nonzero, value of retinal image motion they had before magnifying or minifying spectacles were put in front of their eyes. Both of our findings depended heavily on having much better instrumentation for measuring eye movements than had been used by prior investigators. All prior primate research on adaptation of the VOR had used EOG [electro-oculography], a poor technique for anything beyond crude clinical observations (see Collewijn et al., 1985, for a comparison of EOG with his sensor coil method and also for other references that highlight limitations of this outdated, but unfortunately still used, methodology.)

Time was running out on one of our often repeated claims. In Boston, in Rotterdam and now in College Park, we had found incomplete oculomotor compensation for head movement and had noted that the visual world remained phenomenally clear and stable despite all these perturbations of the retinal image. We had just reported an experiment designed to show explicitly that the goal of oculomotor compensation was some nonzero amount of retinal image slip and went on to claim that this motion might be beneficial for vision because it could prevent fading of the visual scene when a human being was not moving about. Where were the visual psychophysical measurements to support these claims? All we actually had were informal subjective reports of a phenomenally clear and stable world while we were moving. Where were our concurrent visual threshold measurements?

1.9 An Intramural Phone Call

... announced that it was time to "fish or cut bait." This call came from Jack Levinson, a colleague in my department's Program in Sensory and Perceptual Processes. Jack had been asked to organize a symposium on the detection of contrast in the presence of motion for the Annual Meeting of the Optical Society to be held in New Orleans in October, 1983. Jack phoned from across the hall in the Spring, and I agreed to participate with one proviso. Namely, I explained to Jack that we would need his help. It's one thing to say that we must study visual processing in the presence of natural retinal image motion to oculomotorists (or to journal referees with such background) and quite another to make visual psychophysical measurements properly and then get up in front of a group of hard-nosed professional visionaries and tell them how natural image motion affects basic visual processing. Collewijn and I (and our collaborators) were mere oculomotorists, members of a much younger specialty than physiological optics (Helmholtz's term) or visual science (a currently popular label). Oculomotoring is in its intellectual infancy. We don't even have real quantitative models. Our "models", such as they are, are only minor variations of circuit designs taught in the "How to do Electronics" manuals issued by the Armed Forces just after World War II (see Steinman, 1986b, for an evaluation of the value of the linear systems engineering approach introduced into the study of the oculomotor system in the decade following the end of the war). How could we advise a group, working in traditions begun by Helmholtz, Maxwell and Mach, about their own specialized research area. This was a tradition in which real natural scientists discussed real universal problems, not the just-emerging oculomotor traditions where fads launched by the half-baked ideas of the odd assortment of dentists, psychologists, physicians and engineers, who work on eye movements, were discussed. Jack (a genuine, card-carrying physicist) was reassuring. He promised to keep us from making fools of ourselves. He went even further and offered to make the measurements of contrast sensitivity in the presence of natural retinal image motion that we would need for his symposium. A set of these threshold measurements is summarized in Fig. 1.14(A–C).

These graphs show the contrast required for each of 3 subjects (Eileen Kowler, a; the author, b; and Han Collewijn, c) to detect the presence of a sinusoidally-modulated raster in a CRT display as a function of the spatial frequency of the bars. The parameter separating the 3 functions in each graph shows whether the head was supported artificially (triangles) or whether the subjects were oscillating their heads in time to a metronome beating at 1/3 or 4/3 Hz (Xs or diamonds, respectively). These particular subjects and particular paced head movements were used because we knew a lot about the natural retinal image motion they produced (Collewijn et al., 1981). Each subject had a typical contrast sensitivity function (CSF) when the head was supported artificially, namely, they were most sensitive somewhere between 3 and 6 cycles/degree and needed more contrast at lower spatial frequencies. Their high frequency cut-offs lay between 40 and 60 cycles/degree. The intersubject differences, which are apparent in Fig. 1.14(A–C), would be expected because a and c were myopes, who had worn spectacle corrections since childhood. This is known to make them relatively insensitive to high spatial frequencies when compared to b, who had no history of refractive errors.

The interesting results are to be found within each subject's performance, namely, by looking for differences between the CSFs measured with the head stationary and when it was moving. First, we see higher sensitivity for moving low

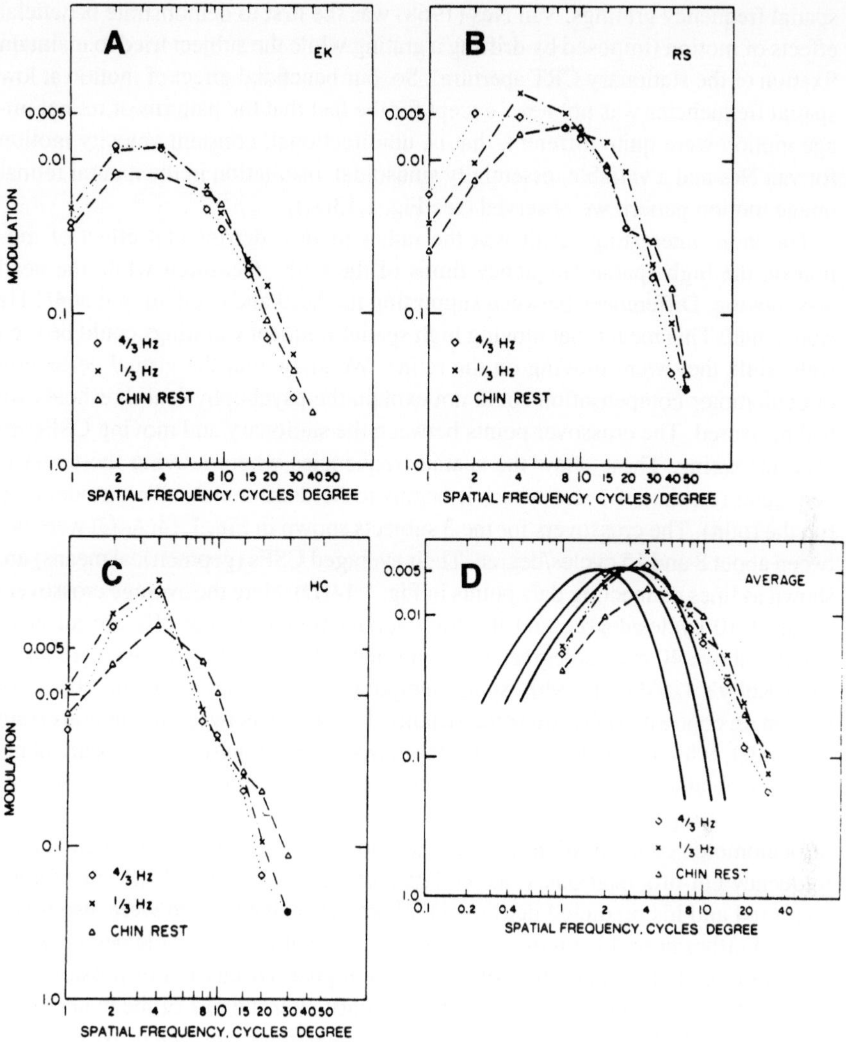

FIGURE 1.14. Threshold grating contrast-modulation settings as fractions of 100% contrast at various spatial frequencies. Three frequencies of head oscillations about its vertical axis were used, namely, near 0 with the head supported by a chin-rest or oscillating at 1/3 and 4/3 Hz. (A) shows the CSFs obtained under these conditions for subject EK, (B) for subject RS and (C) for subject HC. (D) shows their average CSF (the geometric mean) as broken lines with plotting symbols. Values and extrapolations from Kelly's model of the spatio-temporal transfer surface are shown as solid lines. These hypothetical functions are based on unidirectional motion of similar speed imposed on a grating "stabilized" with a Stage III, SRI Eyetracker. See the text for a discussion of these functions (from Steinman, Levinson, Collewijn & van der Steen, 1985).

spatial frequency gratings. Van Nes (1968) was the first to demonstrate beneficial effects of motion (imposed by drifting a grating while the subject tried to maintain fixation of the stationary CRT aperture). So, our beneficial effect of motion at low spatial frequencies was not news except for the fact that the patterns of retinal image motion were quite different, that is, unidirectional, constant velocity motion for van Nes and a variable, essentially sinusoidal, oscillation in the natural retinal image motion pattern we observed (see Fig. 1.13(A)).

The more interesting result was the rather modest detrimental effect of motion on the high spatial frequency limbs of the CSFs measured while the head was moving. Differences between supporting the head and oscillating it at 4/3 Hz were small. This meant that moving high spatial frequency gratings could be seen well while they were moving on the retina. We knew that the virtual perfection of oculomotor compensation could not explain the psychophysical thresholds we had measured. The crossover points between the stationary and moving CSFs are also interesting. These show the spatial frequencies where moving the target is beneficial (on the left) and where it begins to be detrimental, albeit moderately (on the right). The crossovers for the 3 subjects shown in Fig. 1.14(A–C) were between about 8 and 15 cycles/degree. Their averaged CSFs (geometrical means) are shown as lines connecting data points in Fig. 1.14(D). Here the average crossovers occur at 10 cycles/degree and the high spatial frequency cut-offs are all at or slightly above 40 cycles/degree. The solid lines plotted in Fig. 1.14(D) are taken from Kelly's (1979a, 1979b) "spatio-temporal surface model" of the effects of motion on contrast. There are some striking differences between predictions from his model, which was developed on the basis of psychophysical measurements of CSFs obtained after imposing constant velocity motion on a "stabilized" sinusoidal grating, and our results where motion was generated by natural failures of oculomotor compensation during head movement. His predicted high spatial frequency cut-offs (stationary, as well as moving!) were half of the cut-offs we measured and his predicted detrimental effects of image motion were also much larger. Furthermore, his predicted crossovers were at about 1 cycle/degree while ours were seen almost an order of magnitude higher. Discussion of possible reasons for these differences will be postponed until after the rest of the material we collected for the October 1983 Optical Society Meeting is described. We did at lot more than measure CSFs and, as will be seen, this other work was much more interesting than the most-likely reasons for the discrepancies between our results and the predictions of Kelly's (1979a, 1979b) model. We were taking Kelly's model much more seriously than it deserved when we first reported our measurements. Reasons for saying this will be given later.

The CSF measurements were made in College Park just before the annual ARVO meeting in 1983. Only psychophysical thresholds were measured at the time (i.e., no concurrent eye movement recordings were made) because the 1-D, phase-detecting system in College Park was in its final phase of development, which made it a unique, 2-D, phase-detecting instrument. Bandwidth and linearity were also improved in this period (effective bandwidth was raised to 244 Hz and linearity

A 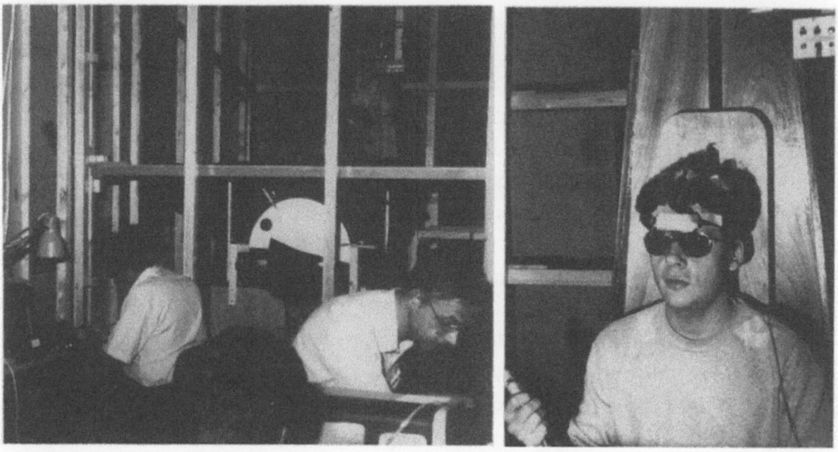 B

FIGURE 1.15. (A) Collewijn's large cube-surface field coil phase-detecting apparatus in Rotterdam. See the text for details. (B) Close-up of a subject in the apparatus shown in (A) wearing binocular eye coils beneath a pair of spectacles with a lowpass filter in front of one eye and a highpass filter in front of the other eye. These filters made it possible to present different stimuli to each eye, allowing psychophysical measurement of stereoacuity while head and eye movements were recorded.

made better than 0.01%). These improvements took several years and the first 2-D work with the completed instrument was published in 1988[2].

The rest of our data collection for the Optical Society Symposium was done in Rotterdam where Collewijn had just completed the new cube-surface field coil, 1-D phase-detecting system illustrated in Fig. 1.15. This was a large set-up, designed specifically for human beings, who could now sit in a comfortable chair near the center of the field coils rather than straddle a narrow wooden beam at the center of the much smaller rabbit apparatus. In the photo reproduced in Fig. 1.15(A), Han Collewijn (wearing spectacles) can be seen on the right. The back of Hans van der Steen's head (like Han, serving as an experimenter in this photo) can be seen in the foreground. I can be seen on the left side, just outside the field coils where I am preparing data sheets for psychophysical measurements of stereoacuity thresholds. The subject (Dusan Poboda) is faintly visible within the coils. Two slide projectors

[2]See Collewijn and colleagues (1988a, 1988b) for a description of the completed 2-D instrument, which can measure, simultaneously, by means of the phase-detection principle, rotations on the N-S and E-W vertical, as well as on the horizontal, meridian. Also, see Collewijn and colleagues (1990) and Kowler and colleagues (1990) for its first uses after the addition of a device, which uses acoustic propagation times, to measure translations of the head to better than 1 mm. This most recent addition finally made it possible to work with nearby targets while the head was free. This new capability paves the way for beginning to try to answer the question that Prof. Volkmann had posed over the telephone 16 years earlier. Also, see Edwards and colleagues (1994) for a complete treatment of the unique 3-D eye/head monitor now in use.

were located behind the large disk with 2 apertures that can be seen in front of and below the subject in Fig. 1.15(A). The output of these projectors passed through either a highpass (red appearing) or a lowpass (green appearing) gelatin filter that permitted separate visual input for each eye (<1% crosstalk) because the subject wore a matching pair of filters in front of his eyes. The filters in front of the slide projectors could be interchanged quickly by rotating the disk. This made it possible to vary the eye that received each half of a particular stereo-pair. Changing the slide shown to each eye had the effect of making the critical detail (a bright thin line surrounded by a haphazard arrangement of bright and dark rectangles) to appear either in front or in back of the plane of the rectangles when the display was fused. These arrangements allowed a forced-choice psychophysical technique to be used (i.e., the subject always responded, saying either "front" or "back" on all trials; he was forced to choose regardless of the confidence he had about the relative depth of the line target, whose disparity was varied randomly from trial to trial). This kind of psychophysical procedure is probably the easiest way of getting a reliable threshold estimate within the 40 minute session-length imposed by wearing silicone annuli. One of the 3 subjects, who is ready to participate in the experiment (i.e., wearing a head coil and binocular eye coils beneath a pair of goggles serving to mount the red and green filters), is shown in Fig. 1.15(B). During the experiment he either sat still or oscillated his head in pace with a metronome beating at 1/3, 2/3 or 4/3 Hz. The head and eye movements of the 3 subjects who served in these experiments are illustrated in Fig. 1.16(A). All showed considerable retinal image motion within each eye and considerable perturbation of vergence (differences between each eye's motion) — expected results in light of our prior work.

Fig. 1.16(B) shows their average vergence speed under the 4 conditions of movement and Fig. 1.16(C) shows their averaged psychophysical performance (i.e., percent correct reports of relative depth as a function of the disparity of the line target). Conventional thresholds (e.g., 75% correct) could not be measured with the disparities available. Our smallest disparity was 11.4 seconds of visual angle (somewhere between 1/2 to 1/3 the diameter of the finest human foveal cone receptor). All 3 subjects got the stereo-depth plane correct more than 80% of the time with this disparity both when they moved their heads and when they sat still. Moving the head with the resulting vergence perturbations had no measurable effect on the subjects' ability to report the relative depth of the line target. In fact, all 3 subjects said that it was easier to see the relative depth of the line when they moved the head then when they sat still. Their subjective impressions, however, did not show up in the likelihood of their making a correct response. Moving and sitting still gave essentially the same results.

To sum up, retinal image motion in each eye and differences in image motion between the eyes had no measurable effects on stereoacuity. The implications of these results for theories of stereovision are discussed in detail in Collewijn and Erkelens (1990) and Collewijn and colleagues (1990). Here, it is sufficient to realize that our results show that the human visual system not only solves what computer scientists and machine visionaries like to call the "correspondence problem", but

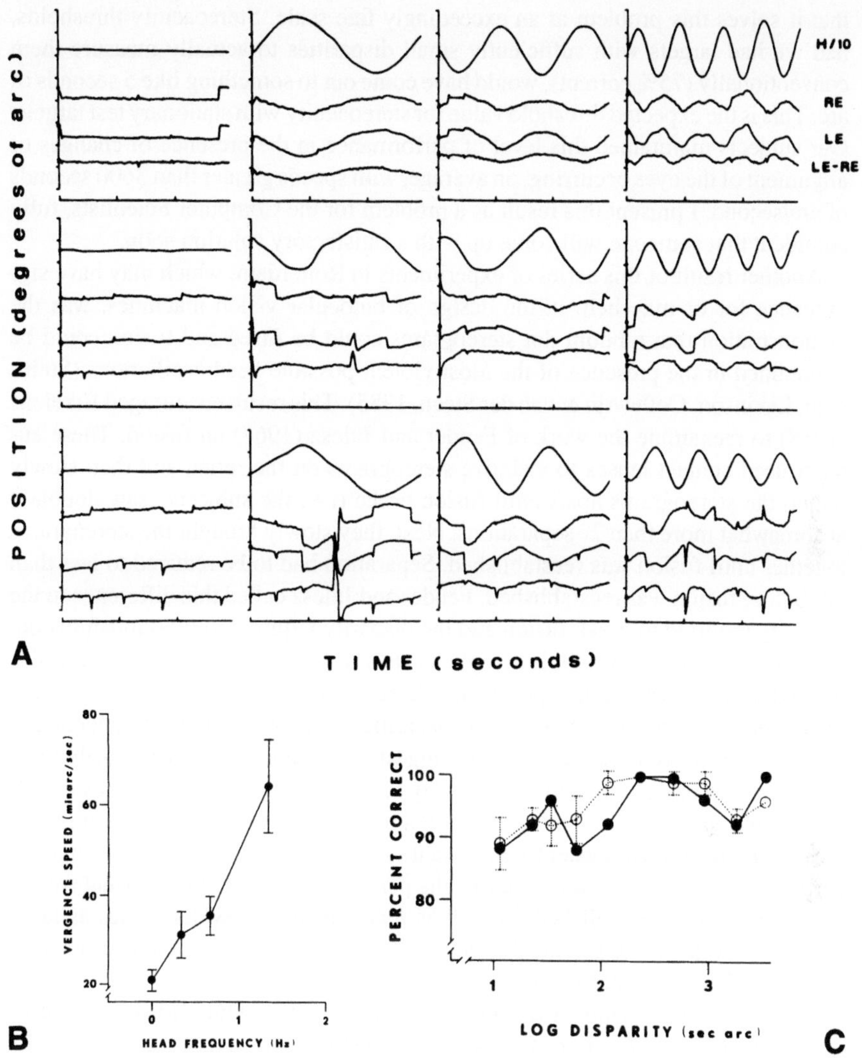

FIGURE 1.16. (A) Head and binocular eye movements of three subjects while psychophysical stereoacuity thresholds were measured during four conditions of head oscillation; sitting still on the left and oscillating about a vertical axis at 1/3, 2/3 and 4/3 Hz in the records to the right. LE-RE is vergence, the eye movement of significance in this experiment. See the text for details (from Steinman et al., 1985.). (B) Mean vergence speed (absolute velocity) as a function of head frequency for the three subjects whose eye and head movements are illustrated in (A). The error bars (standard deviations) show intersubject variability (from Steinman et al., 1985). (C) Accuracy of report (percent correct) as a function of the disparity of the stimuli. Data points show the averaged performance of the three subjects shown in (A) and (B). The closed symbols show accuracy when the subjects sat still. The open symbols show accuracy when they oscillated their heads. The error bars (standard deviations) show the variability associated with moving the head at three frequencies (from Steinman et al., 1985).

that it solves this problem at an exceedingly fine scale. Stereoacuity thresholds, had we had targets with sufficiently small disparities to actually measure them conventionally (75% correct), would have come out to something like 5 seconds of arc. This is the expected threshold value for stereoacuity with stationary test targets. Our subjects maintained this level of performance in the presence of changes of alignment of the eyes occurring, on average, with speeds greater than 3600 seconds of arc/second. I present this result as a problem for the Computer Scientists, fully confident that someone will come up with a satisfactory solution soon.

Another result of this series of experiments in Rotterdam, which may have significance for or may help in the design of binocular vision machines, was the demonstration that random dot stereograms could be fused and fusion could be maintained in the presence of the most violent possible head oscillations (Steinman, Levinson, Collewijn & van der Steen, 1985). This result encouraged Erkelens (1988) to reexamine the work of Fender and Julesz (1967) on fusion. These authors used contact lenses to stabilize stereograms on the retina and then slowly pulled the stereograms apart until fusion broke (i.e., the subjects "saw double") at somewhat more than 2° separations. Next, they slowly brought the stereograms together until fusion was reestablished. Separation had to be reduced to less than 10' before fusion was reestablished. Fender and Julesz called this difference in the disparity required to break fusion and the disparity required to reestablish fusion, "hysteresis". Erkelens was able to replicate these results but found that "hysteresis" depended on the particular experimental design, that is, it depended on the preceding fusional history. Fusion over large disparities was possible if brief, randomized tests were employed. Erkelens' result suggests that "hysteresis" is not likely to be significant for binocular vision in everyday life. Furthermore, Erkelens' results are compatible with our observations of natural binocular eye movements during head movement. It would be easier for the visual system to tolerate the kinds of vergence perturbations we observed (i.e., allow the percept of a single, fused visual world) if fusion could be established, as well as maintained, across disparities as large as 2° (see Collewijn & Erkelens, 1990; Collewijn et al., 1990, for a discussion of "hysteresis" in binocular vision).

Our report at the Optical Society Meeting was successful. No one booed or threw rotten vegetables. In fact, only a few seemed to realize that our novel results mattered a lot for the way the brain does visual processing. Why? Well, first off, no one in the audience, or elsewhere for that matter, was in a position to make the kind of measurements we had made. So, even if the listener believed our results, what could he do experimentally along these lines in his own lab? Also, even if he accepted our results and did theoretical, rather than experimental work, how could he model these results? Retinal images were moving too far and too fast on the retinas to be incorporated readily into contemporary models of visual processing. In time, what seemed at first to be benign acceptance gave way to benign skepticism. Our CSF measurements were the easiest to doubt because we had not at the time been able to make eye movement recordings while we measured psychophysical thresholds. Could it be that the CSFs, measured in the presence of natural retinal image motion, contained an artifact? Could we be sure that the subjects (ourselves) had followed the instructions and made threshold judgments near the center of their head oscillations when its velocity, and consequently, the gratings slip on the retina, was at its highest value? Maybe we cheated. In its more friendly form, one might say that we inadvertently failed to follow instructions and made our judgments when the head, and consequently the retinal image of the grating, slowed down just before and after the head changed direction. Said more succinctly, can we really be sure that we didn't sneak a peek as we turned about? If we had done this, our failure to find marked, adverse effects of image motion with high spatial frequency gratings would be an artifact. The high spatial frequency gratings were actually moving more slowly than we believed. This hypothesis is not as straightforward as it seems at first once it is realized that we would have had to use a different strategy with low spatial frequency gratings. Here, we would have had to make our judgments at the center of our head swings to obtain the enhancement of contrast sensitivity by motion known ever since van Nes (1968). Human beings can, of course, be this devious and this skillful. They might even be capable of behaving this way without realizing they were doing so.

This artifact was ruled out recently (Steinman & Levinson, 1990) by using the eye monitoring instrumentation at College Park to do control experiments that precluded such devious strategems. A dedicated microprocessor was used to calculate head and retinal image speed in real-time during oscillations of the head. The minimum amount of contrast available on the display was made contingent on how fast either the head or retinal image was moving. The experimenter set the maximum contrast available on a given trial when the head moved and used these settings to estimate contrast thresholds. The results of these experiments are summarized in Fig. 1.17. In Fig. 1.17(A), head speed set the lower limit of contrast available (i.e., No speed, no contrast. No contrast, no bars in the display). In Fig. 1.17(B), retinal image speed, rather than head speed, controlled contrast with the same consequences for the display. The experiment in which the lower limit of contrast was controlled by the head allowed very complete threshold measurements because no silicone annulus was worn on the eye and thresholds could be estimated in sessions that lasted 2 hours. Psychophysical sessions in which retinal image speed controlled contrast were much shorter (20–30 minutes). These thresholds are, therefore, necessarily noisier, but they, like the head thresholds, are good enough to rule out the possibility that our prior thresholds were artifactual (compare Figs. 1.17(A,B) with Fig. 1.14(B), which shows the same subject's CSFs when he could sneak a peek at turnabouts). The visual system does actually tolerate appreciable retinal image motion much more than has been observed in experiments in which constant velocity motion was imposed on a "stabilized" grating display.

Now that we were sure that we had no artifact in the main features of the CSFs in our original report, it becomes reasonable to ask why did natural retinal image motion have different effects on contrast sensitivity than imposed image motion? When we published our original work (Steinman et al., 1985), we sought explanation in subtle differences, such as differences in the waveforms of the natural and imposed motions and different light levels in the several experiments.

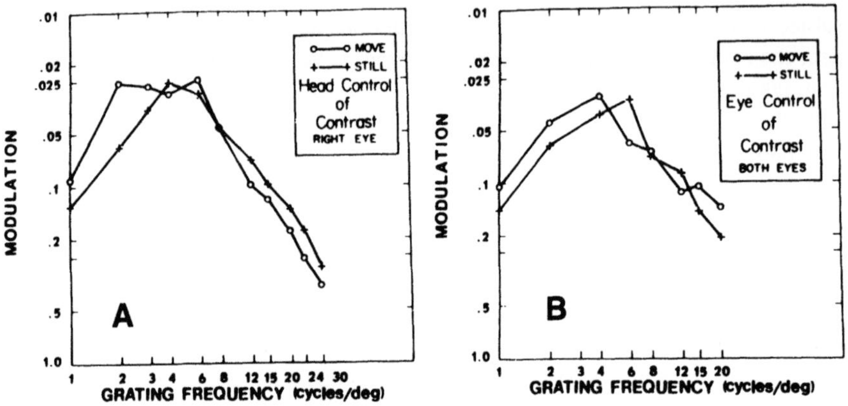

FIGURE 1.17. (A) Threshold grating contrast-modulation settings as fractions of 100% contrast at 11 spatial frequencies. These contrast sensitivity functions were obtained either while Steinman sat still, shown by the filled circles, or while he oscillated his head about its vertical axis, shown by the open circles. When he oscillated his head, the speed with which the head was moving determined the maximum available contrast. See the text for a discussion of these functions (from Steinman & Levinson, 1990). (B) Threshold grating contrast-modulation settings as fractions of 100% contrast at 8 spatial frequencies. These contrast sensitivity functions were obtained either while Steinman sat still, shown by the filled circles, or while he oscillated his head about its vertical axis, shown by the open circles. When he oscillated his head, the speed with which the retinal image was moving in either the right or left eye determined the maximum available contrast. The functions during movement are the mean of the threshold settings made with each eye at a separate session. See the text for a discussion of the significance of these functions and details about how these observations were made (from Steinman & Levinson, 1990).

Looking harder at this question over a period of several years, we now believe that the explanation lies in deficiencies in Kelly's (1979a, 1979b) stabilizing instrument. In short, we do not actually have any meaningful measurements of CSFs with imposed motion to compare with our measurements with natural motion. Kelly used a Stage III SRI Double Purkinje Image Eyetracker to stabilize his gratings prior to imposing constant velocity motion. This instrument cannot be used with sine wave gratings above 12 cycles/degree (only about 1/4 of the way to the normal high spatial frequency cutoff of the human visual system) in the most suitable subject described to date (Kelly himself). It has not been found to be useful above 7 or 8 cycles/degree in other subjects. Only a brief justification of these assertions is possible here and the reader is directed to Steinman and Levinson (1990) for a more complete treatment. I will start with a brief description of Kelly's stabilizing instrument.

The SRI Eyetracker is a very ingenious and complicated, noninvasive, instrument for measuring 2-D eye rotations. In principle, this makes it ideal for psychophysical research because very long periods of data collection, not possible with any invasive method, are possible. Its principle of operation is as follows: The SRI Eyetracker measures changes in the distance between the first Purkinje

Image (the light reflected from the cornea) and the fourth Purkinje Image (the light reflected from the concave surface at the back of the crystalline lens within the eye). When a collimated beam enters the eye, a real image of its source is formed, within the eye, by reflection from the concave surface of the crystalline lens. A virtual image of the source is also formed by the convex outer surface of the cornea. This virtual image from the cornea is also located within the eye in a position near to the virtual image formed by the crystalline lens. These reflecting surfaces, which form the first and fourth Purkinje Images, are at different distances from the center of rotation of the eye. This causes the distance between the first and fourth images to change as the eye rotates. This distance does not change when the eye translates relative to the source, allowing an instrument built on this principle to measure eye rotation free from translational artifacts. The degree, size and location of regions of relative linearity of the eye position output of this instrument depend on anatomical structures of each individual's eye.

The device is electromechanical with voltages proportional to eye position being derived from the X and Y position of mirrors mounted on 2-D servomotors that track the first and very faint fourth images so as to keep them centered on silicon quadrant detectors. The indications of eye position along the horizontal and vertical meridians come from piezo-electric crystals mounted on the tracking servo mirror mountings. These crystals must resolve displacements as small as 1 micrometer or better. There are many sources of noise inherent in such instrumentation, ranging from shot noise in the photodetectors to nonstationary variations of the reflecting properties of a given individual's eyes. It is not possible to determine the noise spectrum of the instrument independently of noise contributed by a particular individual's eyes.

Herein lies the main problem. It is easy to show (as Kelly, 1979a, did), by determining the highest spatial frequency that the instrument can stabilize sufficiently well to make an afterimage with discriminable bars, that inherent noise is such as to limit its use as a stabilizer to gratings of 12 cycles/degree in Kelly's eye, and 7 cycles/degree in his subject's eye. Subsequent work with the latest version of this instrument (Arend & Timberlake, 1986) also has found its use to be restricted to spatial frequencies under 10 cycles/degree. Accepting Kelly's data (1979a) and his model of the effects of motion on contrast sensitivity (Kelly, 1979b), as we did in our original report, is only plausible if one is ready to assume the SRI Eyetracker, which serves as a lowpass filter when used for stabilization, does not distort the shape of the CSFs measured with different amounts of imposed motion with spatial frequencies below 12 or 7 cycles/degree, depending on the subject. This is asking a lot, particularly inasmuch as the actual noise spectrum of the Eyetracker has not, and probably cannot be measured because it includes unpredictable contributions from properties of each individual subject's eye.

To sum up, the noise spectrum of the SRI Eyetracker can distort the CSFs in unknown (and probably unknowable ways) with lower spatial frequency gratings and it is useless across the most interesting range we studied (above 10 cycles/degree). We did not appreciate these limitations when we prepared our original report and the predictions from Kelly's model shown in Fig. 1.14(D) are, in retrospect, meaningless. Once this pointless comparison is put aside, we are left with an intriguing problem. Namely, how does the visual system manage to tolerate (or, perhaps, even put to good use) the relatively large, fast displacements of light distributions on the retinal surface that accompany all natural bodily movements? The new and particularly intriguing fact, here, is that this tolerance for motion extends down to the lowest, most primitive level of visual processing, that is, the discrimination of threshold differences of contrast in a single eye. This discovery is harder to understand than the tolerance we found for vergence perturbations on stereoacuity thresholds or the fusion of random dot stereograms. These situations are far less mysterious because relative positions of relevant target details are preserved when vergence changes. If the important stimulus feature underlying the percept is the relative positions of target details, vergence perturbations are of little or no consequence and should, therefore, be tolerated (see Collewijn & Erkelens, 1990; and Collewijn, Steinman, Erkelens & Regan, 1992, for elaboration of this point). But note, both stereoacuity and stereofusion are binocular accomplishments that require processing at relatively high levels in the visual system after input from the two eyes is combined. The tolerance of appreciable retinal image motion contrast thresholds within a single eye is harder to understand at this point in time.

It is hard to predict, with any degree of certainty, what the future holds for the role of eye movement in visual processing because this role has changed so frequently during the past 136 years. At present it is possible to say with reasonable certainty that eye movement is useful for seeing very sharp edges or small details we wish to inspect for more than 2/10ths of a second. It is also possible to say with reasonable certainty that the retinal image is, fortunately, in no danger of standing still in any natural situation. It is even possible to assert with reasonable certainty that vision continues to function at near capacity levels without virtual perfection of oculomotor control. Finally, it is possible to say with reasonable certainty that the role of eye movement in maintaining a phenomenally clear and stable world can now be studied with techniques that may, one day, make it possible to do more then shake one's head or look upwards with raised, outstretched arms whenever the telephone rings.

Acknowledgments: Supported by grants from the Life Sciences Directorate of the Air Force Office for Scientific Research: AFOSR 91-0124 and F49620-94-1-0333.

1.10 References

Adler, F. H. & Fliegelman, M. (1934). Influence of fixation on the visual acuity. *Archives of Ophthalmology, 12*, 475–483.

Arend, L. E. & Timberlake, G. T. (1986). What is perfect stabilization? do perfectly stabilized images always disappear? *Journal of the Optical Society of America A, 3*, 235–241.

Averill, H. L. & Weymouth, F. W. (1925). Visual perception and the retinal mosaic. ii. the influence of eye movements on the displacement threshold. *Journal of Comparative Psychology*, *5*, 147–176.

Barlow, H. B. (1963). Slippage of contact lenses and other artefacts in relation to fading and regeneration of supposedly stabilized images. *Quarterly Journal of Experimental Psychology*, *15*, 36–51.

de Bie, J. (1985). An afterimage vernier method for assessing the precision of eye movement monitors: results for the scleral search coil technique. *Vision Research*, *25*, 1341–1343.

Boring, E. G. (1942). *Sensation and Perception in the History of Experimental Psychology*. New York: Appleton-Century-Crofts.

Collewijn, H. (1977). Eye and head movements in freely moving rabbits. *Journal of Physiology*, *266*, 471–498.

Collewijn, H. & Erkelens, C. J. (1990). Binocular eye movements and the perception of depth. In E. Kowler (Ed.), *Eye Movements and Their Role in Visual and Cognitive Processes*, Volume 4, Reviews of Oculomotor Research (pp. 213–261). Amsterdam: Elsevier.

Collewijn, H., Erkelens, C. J. & Steinman, R. M. (1988a). Binocular co-ordination of human horizontal saccadic eye movements. *Journal of Physiology*, *404*, 157–182.

Collewijn, H., Erkelens, C. J. & Steinman, R. M. (1988b). Binocular co-ordination of human vertical saccadic eye movements. *Journal of Physiology*, *404*, 183–197.

Collewijn, H., van der Mark, F. & Jansen, T. C. (1975). Precise recording of human eye movements. *Vision Research*, *15*, 447–450.

Collewijn, H., Martins, A. J. & Steinman, R. M. (1981). Natural retinal image motion: origin and change. *Annals of the New York Academy of Science*, *374*, 312–329.

Collewijn, H., Martins, A. J. & Steinman, R. M. (1983). Compensatory eye movements during active and passive head movements: fast adaptation to changes in visual magnification. *Journal of Physiology*, *340*, 259–286.

Collewijn, H., van der Steen, J. & Steinman, R. M. (1985). Human eye movements associated with blinks and prolonged eyelid closure. *Journal of Neurophysiology*, *54*, 11–27.

Collewijn, H., Steinman, R. M., Erkelens, C. J., Pizlo, Z. & van der Steen, J. (1990). The effect of freeing the head on eye movement characteristics during 3-d shifts of gaze and tracking. In A. Berthoz, P. P. Vidal & W. Graf (Eds.), *The Head-Neck Sensory Motor System: Evolution Development Disorders and Neuronal Mechanisms* (pp. 412–418). London: Oxford University Press.

Collewijn, H., Steinman, R. M., Erkelens, C. J. & Regan, D. (1992). Binocular fusion, stereopsis and stereoacuity with a moving head. In D. Regan (Ed.), *Vision and Visual Dysfunction*, Volume 10A: Binocular Vision (pp. 121–136). London: MacMillan.

Ditchburn, R. W. (1973). *Eye Movements and Visual Perception*. Oxford: Claredon Press.

Edwards, M. E., Pizlo, Z., Erkelens, C. J., Collewijn, H., Epelboim, J., Kowler, E., Stepanov, M. & Steinman, R. M. (1994). The Maryland revolving-field monitor: Theory of the instrument and processing its data. Technical Report CAR-TR-711, Center for Automation Research, University of Maryland.

Epelboim, J., Collewijn, H., Edwards, M. E., Erkelens, C. J., Kowler, E., Pizlo, Z. & Steinman, R. M. (1993). Coordination of eyes head and hand in a natural 3-d tapping task. *Investigative Ophthalmology and Visual Science Supplement*, *34*, 1502.

Epelboim, J., Collewijn, H., Edwards, M. E., Erkelens, C. J., Kowler, E., Pizlo, Z. & Steinman, R. M. (1994a). Coordinated movements of the arm and head increase gaze-shift velocity. *Investigative Ophthalmology and Visual Science Supplement, 35*, 1550.

Epelboim, J., Collewijn, H., Kowler, E., Erkelens, C. J., Edwards, M. E., Pizlo, Z. & Steinman, R. M. (1994b). Natural oculomotor performance in looking and tapping tasks. In *Proceedings of the Sixteenth Annual Conference of the Cognitive Science Society* (pp. 272–277). Hillsdale, NJ: Erlbaum.

Erkelens, C. J. (1988). Fusional limits for a large random-dot stereogram. *Vision Research, 28*, 345–353.

Erkelens, C. J., van der Steen, J., Steinman, R. M. & Collewijn, H. (1989a). Ocular vergence under natural conditions. i. continuous changes of target distance along the median plane. *Proceedings of the Royal Society, London B, 236*, 417–440.

Erkelens, C. J., Steinman, R. M. & Collewijn, H. (1989b). Ocular vergence changes under natural conditions ii. gaze shifts between real targets differing in distance and direction. *Proceedings of the Royal Society, London B, 236*, 441–465.

Falk, J. L. (1956). Theories of visual acuity and their physiological bases. *Psychological Bulletin, 53*, 109–133.

Fender, D. & Julesz, B. (1967). Extension of Panum's fusional area in binocularly stabilized vision. *Journal of the Optical Society of America, 57*, 819–830.

Festinger, L. (1971). Eye movements and perception. In P. Bach y Rita & C. C. Collins (Eds.), *The Control of Eye Movements* (pp. 259–273). New York: Academic Press.

Fritsch, G. (1908). *Über Bau und Bedeutung der Area Centralis der Menschen.* Berlin.

Gerrits, H. J. M. & Vendrik, A. J. H. (1970). Artificial movements of a stabilized image. *Vision Research, 10*, 1443–1456.

Gerrits, H. J. M. & Vendrik, A. J. H. (1972). Eye movements necessary for continuous perception during stabilization of retinal images. *Bibl. Ophthalmol., 82*, 339–347.

Gerrits, H. J. M. & Vendrik, A. J. H. (1974). The influence of stimulus movements on perception in parafoveal stabilized vision. *Vision Research, 14*, 175–180.

Hartline, H. K. (1938). The response of single optic nerve fibers of the vertebrate eye to illumination of the retina. *American Journal of Physiology, 121*, 400–415.

Hartline, H. K. (1940). The receptive field of the optic nerve fibers. *American Journal of Physiology, 130*, 690–699.

Hartmann, R. & Klinke, R. (1976). A method for measuring the angle of rotation (movements of body head eye in human subjects and experimental animals. *Pflugers Archives von Physiologie, 362 Suppl.*, R52.

Hartridge, H. (1922). Visual acuity and the resolving power of the eye. *Journal of Physiology, 57*, 52–67.

Hebb, D. O. (1949). *The Organization of Behavior*. New York: John Wiley and Sons.

Hecht, S. (1927). A quantitative basis for the relation between visual acuity and illumination. *Proceedings of the National Academy of Science, 13*, 569–574.

Hecht, S. (1928). The relation between visual acuity and illumination. *Journal of General Physiology, 11*, 255–281.

Hecht, S. & Mintz, E. (1939). The visibility of single lines at various illuminations and the retinal basis of visual resolution. *Journal of General Physiology, 22*, 593–612.

von Helmholtz, H. L. F. (1866). *Handbuch der Physiologischen Optik*, Volume 2. Hamburg-Leipzig: Voss.

Hering, E. (1899). Über die grenzen der sehschärfe. *Berichte der mathematisch-physicalischen Klasse der Königlichen Gesellschaft der Wissenschaft zu Leipzig* (pp. 16–24).

Hering, E. (1920). *Grundzuge der Lehre vom Lichtsinn*. Berlin: Springer.

Hochberg, J. (1964). *Perception*. Old Tappan, NJ: Prentice-Hall.

Hubel, D. H. & Wiesel, T. N. (1962). Receptive fields binocular interaction and functional architecture in the cat's visual cortex. *Journal of Physiology*, *160*, 106–126.

Hubel, D. H. & Wiesel, T. N. (1963). Receptive fields of cells in striate cortex of very young visually inexperienced kittens. *Journal of Neurophysiology*, *26*, 994–1011.

Kelly, D. H. (1979a). Motion and vision I. stabilized images of stationary targets. *Journal of the Optical Society of America*, *69*, 1266–1274.

Kelly, D. H. (1979b). Motion and vision II. stabilized spatio-temporal transfer surface. *Journal of the Optical Society of America*, *69*, 1340–1349.

Kowler, E., Pizlo, Z., Zhu, G. L., Erkelens, C. J., Steinman, R. M. & Collewijn, H. (1990). Coordination of head and eyes during the performance of natural(and unnatural) visual tasks. In A. Berthoz, P. P. Vidal & W. Graf (Eds.), *The Head-Neck Sensory Motor System: Evolution Development Disorders and Neuronal Mechanisms* (pp. 419–426). London: Oxford University Press.

Kowler, E. & Steinman, R. M. (1980). Small saccades serve no useful purpose: reply to a letter by R. W. Ditchburn. *Vision Research*, *20*, 273–276.

Krauskopf, J. (1957). Effect of image motion on contrast thresholds for maintained vision. *Journal of the Optical Society of America*, *47*, 740–745.

Krauskopf, J. (1962). Effect of target oscillation on contrast resolution. *Journal of the Optical Society of America*, *52*, 1306.

Krauskopf, J. (1963). Effect of retinal image stabilization on the appearance of heterochromatic targets. *Journal of the Optical Society of America*, *53*, 741–744.

Le Grand, Y. (1967). *Form and Space Vision*. Bloomington: Indiana University Press. Revised edition, translated by M. Millodot and G. Heath.

Lotze, R. H. (1852/1966). *Medicinische Psychologie*. Amsterdam: E. J. Bonet.

Marshall, W. H. & Talbot, S. A. (1942). Recent evidence for neural mechanisms in vision leading to a general theory of sensory acuity. *Biological Symposium*, *7*, 117–164.

Murphy, B. J., Haddad, G. M. & Steinman, R. M. (1974). Simple forms and fluctuations of the line of sight: Implications for motor theories of form processing. *Perception & Psychophysics*, *16*, 557–563.

van Nes, F. L. (1968). Enhanced visibility by regular motion of the retinal image. *Journal of the Optical Society of America*, *57*, 401–406.

Polyak, S. L. (1941). *The Retina*. Chicago: University of Chicago Press.

Ratliff, F. & Riggs, L. A. (1950). Involuntary motions of the eye during monocular fixation. *Journal of Experimental Psychology*, *40*, 687–701.

Riggs, L. A. (1965). Visual acuity. In C. H. Graham (Ed.), *Vision and Visual Perception* (pp. 321–349). New York: John Wiley and Sons.

Riggs, L. A., Ratliff, F., Cornsweet, J. C. & Cornsweet, T. N. (1953). The disappearance of steadily-fixated objects. *Journal of the Optical Society of America, 43*, 495–501.

Rubens, S. R. (1945). Cube-surface coil for producing a uniform magnetic field. *Review of Scientific Instrumentation, 16*, 243–245.

Skavenski, A. A., Hansen, R. H., Steinman, R. M. & Winterson, B. J. (1979). Quality of retinal image stabilization during small natural and artificial body rotations in man. *Vision Research, 19*, 365–375.

Sperling, G. (1990). Comparison of perception in the moving and stationary eye. In E. Kowler (Ed.), *Eye Movements and Their Role in Visual and Cognitive Processes*, Volume 4, Reviews of Oculomotor Research (pp. 307–351). Amsterdam: Elsevier.

Steinman, R. M. (1975). Oculomotor effects on vision. In P. Bach y Rita & G. Lennerstrand (Eds.), *Basic Mechanisms of Ocular Motility and Their Clinical Implications* (pp. 395–416). Oxford-New York: Pergamon Press.

Steinman, R. M. (1976). Role of eye movements in maintaining a phenomenally clear and stable world. In R. A. Monty & J. W. Senders (Eds.), *Eye Movements and Psychological Processes* (pp. 121–149). Hillsdale, New Jersey: Lawrence Erlbaum Associates.

Steinman, R. M. (1986a). Eye movement. *Vision Research, 26*, 1389–1400.

Steinman, R. M. (1986b). The need for an eclectic rather than a systems approach to the study of the primate oculomotor system. *Vision Research, 26*, 101–112.

Steinman, R. M. & Collewijn, H. (1978). How our two eyes are held steady. *Journal of the Optical Society of America, 68*, 1359.

Steinman, R. M. & Collewijn, H. (1980). Binocular retinal image motion during natural active head rotation. *Vision Research, 20*, 415–429.

Steinman, R. M., Cushman, W. B. & Martins, A. J. (1982). The precision of gaze. *Human Neurobiology, 1*, 97–109.

Steinman, R. M., Haddad, G. M., Skavenski, A. A. & Wyman, D. (1973). Miniature eye movements. *Science, 181*, 810–819.

Steinman, R. M. & Levinson, J. Z. (1990). The role of eye movements in the detection of contrast and spatial detail. In E. Kowler (Ed.), *Eye Movements and their Role in Visual and Cognitive Processes* (pp. 115–212). Amsterdam: Elsevier.

Steinman, R. M., Levinson, J. Z., Collewijn, H. & van der Steen, J. (1985). Vision in the presence of known natural retinal image motion. *Journal of the Optical Society of America A, 2*, 226–233.

Talbot, S. A. & Marshall, W. H. (1941). Physiological studies of neural mechanisms of visual localization and discrimination. *American Journal of Ophthalmology, 24*, 1255–1264.

Volkmann, F. C. (1986). Human visual suppression. *Vision Research, 26*, 1401–1416.

Walls, G. L. (1943). Factors in human visual resolution. *Journal of the Optical Society of America, 33*, 487–505.

Westheimer, G. (1981). Visual hyperacuity. *Progess in Sensory Physiology, 1*, 1–30.

Weymouth, F. W., Andersen, E. E. & Averill, H. L. (1923). Retinal mean local sign; a new view of the relation of the retinal mosaic to visual perception. *American Journal of Physiology, 63*, 410–411.

Wilcox, W. W. & Purdy, D. M. (1933). Visual acuity and its physiological basis. *British Journal of Psychology, 23*, 233–261.

Wilson, V. J. & Melvill Jones, G. (1979). *Mammalian Vestibular Physiology*. New York: Plenum Press.

Wundt, W. (1910). *Grundzüge der Physiologischen Psychologie* (5th Ed.), Volume 2. Leipzig: W. Engelmann.

Yarbus, A. L. (1957a). A new method for studying the activity of various parts of the retina. *Biophysics, 2*, 165–167.

Yarbus, A. L. (1957b). The perception of an image fixed with respect to the retina. *Biophysics, 2*, 683–690.

Yarbus, A. L. (1967). *Eye Movements and Vision*. New York: Plenum Press. Translated by B. Haigh and L. A. Riggs.

2

Cogito Ergo Moveo: Cognitive Control of Eye Movement

Eileen Kowler[1]

ABSTRACT
Eye movements serve the needs of vision. They are used to bring images of eccentric targets to the central fovea, where visual acuity is best, and to keep images sufficiently stable on the retina to allow accurate visual processing. Most models assume that eye movements are driven by low-level sensory signals, such as retinal image position or retinal motion. This chapter shows that eye movements are driven by a higher-level signal, incorporating cognitive states, such as selective attention and expectations. By relying on high-level signals, eye movements become inextricably tied to concurrent cognitive activities, helping ensure that the eye will be directed to objects of current interest with little or no additional effort on the part of the observer. Cognitive control of eye movements, which is so prominent in human beings, may also be a useful principle to apply to the design of robots.

2.1 Introduction

The task of recognizing objects in natural visual environments is made enormously difficult by the existence of the fovea: the small (1.5 deg diameter), central, retinal region where visual acuity is best. The visual limitations created by the existence of the fovea make an effective pattern of eye movements essential if we are to be able to process visual information over extended regions of space. Eye movements are needed both to shift gaze to objects of interest, bringing the selected portion of the retinal image to the fovea, and to maintain stable gaze on selected targets as they, or as we, move about. Mobile artificial visual systems similarly plagued with a limited area of high-quality vision must also have appropriate means of controlling the movements of sensors in order to recognize objects in the surrounding visual environment.

Research on human eye movements over the past 40 years or so — going back to the beginning of the modern era of accurate instrumentation for eye movement recording — has documented an impressive array of oculomotor skills, which at

[1]Department of Psychology, Rutgers University

first glance, would seem to be more than adequate to accomplish the tasks confronting us when we inspect the visual world. For example, when we try to maintain a stable line of sight on a stationary target, the standard deviation of eye position is only 2-3′ when the head is artificially supported, and increases to about 30′ when head movements are permitted (Steinman, Cushman & Martins, 1982). These values are small enough to keep the target image well within the foveal floor. If the target is set in motion at a constant velocity, smooth eye movements can follow along quite accurately, with eye velocity reaching values that can be as high as 90% or more of the velocity of the target (Collewijn & Tamminga, 1984; Kowler, Murphy & Steinman, 1978). If the target should change its position abruptly, or if we should take an interest in another object, a single saccadic eye movement can take the line of sight to the new position of the target with considerable accuracy (Collewijn, Erkelens & Steinman, 1988; Kowler & Blaser, 1995; Lemij & Collewijn, 1990). Saccades are not only accurate, but are also precise: Standard deviations of saccade offset error are only about 5% of the size of the saccade (Aitsebaomo & Bedell, 1992; Kowler & Blaser, 1995).

The oculomotor skills described above have been demonstrated in laboratory environments, usually containing a simple target (often a point of light) seen against a dark, or otherwise homogeneous, visual background while the head is stabilized by a biteboard or a chinrest. (Head support is required by most oculomotor instrumentation with sensitivity better than about 0.5 degree; for an exception to the general rule, see Steinman, this volume, Chapter 1.) Despite the obviously unnatural aspects of these testing conditions, the high degree of oculomotor skill that has been demonstrated would seem to be adequate, if not optimal, in more natural settings for providing a stable retinal platform for the subsequent analysis needed to recognize visual patterns. I say "subsequent analysis" because, according to most popular oculomotor models, oculomotor skill does not depend on high-level visual analysis, but instead precedes it. The control of eye movements is assumed to be a reflexive process based on fairly primitive signals — such as retinal position, retinal velocity, or vestibular signals — that are coded by low-level sensory mechanisms and transformed, more or less automatically, into the appropriate oculomotor commands (for a review of such approaches, see Carpenter, 1991; Hallett, 1986; Kowler, 1990). These commands are designed to minimize the putative "error" signals (e.g., retinal slip velocity or retinal position error) and create the retinal image conditions that are presumably ideal for vision. This is an attractive and popular story, but problems arise as soon as we begin to ask whether these mechanisms would be able to explain the performance of intelligent, mobile beings exploring richly patterned visual environments. When we ask such questions, we discover that effective oculomotor performance is not achieved by reflexive responses to sensory "error" signals. Instead, effective oculomotor responses require the involvement of cognitive processes, most notably, selective attention and expectations about the future motion or position of targets.

This chapter considers four examples of the cognitive control of eye movements. The first is the selection of the target for smooth eye movements from a crowded visual environment. The second is the effect of cognitive expectations about the future motion of targets on smooth eye movements. The third is the role of cognitive planning in the programming of sequences of saccadic eye movements. The fourth is the selection of the target for saccades. These examples will all show that cognitive processes play important and unavoidable roles in the control of eye movements, both in the selection of the input and in the formulation of the oculomotor command. Their contribution is not tangential or ephemeral — that is, cognitive processes do not simply make minor adjustments to already-formed reflexive responses. If anything, the story is the reverse: sensory cues operate on a system that, to use Lashley's (1951) expression, is already "actively excited and organized" to respond in particular ways.

2.2 Example 1: Selection of the Target for Smooth Eye Movements

2.2.1 Smooth eye movements in the presence of visual backgrounds

Smooth eye movements are not under voluntary control. By this I mean that in the presence of nothing but stationary objects, observers (with the exception of a few very rare individuals) are not able to move the eye smoothly across the visual field. Similarly, if the visual field contains nothing but smoothly moving objects, the observer cannot keep the eye stationary. It is dragged along smoothly in the direction of the stimulus motion. This observation implies that whether the eye is relatively stationary or moving smoothly in one or another direction depends not on voluntary choice, but rather on what sort of stimulus motion is present in the visual field. This is consistent with traditional reflexive models of smooth eye movement in which stimulus motion is detected, coded, and automatically transformed into the appropriate smooth following commands.

How would such reflexive, smooth-following mechanisms behave in the typical visual environment, in which a variety of stationary and smoothly moving targets are present at the same time? There are several possibilities. One is that the stimulus motion signals originating from the most intense target, or from the target that happens to fall on the fovea, might dominate and constitute the effective stimulus for smooth eye movements. Alternatively, all the available motion signals might be pooled, causing the eye to track at a direction and speed determined by the average of all the signals. Both of these schemes are consistent with models, described above, in which stimulus motion is coded and automatically transformed into the smooth oculomotor command because for both schemes the smooth response of the eye is determined by nothing but the properties of the stimulus.

Neither of these schemes is correct. Observers do not track the motion of the strongest target, nor do they track the average of the available motions. They track the target they select.

Selective capacity was studied by Kowler, van der Steen, Tamminga and Collewijn (1984b). In this experiment subjects were confronted with two identical, full-field patterns of randomly positioned dots. One field was stationary and the other moved to the left at 1 deg/sec. The subjects were asked to maintain a steady line of sight on either one of the two fields while ignoring the other, superimposed field. The two fields were the same intensity and density (either 1 or 8 dots/deg^2) so that one field would not constitute a stronger stimulus than the other. The rest of the stimulus attributes were chosen to make the task as difficult as possible and, therefore, provide a good test of selective capacity. For example, cues, such as differences in depth planes of the fields, which might have aided the selection process, were absent. Also, the velocity of the moving field was set low enough so that dots would be seen clearly (Murphy, 1978; Westheimer & McKee, 1975). And, the denser field used (8 dot/deg^2) was sufficiently dense so that subjects could not isolate a single dot moving against a blank portion of the background. In the dense stimulus, dots from one field were continually passing across dots from the other.

Fig. 2.1(A) shows that smooth eye movements were as effective in maintaining the line of sight on the chosen field regardless of whether the field was presented alone (left-hand eye movement records) or with the superimposed background field (right-hand records). These records are representative of overall performance. The overall effect of the background on eye velocity amounted only to about 2–4% of the velocity of the moving field (Fig. 2.1(B–C)). Similar results were recently reported by Niemann, Ilg and Hoffman (1994).

How was the influence of the background field prevented? It was not done by the subjects choosing to track the dots at a particular location in the visual field because dots from each display were everywhere. (Too bad: This would have been an easy selection rule to incorporate into oculomotor models because sensory velocity signals could be tagged according to retinal location.) The influence of the background field could not have been prevented by selecting one or another stimulus velocity because the velocity the subjects perceived was very different from the velocity of the eye. Subjects perceived vivid induced motion of the stationary field when the moving background was present, but induced motion had no effect on the smooth eye movements, as can be seen in the upper right-hand graph of Fig. 2.1(A) (see also Mack, Fendrich & Wong, 1982). The remaining alternative is that the subjects chose what to track, and not how fast to move the eye. For such a selection process to work, the two fields must first be segregated into distinct "objects" (a process not yet understood with transparent fields such as these that move in the same apparent depth plane). Then, after the segregation occurs, motion signals from each field must be appropriately tagged so that only the selected set is permitted to reach the smooth oculomotor circuitry.

Understanding segregation and selection will not prove to be an easy task. To appreciate only one of the challenges, realize that the voluntary component of the selection process, the one that decides which object to track, is doing so based on motion information that is highly sensitive to the relative motion of the different fields (recall the strong impression of induced motion noted above). At the same

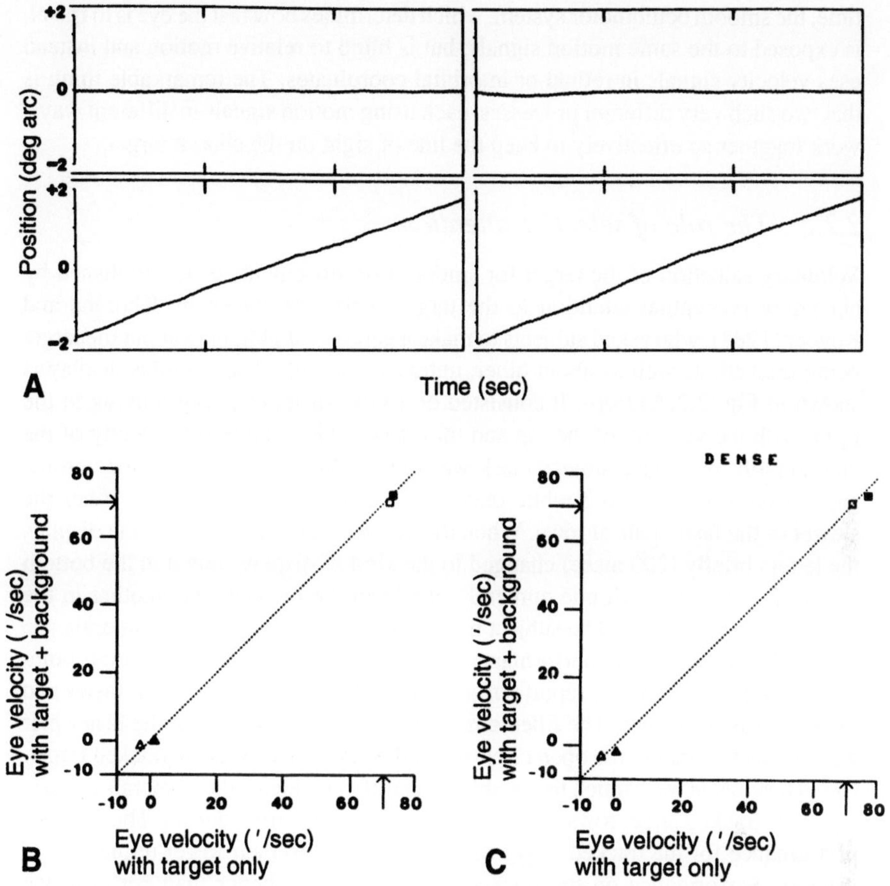

FIGURE 2.1. (A) Representative records of horizontal eye movements for subject RS under instructions to maintain the line of sight on the stationary (top 2 graphs) or moving (bottom 2 graphs) field of random dots. In the left-hand graphs, only one field was present; in the right-hand graphs, both were presented superimposed. Tic marks on the x-axis separate 1 sec intervals. Upward deflections of the eye trace indicate movements to the left. (B) Mean 21 msec eye velocities for subjects HC (open symbols) and RS (solid symbols) under the instruction to maintain the line of sight on the random dot field presented either alone (abscissa) or with the superimposed background field (ordinate). Triangles show eye velocity when the stationary field was the target, squares when the moving field was the target. The density of the dots was 1 dot/deg^2. Standard errors were smaller than the plotting symbols. Negative values on the axes indicate rightward velocities. The arrow indicates the velocity of the moving field. Velocities falling on the dotted diagonal line indicate no effect of the background. Velocities falling above the line, when the stationary field was the target, indicate smooth eye movements in the direction of the moving background. Velocities below the line, when the moving field was the target, indicate smooth eye movements slowed by the stationary background. (C) Same as (B) except that the density of the dots was increased to 8 dots/deg^2 (from Kowler et al., 1984b).

time, the smooth oculomotor system, which determines how fast the eye is to travel, is exposed to the same motion signals, but is blind to relative motion and instead uses velocity signals in retinal or in orbital coordinates. The remarkable thing is that two such very different processes, each using motion signals in different ways, work together so effectively to keep the line of sight on the chosen target.

2.2.2 *The role of selective attention*

Voluntary selection of the target for smooth eye movements is accomplished by allocating perceptual attention to the target. This was shown by Khurana and Kowler (1987), who asked subjects to make a perceptual judgment about the target being tracked, as well as about other, untracked stimuli. The stimulus display is shown in Fig. 2.2(A) (top). It consisted of 4 rows of letters, each moving to the right, with the velocity of the top and third rows twice that of the velocity of the 2nd and 4th rows. The subject's task was to keep the line of sight in the vertical gap between rows 2 and 3 while matching horizontal eye velocity to either the slower or the faster pair of rows. When the eye was near the middle of the display, the letters briefly (200 msec) changed to the kind of display shown in the bottom of Fig. 2.2(A), in which one numeral appears in the tracked, and another in the untracked, pair of rows. The subject had to identify and locate both numerals.

Fig. 2.2(B) shows the performance on the perceptual task. Each datum point shows the percent correct reports of identity and location for both the faster and the slower pair of rows. The filled circle shows performance when the faster pair was the tracked target, the open circle when the slower pair was the tracked target. Clearly, subjects were much better able to identify and locate the numerals appearing in the tracked target rows than in the untracked, background rows. The superior performance for the tracked target rows was not due to their lower retinal image velocity. Performance on the tracked target rows was better than performance on the untracked background rows even when retinal image speeds of target and background were equal (see Fig. 2.1(C)).

Finding superior perceptual performance for the tracked target meant that the same selective attentional decision serves both perceptual and motor systems. Even strenuous effort proved unable to dissociate two "attentions", one perceptual and the other motor. When the subjects — both experienced and highly-motivated — tried as best they could to track one set of rows and attend to the other, perceptual performance for the untracked rows did improve slightly, but at the cost of a shift in eye velocity toward the velocity of the untracked rows. In other words, they could not track one thing and attend to another. The best they could do in response to the instructions was to shift a bit of attention over to the untracked rows. Recently, links analogous to those Khurana and Kowler (1987) found between smooth pursuit and attention have been found to hold for saccades and attention (Kowler, Anderson, Dosher & Blaser, 1995).

These results have two consequences for the role of selective attention in active vision: First, it is known from numerous studies done over the past few years that instructions to attend to one or another location in the visual field can improve

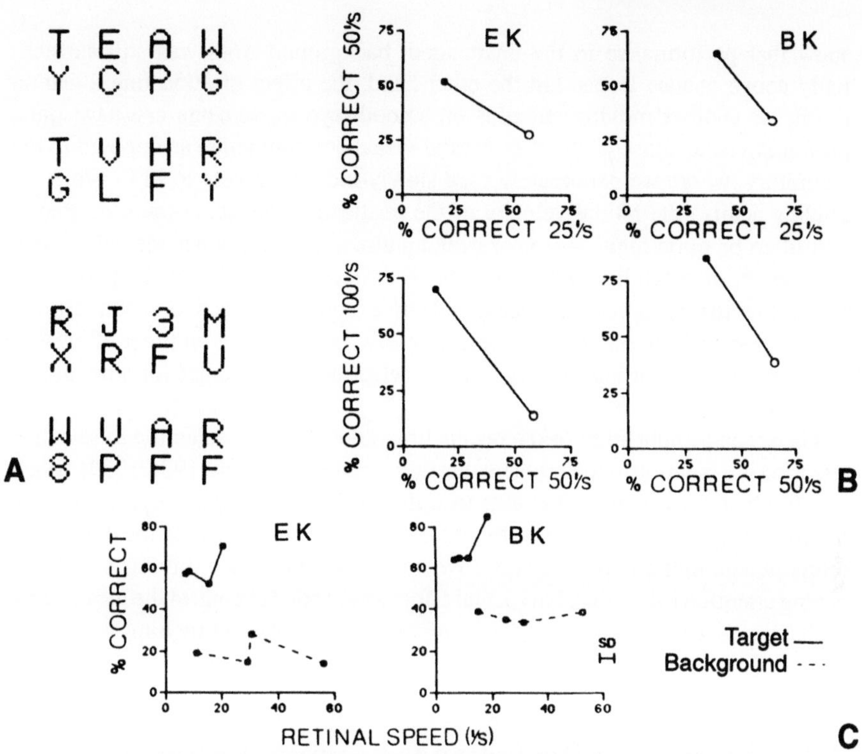

FIGURE 2.2. (A) The stimulus used in an experiment that measured smooth eye movements and selective perceptual attention at the same time. An array of 16 characters began moving horizontally at the beginning of the trial. The velocity of the characters in row 1 (top row) was the same as the velocity of characters in row 3. Similarly, the velocity of row 2 matched that of row 4. (Velocities were as follows: When one pair moved at 25′/sec, the other moved at 50′/sec; when one pair moved at 50′/sec, the other moved at 100′/sec.) The subject kept her line of sight in the vertical gap between rows 2 and 3 and tried to match horizontal eye velocity to one of the row-pairs (called the "target" pair). When the line of sight reached the approximate center of the display, the characters were replaced briefly (200 msec) by an array such as that shown in the bottom portion of (A). Note that one numeral is present in each pair of rows. In this example, a "3" is in row 1 and an "8" in row 4. Subjects had to identify both numerals and report the row in which they were located. (B) Visual search performance. Percent correct reports for the slower pair of rows is shown on the abscissa, for the faster pair on the ordinate. The open symbol shows performance when the slower pair was the target, the filled symbol when the faster pair was the target. Performance was always better for the target rows. (C) The same data points in (B) plotted as a function of measured retinal speed. Performance was always better for the target rows and retinal speed was largely irrelevant (from Khurana & Kowler, 1987).

the perceptibility of objects at that location at the expense of other locations — an example of just such an improvement was illustrated above. But the improvements are limited in degree. Attention, by itself, does not have large effects on visual thresholds or visual resolution. The data in Fig. 2.2(B–C), for example, show that performance in the unattended, background rows was still substantially above chance levels. On the other hand, the effect of allocating attention to one or another moving stimulus on smooth eye movements can have quite profound visual consequences. In natural visual environments (as opposed to the laboratory, where we deliberately used slowly-moving targets so as to avoid degrading acuity) the retinal velocity of the unattended, hence, untracked, objects will often be quite high — tens or even hundreds of degrees per second — and, as a result, detectability and resolution of untracked stimuli will be poor. This means that the main consequence of selective attention on active vision will not be its direct effects on stimulus perceptibility, but rather its indirect effects on visibility, exerted through the attentional selection of the target for smooth eye movements.

The second implication of the results linking oculomotor selection to selective attention is a solution to the problem that perturbed Ernst Mach (1906/1959). Mach was fascinated by how he was able to walk forward, keeping his eye steadfastly fixed on his goal ahead of him, without the eye being dragged off by the flow of the retinal image motion on either side. We now know that Mach did this simply by paying attention to his goal. No special effort was needed to control the movements of the eye. It was sufficient to take an interest in what was before him.

2.3 Example 2: Predicting the Future Position of Targets

2.3.1 *The effect of expectations on smooth eye movements*

As described at the beginning of Section 2.2.1, the traditional view of smooth eye movements is that they are evoked by the smooth motion of the target across the retina. According to this traditional view, the retinal velocity of the selected target is encoded by the visual system and used to compute a smooth pursuit response that allows the eye to move at a velocity nearly matching that of the target, thus reducing retinal image motion to values low enough to support clear vision. Let us now turn to the processes that use the retinal velocity signals to determine the smooth oculomotor command.

In recent years it has been popular to revive an idea of Rashbass (1961), which was suggested earlier by Craik (1947), that the signal driving smooth eye movements is not retinal velocity by itself, but includes a signal representing the current velocity of the eye in the orbit. The combination of retinal image velocity and orbital eye velocity produces a signal that represents the velocity of the eye with respect to the head. This model has been popular because it can account for several oculomotor phenomena, most notably, the ability of some subjects to generate directed smooth eye movements with retinally-stabilized stimuli, such as an afterimage. (The pursuit of stabilized targets is actually more complex because smooth eye movements with afterimages and other retinally-stabilized targets are subject to large individual differences that are not explained simply by the addition of a positive feedback signal; see Cushman, Tangney, Steinman & Ferguson, 1984.)

Are these reflexive models of smooth eye movement, in which responses are determined by the motion of targets in retinal or in orbital coordinates, adequate to guarantee retinal image velocities low enough to support clear vision, once the target has been selected? The research to be described below shows that they are not. Smooth pursuit eye movements follow the target accurately, but only if the target is moving in a simple, repetitive pattern, or moves in some pattern that can be predicted by the subject with reasonable accuracy. This suggests that knowledge of the upcoming target motion, and not simply the presence of a moving stimulus, is necessary to obtain an accurate response.

A vivid example of how smooth pursuit eye movements take advantage of knowledge of future target motion is contained in Fig. 2.3. This record was made while a subject was tracking a light-emitting diode moved by the experimenter. The top trace shows the target and the bottom trace the motion of the eye. The figure shows that the eye started moving smoothly several hundred milliseconds before the target began to move. The eye reached a velocity of about 50 deg/sec by the time the target started to move.

We have called the pursuit response observed before the onset of expected target motion "anticipatory smooth eye movement" (Kowler & Steinman, 1979a,b). Anticipatory smooth eye movements, such as those shown in Fig. 2.3, were observed

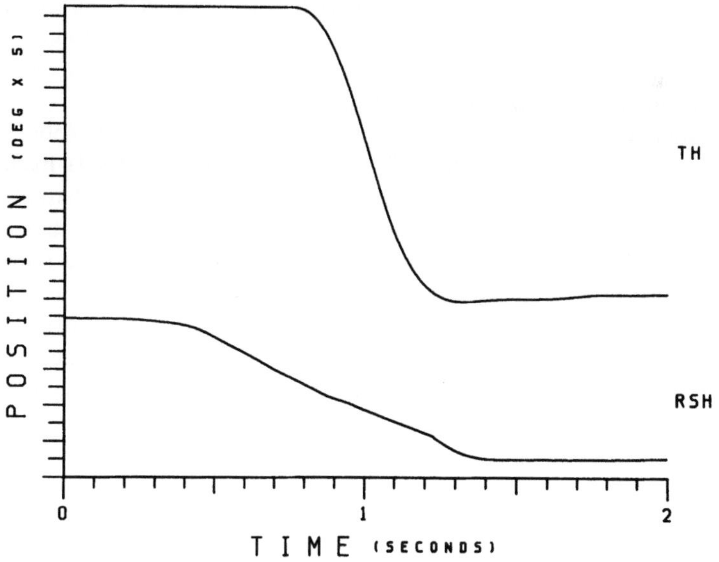

FIGURE 2.3. Movements of the right eye (RSH) of a subject tracking a single point target (TH) moved by an experimenter across the subject's visual field. The subject knew the direction of target motion but did not know when the target was to begin moving relative to the onset of the trial.

in the classical work of Raymond Dodge and colleagues more than 50 years ago (see Dodge, Travis & Fox, 1930). More recent investigators, confirming another one of Dodge's observations, noticed "anticipatory reversals" (Dodge's term) during the tracking of sinusoidal motion. These were episodes in which the eye would change direction ahead of the target (e.g., Stark, Vossius & Young, 1962; Westheimer, 1954; Winterson & Steinman, 1978). There were also several reports that pursuit of repetitive, predictable motions was far more accurate than pursuit of random motions, a result that not only implicated predictive processes in the control of pursuit, but also showed that linear systems models, in which responses to complex patterns of motion can be predicted from the response to sinusoidal motion, could not successfully explain pursuit (Collewijn & Tamminga, 1984; Dallos & Jones, 1963; Pavel, 1990; Stark et al., 1962).

Anticipatory smooth eye movements show that the motion of the target cannot be the only thing that drives the motion of the eye. In the past, modelers have tried to avoid the contribution of anticipation by assuming that anticipatory smooth eye movements were no more than learned habits formed after many cycles of tracking the same, periodic motion (e.g., Dallos & Jones, 1963; Westheimer, 1954). This attractive idea was not supported by several observations, including Dallos and Jones' (1963) own report of anticipatory reversals during the very first cycle of tracking — hardly leaving much time for learning. More recently, the habit hypothesis was contradicted by observations of anticipatory smooth eye movements preceding aperiodic (Kowler & Steinman, 1979a,b) and even randomly selected target motions (Kowler, Martins & Pavel, 1984a; Kowler & Steinman, 1981). When targets move in randomly selected directions, the eye moves in the direction determined by the subject's guess about the future direction of the target motion (Kowler & Steinman, 1981). The guess is determined by which target motions were seen or tracked in the recent past (Kowler et al., 1984a).

An example of anticipatory pursuit when targets move at randomly selected velocities is shown in Fig. 2.4, taken from Kowler and McKee (1987). Each graph shows average eye velocity as a function of time, with target motion starting at time 0 on the abscissa. There were 5 possible target velocities in each set, indicated by the horizontal lines on the graphs, and the subject did not know ahead of time which would be presented on any given trial. The target began to move at time "0" on the abscissa and stopped moving 1 second later. Notice that the eye started to move before the target and reached the approximate velocity of the target quite quickly (in about 250 msec). The interesting thing is that the initial pursuit responses were almost the same for all the targets in each set, regardless of which target velocity had actually been presented. These initial responses tended to be biased towards the mean velocity of the target set, that is, responses to the slower target velocities in the set were too fast, and responses to the faster target velocities were too slow. The eye velocity traces did not fully sort themselves out until several hundred milliseconds after the target started to move. (In this example, target direction was always leftward, but the outcome was the same when direction was selected randomly.)

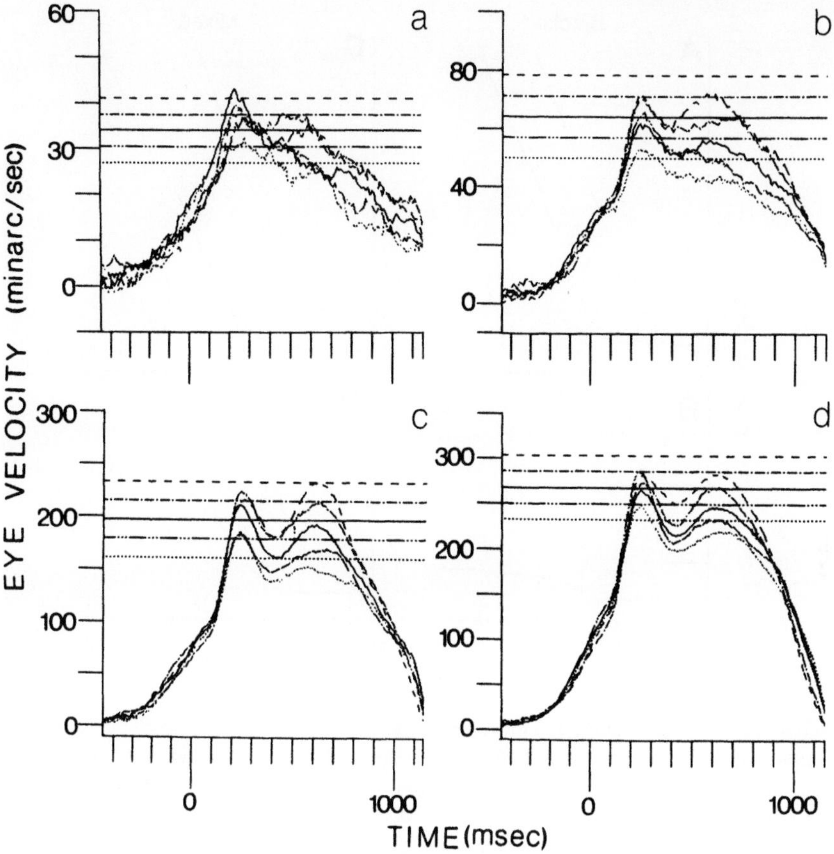

FIGURE 2.4. Mean 100 msec eye velocity for subject EK pursuing leftward target motions in 4 different sets of constant velocity target motion (A–D). Eye velocity is shown as a function of the midpoint of successive 100 msec intervals whose onsets are separated by 10 msec. Target motion began at 0 msec and ended at 1000 msec. Velocities less that 0′/sec indicate rightward motion. Scales are different on the ordinate of each graph. The horizontal lines indicate the velocities of the targets (from Kowler & McKee, 1987).

This example shows that the initial few hundred milliseconds of pursuit of randomly chosen stimuli is based primarily on the set of stimuli presented during previous trials of the experimental session. The contribution of the immediate stimulus motion to the initial pursuit is, by contrast, relatively modest. The modest contribution of the immediate stimulus motion to pursuit is not due to any deficiencies in sensory motion systems. Human beings can distinguish differences in target velocity of only about 5% with very brief (100–200 msec) exposures (Kowler & McKee, 1987; McKee, 1981), showing that precise information about stimulus motion is available in the brain after only about 100 msec. Smooth pursuit does not reach this level of precise discrimination until 600 or 700 msec after the

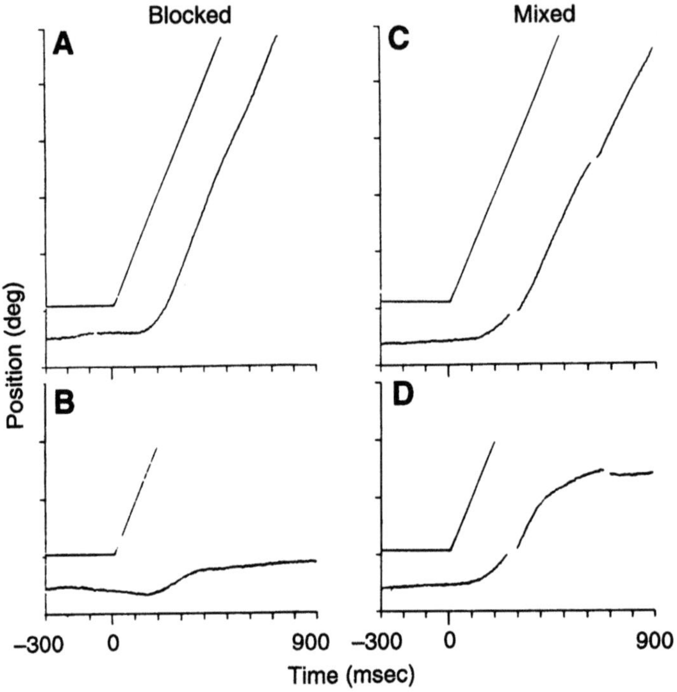

FIGURE 2.5. Representative eye movements (bottom traces) during smooth pursuit of 9.5 deg/sec target motion (top traces). On the left (Blocked) the duration was set to a constant value of either 1 sec (top graph) or 200 msec (bottom graph). On the right (Mixed) duration was selected randomly to be either 1 sec or 200 msec on each trial. Note the brisk initial pursuit of the longer duration and the poor initial pursuit of the shorter duration in the Blocked condition. The initial response took on a value roughly in between these 2 extremes when durations were randomly mixed. The gaps in the eye traces indicate when saccades occurred; the eye traces were shifted by amounts about equal to the size of these saccades.

target starts to move (Kowler & McKee, 1987) — well after pursuit reaches the approximate velocity of the target. Evidently, smooth eye movements do not rely on precise information about the immediate target motion to launch the brisk initial pursuit response. The initial response depends primarily on the expected velocity of the target, which is determined by the target motions tracked in the recent past.

Fig. 2.5 shows another example of anticipatory eye movements with randomly selected target motions. This figure contains representative records showing how pursuit velocity near the onset of target motion depends on how long the target is expected to continue moving (Kowler & McKee, 1987; Kowler, Steinman, He & Pizlo, 1989). In the left-hand graphs, the duration of target motion was known in advance. Brief durations (200 msec) led to a slow initial pursuit response (lower-left), while long (1 sec) durations led to a fairly brisk initial response (upper-left). This result, illustrating the effect of expected duration, shows that pursuit is not launched solely by the initial sweep of the target across the retina, but requires the expectation that target motion will continue into the future.

In the graphs on the right of Fig. 2.5, the short- and long-duration target motions were randomly mixed. This led to higher pursuit velocities for the brief motions and lower pursuit velocities for the long duration motions. So, randomizing duration did not remove the effect of expectations. It simply encouraged the development of a response intermediate between the one deemed most appropriate for each of the 2 durations in the stimulus set — a compromise that took past history into account in an attempt to avoid large errors during the initial part of pursuit.

These studies of pursuit with randomly chosen velocities and randomly chosen durations show that smooth eye velocity near the onset of pursuit depends on information derived from the past history of target motions. This appears to be a useful state of affairs because, left to itself, the smooth oculomotor subsystem evidently cannot incorporate new information about target motion fast enough to program an accurate, timely response that would bring the line of sight to the target before the target has moved to a very different location.

2.3.2 *Past history vs. cognitive expectations of future target motion*

It seems reasonable to use the past history of target motions to generate predictive smooth eye movements because there is little else on which to base predictions, at least in the laboratory. In the laboratory, where target motions are selected at random and any cues about the likely pattern of target motion are carefully removed, there is no information allowing prediction of future motion other than the pattern of motions presented in the past. The dependence on past history makes smooth eye movements prime candidates for adaptive models, in which various parameters of the smooth oculomotor system are continually adjusted on the basis of past stimuli or past performance to produce an optimal future response (Pavel, 1990).

Despite the value of taking past history into account, there may be more involved in the generation of smooth eye movements than just the adaptive responses made on the basis of prior events. The natural world, in contrast to the laboratory, presents a host of cues about the likely future motion of targets. For example, we can easily predict the motion of targets we control ourselves, such as the motion of objects we hold in our hands, or the motion of stationary objects in the environment relative to ourselves as we move about. In natural environments, the motion of objects we do not control often can be predicted on the basis of situational cues. For example, it is easy to predict the future motion of an animal based on the direction it is facing. Are smooth eye movements able to take advantage of such cues? In other words, are smooth eye movements generated on the basis of genuinely intelligent processes that take into account various sorts of symbolic cues about future motion? Or, are smooth eye movements blind to such cues and instead constrained to repeat successful pursuit responses made in the immediate past, even in the face of new information signaling that the pattern of target motion is about to change?

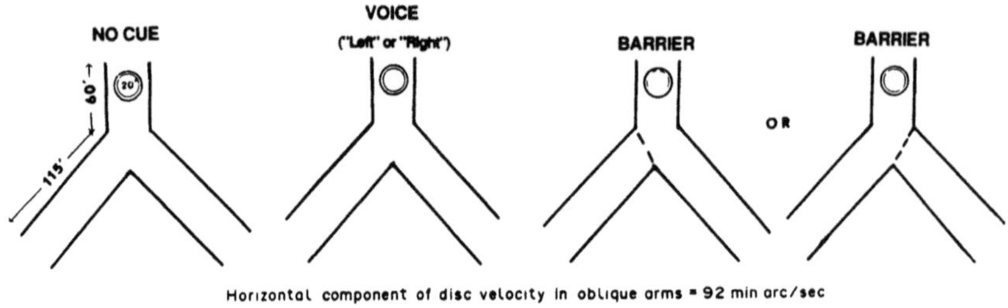

FIGURE 2.6. The stimulus display in an experiment comparing habits to cognitive expectations. It consisted of a stationary, inverted Y-shaped tube and an annulus which served as the moving target. The velocity of the target was 130′/sec. The target moved down the tube and continued at the same velocity down either the right-hand or left-hand oblique branch of the Y (horizontal component of velocity when the target was in either branch of the Y was 92′/sec). The target was equally likely to travel down either branch. The branch in which the target moved was either undisclosed before each trial (No Cue), disclosed by a Voice cue, or disclosed by a visible Barrier cue blocking access to either the left-hand or the right-hand branch (from Kowler, 1989).

To answer this question, target motions of the recent past were put in conflict with cues signaling a new pattern of motion in the future. The conflict was created by using the stimulus shown in Fig. 2.6, taken from Kowler (1989). It consists of a disc that moved downward inside an inverted Y-shaped tube. When the disc reached the junction of the oblique branches of the Y, it could travel down either the right-hand or the left-hand branch. The branch was chosen at random before each trial. In one experimental condition ("no cue"), the branch was not disclosed to the subject before the trial. In other conditions the branch was disclosed before each trial, either by a speech-synthesized voice saying "right" or "left", or by a visual barrier blocking access to one branch. What is of most interest is the velocity of the horizontal anticipatory smooth eye movements measured just before the target entered either branch.

Fig. 2.7 shows the average velocity of the eye as a function of time when there was no cue, a voice cue, or a barrier cue. The moving target-disc entered the oblique branch at time-0 on the abscissa of each graph. When there was no cue (Fig. 2.7(A)), anticipatory smooth eye movements depended on the motion in the prior trial — they were rightward when prior motion was to the right and leftward when prior motion was to the left.

Effects of the past on smooth eye movements were overridden by the symbolic cues (Fig. 2.7(B–C)). The eye began to travel in the direction of the expected motion at least 300 msec before the target entered either branch of the Y. Anticipatory smooth eye velocity was about 1/3 of the velocity of the disc by the time the disc entered the oblique branch.

The stronger influence of the symbolic cues, relative to the influence of prior trials, is also shown in Fig. 2.8. It shows the average velocity of the purely anticipatory portion of the pursuit response measured during the 200 msec interval before the start of horizontal motion. The cues produced a change in anticipatory

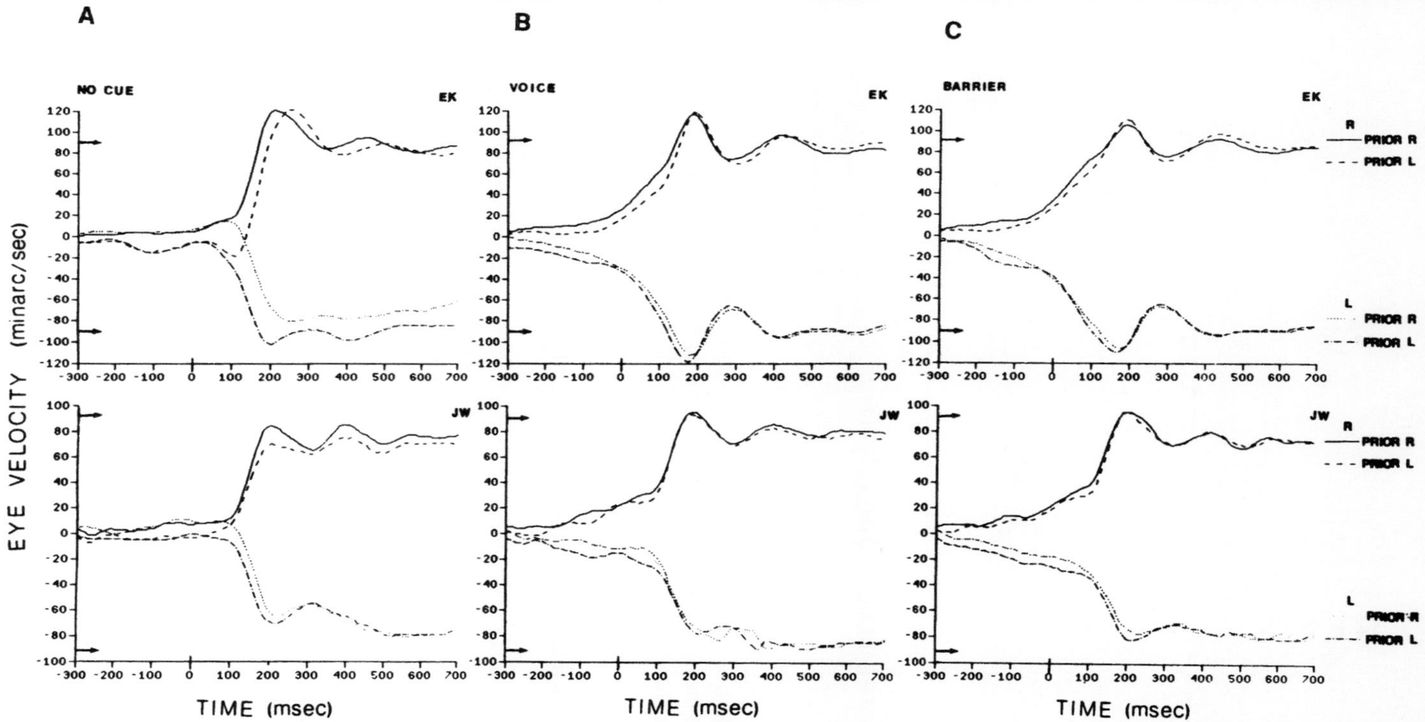

FIGURE 2.7. Mean horizontal eye velocity during successive 100 msec intervals (onsets 10 msec apart) as a function of the midpoint of the interval when either (A) No Cue, (B) a Voice Cue, or (C) a Barrier Cue about the direction of future horizontal target motion was given. Top graphs, EK; bottom graphs, naive subject JW. Time 0 is the start of horizontal target motion (the first entry of the moving target into the oblique branch of the Y-shaped tube). Arrows on the ordinate show horizontal target velocity; negative values denote leftward motion. The top pair of functions in each graph show eye velocity when the eye moved down the left-hand branch. One function in each pair shows eye velocity when the target motion in the prior trial was to the right; the other when motion in the prior trial was to the left. Each mean is based on 80–100 observations. Standard errors were 1-2′/sec and as high as 3′/sec (5′/sec with No Cue) only during the interval (0–200 msec) of most rapid eye acceleration (from Kowler, 1989).

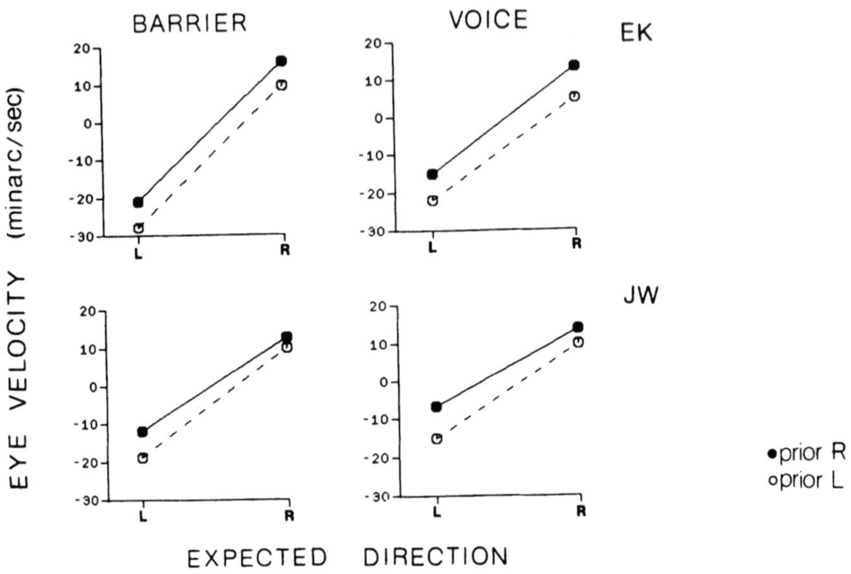

FIGURE 2.8. Mean horizontal eye velocity during the 200 msec interval before the start of horizontal target motion (first entry of the moving target into the oblique branch of the Y-shaped tube) for expected motion to the left and to the right with the Barrier and Voice cues. Top, subject EK; bottom, subject JW. Solid symbols show eye velocity when the prior target motion was to the right; open symbols when it was to the left. Means are based on 80–100 observations; standard errors are smaller than the plotting symbols (from Kowler, 1989).

eye velocity of as much as 40′/sec while the past produced changes of only about 5′/sec.

These results show that cognitive expectations drive anticipatory pursuit. How might expectations exert their influence? It seems implausible that cognitive expectations should operate by means of a separate neural pathway, existing alongside, and in continual conflict with, the "reflexive" pathways that carry the immediate retinal motion signals. A better arrangement would have smooth eye movements driven by a single representation of target motion. This representation would include not only the current motion of the target, but its motion path projected several hundred milliseconds into the future. A representation of this sort is ideal for motor control because it removes potential conflicts between retinal signals and expectations. It does, on the other hand, present interesting problems for those trying to understand the neural coding of visual motion. Not only is it necessary to study neural responses to motion that hasn't yet occurred, but it is, once again, necessary to distinguish the effect of motion signals on perception from the effects of motion signals on oculomotor responses. This distinction is compelled by this experiment because the eye began to travel in the direction of expected motion well in advance, but the subjects, nevertheless, perceived the path of the moving target accurately.

The research summarized here, as well as many prior experiments (see Kowler, 1990, & Pavel, 1990, for reviews), show that the contribution of expectations that accurately reflect the state of affairs in the physical world is essential for accurate pursuit. Pursuit of random motions is notoriously poor. Pursuit is best when the subject knows what sort of target motion to expect, either by means of appropriate environmental cues or by having enough time to figure out the future path of the target based on its motion in the recent past. If smooth eye movements are to confine retinal image velocities to levels that allow clear vision in the natural world, then the contribution of expectations is essential. The results summarized here show that our knowledge about the likely future motion of a target is sufficient to evoke the appropriate predictive response. No special types of target motions, or elaborate repetitive learning experiences are required. Cognitive expectations are an important component of the normal operation of pursuit. There is no evidence to support a separate, low-level, purely reflexive, smooth oculomotor subsystem in human beings.

The next 2 examples of the cognitive control of eye movement deal with saccades: the rapid, voluntary, eye movements we make to shift gaze between stationary targets.

2.4 Example 3: Planning Sequences of Saccades

Inspection of the visual world requires saccadic eye movements to bring eccentric visual details to the fovea, where visual acuity is best. Intuitively, we all have an idea about how this sort of thing proceeds, namely, we select something of interest to look at and move the line of sight (sometimes, rotating the head as well as the eyes) to the chosen target.

Oculomotor researchers studying this process have concentrated on a simple and convenient laboratory version of the natural task. In the laboratory version, a target point jumps from one location to another and the subject's task is to make a saccade to follow the target when it jumps. Most models of saccades have concentrated on the way that the assumed error signal — the eccentric position of the target on the retina — is used to program a saccade of the appropriate size and direction to bring the line of sight to the target. Natural scanning, however, involves processes that are not revealed by this conventional laboratory target "step-tracking" task.

In natural scanning, targets are stationary objects presented in the midst of irrelevant visual backgrounds. Instead of a single saccade to a target, natural scanning consists of continual sequences of saccades made to look from one stationary target to the next.

Zingale and Kowler (1987) studied the properties of patterns of saccadic sequences. Their stimulus was a 2-dimensional array of single points located at the vertices of a small (90′ on a side) imaginary pentagon. On each trial, anywhere from 2 to 5 points would be presented. Subjects had to look at one of the points

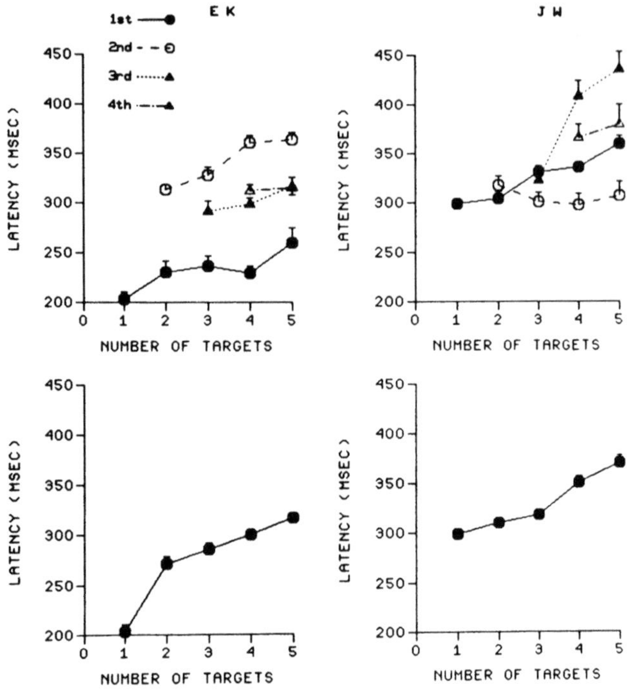

FIGURE 2.9. Top: Mean latency of the first through fourth saccades in the sequence as a function of the number of targets in a sequence for subject EK and naive subject JW. Bottom: Mean latency averaged over all saccades in the sequence as a function of the number of targets in the sequence. Vertical bars represent 1 SE (from Zingale & Kowler, 1987).

to start and, at a signal, begin to make a sequence of saccades to look from one visible target point to the next.

Fig. 2.9 shows that the latency of the first saccade of the sequence, and the time between subsequent saccades, depended on how many targets the subject was going to have to scan. This sort of pattern is by no means unique to saccades. Typewriter keypresses and spoken syllables have the same characteristic, as was demonstrated by Sternberg and colleagues (1978a, 1978b) in a study that provoked our study of saccadic sequences. We also found that the time between successive saccades in the sequence depended on the ordinal position of the target in the sequence (Fig. 2.10) — a property once again shared with typing and speech.

These results show that the timing patterns of saccadic sequences depend not only on the movement that is about to be programmed, but on the entire sequence of movements to be made. Sternberg and colleagues suggested that the increase in latency and in inter-response times with sequence length implies that motor programs for the entire sequence are prepared in advance, stored, and retrieved, as needed, during the execution of the sequence. They developed this model for typewriting and speech, but it could apply to saccades as well.

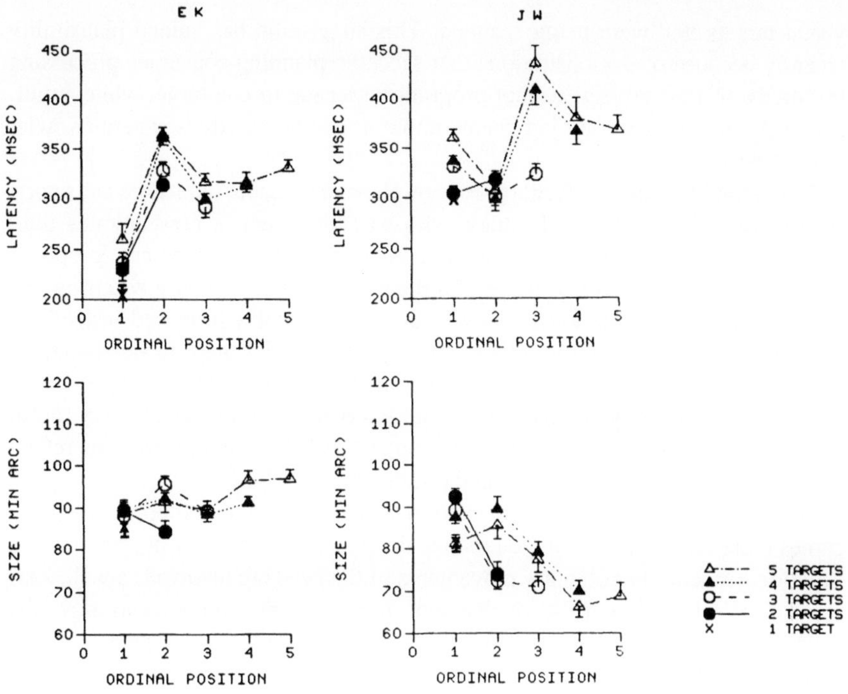

FIGURE 2.10. Mean latency (top) and mean size (bottom) as a function of the ordinal position of the saccade in the sequence for each of 5 sequence lengths. Vertical bars represent 1 SE. The distance between successive targets was 90′ (from Zingale & Kowler, 1987).

Stored programs for saccades can be modified during the execution of the sequence. Zingale and Kowler showed this by removing all the target points just as the signal to begin execution of the sequence was given. Under these conditions, in which performance was based solely on remembered saccadic programs and remembered target positions, sequence length and ordinal position continued to determine the timing of saccades. Saccades sizes, however, were 20% larger than the separation of the targets. This shows that the visible targets do play a role in fine-tuning saccade size during sequence execution, perhaps by modifying parameters of pre-determined saccadic programs (see Rosenbaum, Inhoff & Gordon, 1984, who discuss the modification of pre-planned programs for a finger-tapping task).

What are the advantages of pre-planning sequences of saccades? The answer to this question is not obvious, given that visual targets are always available and it would seem reasonable to plan saccades only as needed, rather than in advance. Zingale and Kowler (1987) made two suggestions.

The first was that pre-planned sequences might be of value in freeing processing resources from the need to continually program saccades. This would make processing resources available for other tasks, such as making decisions about the visual targets that were being scanned. This suggestion has gained plausibility recently because of demonstrations that saccadic planning consumes processing resources, that is, subjects cannot program a saccade to one target while simultaneously making accurate judgments about a target located elsewhere (Kowler et al., 1995).

The second suggested advantage for pre-planned saccadic sequences calls upon an idea of Lashley (1951). Lashley, who was interested in how animals integrate concurrent motor actions into effective patterns of movement, proposed that integration would not be successful if each individual response was triggered independently by a different sensory error signal. Lashley proposed instead the existence of a single internal clock, and a single spatial map, to be shared by concurrent voluntary motor activities. The temporal patterning of sequences of saccades is sufficiently similar to the temporal patterns of other voluntary motor responses (Sternberg et al., 1978a,b) to suggest that the temporal patterns reflect the reliance of all these activities on the same internal clock.

The idea that concurrent motor activities share the same temporal clock becomes more reasonable when you realize that saccades are rarely programmed in isolation. Concurrent voluntary movements of the head are involved as well. Zingale and Kowler inferred the involvement of concurrent commands to move the head when they noticed that the attempt to execute a very rapid sequence of saccades around the points of the pentagon produced striking discomfort of the neck muscles, as the subject fought to keep the head on the biteboard while rapidly looking from one point to the next. This result agrees with other observations that electrical activity in the neck muscles occurs during saccades, even when the head is artificially supported (Andre-Deshays, Berthoz & Revel, 1988; Berthoz & Grantyn, 1986). Others have noticed that allowing head motions during scanning improves performance by decreasing intersaccadic intervals (Kowler et al., 1992) and by increasing the velocity of saccades (Collewijn et al., 1992). All of these results suggest that the instruction to "look at an eccentric target" is, in fact, a more general instruction to orient to the target, moving both head and eyes at the same time. Perhaps, the temporal structure of the saccadic sequences Zingale and Kowler observed are glimpses into the works of the internal clock that sets up the rhythmic patterns used to coordinate the concurrent motions of the head and eye made while we scan natural environments.

2.5 Example 4: Saccades to Selected Targets in the Presence of Irrelevant Visual Backgrounds

The capacity to maintain the line of sight on a chosen target with smooth eye movements, without influence of irrelevant, unattended background stimuli, was the first example of the cognitive control of eye movements illustrated in this chapter. Saccades need this capacity as well, if we are to succeed in aiming the line of sight at targets of interest.

There has been some controversy about the selective capacity of saccades in the oculomotor literature. This controversy is due to studies in which failures of selection were inferred from experiments that required subjects to make saccades as quickly as possible, even if this meant they had not yet ascertained the true location of the target. These studies showed that the line of sight would often be displaced in the direction of irrelevant background stimuli. This led to suggestions that there are automatic sensorimotor "averaging" processes that take the line of sight to the "center of gravity" of an entire stimulus configuration (Coeffe & O'Regan, 1987; Findlay, 1982; Ottes, van Gisbergen & Eggermont, 1985). He and Kowler (1989) questioned the existence of such "centering tendencies" by doing an experiment in which the probability of a target appearing in one or another location was varied. An irrelevant, nontarget stimulus always appeared in the other location. He and Kowler (1989) found that the endpoint of the short-latency saccades was displaced toward the more probable location (Fig. 2.11, top graphs), a result suggesting that observed "centering" tendencies do not represent automatic sensorimotor averaging, but instead represent a high-level search strategy used to program saccades when the true location of the target is still unknown.

Subsequently, He and Kowler (1991) found that subjects were able to aim saccades accurately to designated places within an eccentric target form (a triangle), provided that there was no uncertainty on the part of the subject about where he was supposed to look. Such saccades were as accurate and as precise as saccades directed to single points at the same eccentricities. Saccades that were directed toward the form as a whole, rather than to a particular place within it, did tend to land in the same location each time, which was near the center-of-gravity.

These results shed some light on the process by which a saccadic endpoint is computed when the stimulus is a spatially-extended form, rather than a single point of light. It would seem that "averaging" of sorts might occur, but not an automatic averaging of all the elements in the visual field, as prior workers (see above) have suggested. Rather, "averaging" would operate only on those visual signals that were selected (i.e., attended) by the observer. It would seem that the visual position signals reaching the saccadic circuitry, like those reaching the smooth oculomotor circuitry, must first be subjected to a relative weighting by means of selective attention. Perhaps the same attentional decision that serves perception and smooth eye movements also serves to determine the target region for saccades.

2.6 Summary and Conclusions

The research described in this chapter has shown that:

(1) The target for both smooth eye movements and saccades is determined by spatially-selective attention that decides which signals are to reach lower-level oculomotor circuitry. The selection operates on representations of objects in the environment, rather than on simpler, isolated sensory cues — that is, you choose

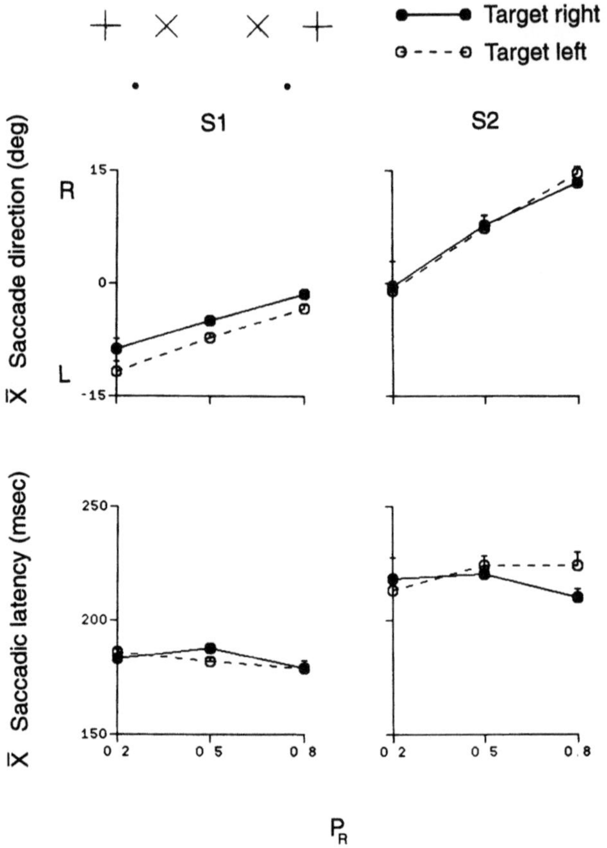

FIGURE 2.11. Mean saccadic direction (top) and latency (bottom) as a function of the probability (Pr) of the target appearing on the right for 2 naive subjects (S1 and S2). The target was either on the right (solid lines) or on the left (dotted lines). Standard errors were smaller than the plotting symbols except where noted by vertical bars (from He & Kowler, 1989).

what to track, not a particular stimulus velocity or eye velocity. How these representations of objects are then decomposed into the isolated velocity or position signals needed to compute the appropriate oculomotor commands is a major research problem, perhaps one as challenging as determining how the object representations were formed in the first place.

(2) The motion signals that guide smooth eye movements include projections of the expected target motion several hundred milliseconds into the future. In a somewhat analogous way, programs for saccadic eye movements also depend on an internal representation of future events, specifically, on a stored plan for an entire sequence of saccades. These phenomena show that effective oculomotor programs are based on representations of the visual world that encompass not only its current state, but its predicted state well into the future.

Each of these points shows that cognitive processes are inextricably tied to the oculomotor programming — they are not switched in and out, leaving the observer fluctuating between "reflexive" and "intelligent" modes of oculomotor control. The main advantage of cognitive control would appear to be the automatic and effortless linkage of oculomotor activity with other, concurrent visual, motor and cognitive events. For example, no special effort is required to select a target for smooth or saccadic eye movements. The line of sight will be drawn toward the target of current interest. Similarly, the knowledge we have about a target's future motion, or the plans we make about how we wish to move about in the environment, will find their way into the oculomotor circuitry without us having to activate any special mode of cognitive control to override low-level reflexes. This link between oculomotor commands and ongoing cognitive processes lets us avoid the need for special effort when we wish to choose targets and program appropriate eye movements. It also ensures that the line of sight is most likely to land on, and remain on, the target of interest.

Perhaps cognitive control will prove to be useful only in multi-purpose beings such as ourselves, who must see and think as we move eyes, head and limbs about in the environment, with all of these activities having to proceed in a coordinated fashion. I suspect, however, that the problem of coordinating diverse actions is as pressing in robots, who must, as we must, select targets of interest from cluttered backgrounds, and program motor responses early enough so that sensors or arms arrive at the selected targets before their location has changed. There are at least two lessons that may be learned from the way human beings accomplish these goals that may prove valuable in designing robots.

The first is that effective motor performance is not achieved by trying to expand sensory capacities or by trying to reduce motor programming time, but rather by taking advantage of the ongoing intelligent processes of selection and prediction — the same processes that are occurring as we (or the robots) are making decisions about the objects in the environment.

The second lesson is that it may be valuable to dissociate the visual representations that guide eye movements from those that determine percepts. We saw two examples of the dissociation in this chapter. One was the insensitivity of smooth eye movements to perceived induced motion. The second was the difference between the perceived path of a moving target and the predicted path that is used to guide the smooth eye movements. Similar dissociations between the visual representations used to organize perception and to guide eye movements have been reported before in the comparison of perceptual and motor localization of the position of a target (Hansen, 1979; Hansen & Skavenski, 1977; Skavenski, 1990; Sperling, 1990) and, also, in the comparison of the stimuli driving binocular vergence eye movements and the stimuli responsible for the perception of motion-in-depth (Collewijn & Erkelens, 1990; Erkelens & Collewijn, 1985). In both of these cases, percepts depended on the position or the motion of targets relative to visual backgrounds, while eye movements depended on the position or the motion of targets relative to the observer. Distinguishing two kinds of visual representations, one for perception and the other for motor control, perhaps at very early stages of processing, may prove to be useful in designing a robot who, like us, must confront the dual challenge of controlling its own movements effectively while at the same time recognizing objects in the environment.

Acknowledgments: The preparation of this chapter was supported by grant 88-0171 and 91-0342 from the Air Force Office of Scientific Research.

2.7 References

Aitsebaomo, A. P. & Bedell, H. E. (1992). Psychophysical and saccadic information about direction for briefly presented visual targets. *Vision Research*, *32*, 1729–1737.

Andre-Deshays, C., Berthoz, A. & Revel, M. (1988). Eye-head coupling in humans. I. Simultaneous recording of isolated motor units in dorsal neck muscles and horizontal eye movements. *Experimental Brain Research*, *69*, 399–406.

Berthoz, A. & Grantyn, A. (1986). Neuronal mechanisms underlying eye-head coordination. *Progress in Brain Research*, *64*, 325–343.

Carpenter, R. H. S. (1991). *Eye Movements*, Volume 8 of *Vision and Visual Dysfunction*. London: MacMillan.

Coeffe, C. & O'Regan, J. K. (1987). Reducing the influence of non-target stimuli on saccade accuracy: Predictability and latency effects. *Vision Research*, *27*, 227–240.

Collewijn, H. & Erkelens, C. J. (1990). Binocular eye movements and the perception of depth. In E. Kowler (Ed.), *Eye Movements and Their Role in Visual and Cognitive Processes* (pp. 213–261). Amsterdam: Elsevier.

Collewijn, H., Erkelens, C. J. & Steinman, R. M. (1988). Binocular coordination of horizontal saccadic eye movements. *Journal of Physiology*, *404*, 157–182.

Collewijn, H., Steinman, R. M., Erkelens, C. J., Pizlo, Z. & van der Steen, J. (1992). The effect of freeing the head on eye movement characteristics during 3-D shifts of gaze and tracking. In A. Berthoz, P. P. Vidal & W. Graf (Eds.), *The Head-Neck Sensory Motor System* (pp. 412–418). New York: Oxford University Press.

Collewijn, H. & Tamminga, E. P. (1984). Human smooth and saccadic eye movements during voluntary pursuit of different target motions on different backgrounds. *Journal of Physiology*, *351*, 109–129.

Craik, K. J. W. (1947). Theory of the human operator in control systems. *British Journal of Psychology*, *38*, 56–61.

Cushman, W. B., Tangney, J. F., Steinman, R. M. & Ferguson, J. L. (1984). Characteristics of smooth eye movements with stabilized targets. *Vision Research*, *24*, 1003–1009.

Dallos, P. J. & Jones, R. W. (1963). Learning behavior of the eye fixation control system. *IEEE Transactions on Automation and Control*, *AC-8*, 218–227.

Dodge, R., Travis, R. C. & Fox, J. C. (1930). Optic nystagmus. III. Characteristics of the slow phase. *Archives of Neurology and Psychiatry*, *24*, 21–34.

Erkelens, C. J. & Collewijn, H. (1985). Eye movements and stereopsis during dichoptic viewing of moving random dot stereograms. *Vision Research*, *25*, 1689–1700.

Findlay, J. M. (1982). Global visual processing for saccadic eye movements. *Vision Research*, *22*, 1033–1046.

Hallett, P. E. (1986). Eye movements. In K. Boff, L. Kaufman & J. Thomas (Eds.), *Handbook of Perception and Human Performance*, Volume 1 chapter 10. New York: Wiley.

Hansen, R. H. (1979). Spatial localization during pursuit eye movements. *Vision Research, 19*, 1213–1221.

Hansen, R. H. & Skavenski, A. A. (1977). Accuracy of eye position information for motor control. *Vision Research, 17*, 919–926.

He, P. & Kowler, E. (1989). The role of location probability in the programming of saccades: Implications for "center-of-gravity" tendencies. *Vision Research, 29*, 1165–1181.

He, P. & Kowler, E. (1991). Saccadic localization of eccentric forms. *Journal of the Optical Society of America A, 8*, 440–449.

Khurana, B. & Kowler, E. (1987). Shared attentional control of smooth eye movement and perception. *Vision Research, 27*, 1603–1618.

Kowler, E. (1989). Cognitive expectations, not habits, control anticipatory smooth oculomotor pursuit. *Vision Research, 29*, 1049–1057.

Kowler, E. (1990). The role of visual and cognitive processes in the control of eye movement. In E. Kowler (Ed.), *Eye Movements and Their Role in Visual and Cognitive Processes* (pp. 1–70). Amsterdam: Elsevier.

Kowler, E., Anderson, E., Dosher, B. & Blaser, E. (1995). The role of attention in the programming of saccades. *Vision Research*, in press.

Kowler, E. & Blaser, E. (1995). The accuracy and precision of saccades to small and large targets. *Vision Research*, in press.

Kowler, E., Martins, A. J. & Pavel, M. (1984a). The effect of expectations on slow oculomotor control. IV. Anticipatory smooth eye movements depend on prior target motions. *Vision Research, 24*, 197–210.

Kowler, E. & McKee, S. P. (1987). Sensitivity of smooth eye movement to small differences in target velocity. *Vision Research, 27*, 993–1015.

Kowler, E., Murphy, B. J. & Steinman, R. M. (1978). Velocity matching during smooth pursuit of different targets on different backgrounds. *Vision Research, 18*, 603–605.

Kowler, E., Pizlo, Z., Zhu, G. L., Erkelens, C. J., Steinman, R. M. & Collewijn, H. (1992). Coordination of head and eyes during the performance of natural (and unnatural) visual tasks. In A. Berthoz, P. P. Vidal & W. Graf (Eds.), *The Head-Neck Sensory Motor System* (pp. 419–426). New York: Oxford University Press.

Kowler, E., van der Steen, J., Tamminga, E. P. & Collewijn, H. (1984b). Voluntary selection of the target for smooth eye movement in the presence of superimposed, full-field stationary and moving stimuli. *Vision Research, 24*, 1789–1798.

Kowler, E. & Steinman, R. M. (1979a). The effect of expectations on slow oculomotor control. I. Periodic target steps. *Vision Research, 19*, 619–632.

Kowler, E. & Steinman, R. M. (1979b). The effect of expectations on slow oculomotor control. II. Single target displacements. *Vision Research, 19*, 633–646.

Kowler, E. & Steinman, R. M. (1981). The effect of expectations on slow oculomotor control. III. Guessing unpredictable target displacements. *Vision Research, 21*, 191–203.

Kowler, E., Steinman, R. M., He, P. & Pizlo, Z. (1989). Smooth pursuit depends on the expected duration of target motion. *Investigative Ophthalmology & Visual Science (Supplement), 30*, 182.

Lashley, K. S. (1951). The problem of serial order in behavior. In W. A. Jeffress (Ed.), *Cerebral Mechanisms in Behavior: The Hixon Symposium* (pp. 112–136). New York: Wiley.

Lemij, H. G. & Collewijn, H. (1990). Difference in accuracy of human saccades between stationary and jumping targets. *Vision Research*, in press.

Mach, E. (1906/1959). *Analysis of Sensations*. New York: Dover.

Mack, A., Fendrich, R. & Wong, E. (1982). Is perceived motion a stimulus for smooth pursuit? *Vision Research, 22*, 77–88.

McKee, S. P. (1981). A local mechanism for differential velocity detection. *Vision Research, 21*, 491–500.

Murphy, B. J. (1978). Pattern thresholds for moving and stationary gratings during smooth eye movement. *Vision Research, 18*, 521–530.

Niemann, T., Ilg, U. J. & Hoffman, K.-P. (1994). Eye movements elicited by transparent stimuli. *Experimental Brain Research, 98*, 314–322.

Ottes, F. P., van Gisbergen, J. A. M. & Eggermont, J. J. (1985). Latency dependence of colour-based target vs. nontarget discrimination by the saccade system. *Vision Research, 25*, 849–862.

Pavel, M. (1990). Predictive control of eye movement. In E. Kowler (Ed.), *Eye Movements and Their Role in Visual and Cognitive Processes* (pp. 71–114). Amsterdam: Elsevier.

Rashbass, C. (1961). The relationship between saccadic and smooth tracking eye movements. *Journal of Physiology, 159*, 326–338.

Rosenbaum, D. A., Inhoff, A. W. & Gordon, A. M. (1984). Choosing between movement sequences: A hierarchical editor model. *Journal of Experimental Psychology: General, 113*, 372–393.

Skavenski, A. A. (1990). Eye movement and visual localization of objects in space. In E. Kowler (Ed.), *Eye Movements and Their Role in Visual and Cognitive Processes* (pp. 263–287). Amsterdam: Elsevier.

Sperling, G. (1990). Comparison of perception in the moving and stationary eye. In E. Kowler (Ed.), *Eye Movements and Their Role in Visual and Cognitive Processes* (pp. 307–351). Amsterdam: Elsevier.

Stark, L., Vossius, G. & Young, L. R. (1962). Predictive control of eye tracking movements. *IRE Transactions on Human Factors and Electronics, HFE-3*, 52–57.

Steinman, R. M., Cushman, W. B. & Martins, A. J. (1982). The precision of gaze. *Human Neurobiology, 1*, 97–109.

Sternberg, S., Monsell, S., Knoll, R. & Wright, C. (1978a). The latency and duration of rapid movement sequences: Comparisons of speech and typewriting. In G. E. Stelmach (Ed.), *Information Processing in Motor Control and Learning* (pp. 117–152). New York: Academic Press.

Sternberg, S., Wright, C., Knoll, R. & Monsell, S. (1978b). Motor programs in rapid speech: Additional evidence. In R. A. Cole (Ed.), *The Perception and Production of Fluent Speech* (pp. 507–534). Hillsdale, New Jersey: Erlbaum.

Westheimer, G. (1954). Eye movement response to a horizontally moving visual stimulus. *Archives of Ophthalmology, 52*, 932–941.

Westheimer, G. & McKee, S. P. (1975). Visual acuity in the presence of retinal image motion. *Journal of the Optical Society of America, 65*, 847–850.

Winterson, B. J. & Steinman, R. M. (1978). The effect of luminance on human smooth pursuit of perifoveal and foveal targets. *Vision Research, 18*, 1165–1172.

Zingale, C. M. & Kowler, E. (1987). Planning sequences of saccades. *Vision Research, 27*, 1327–1341.

Part II

Solving Visual Problems

If a visual system can tolerate movement, can it also benefit from it? This section is concerned with methods and algorithms that make use of the motion of an eye or camera to improve vision qualitatively or quantitatively in somewhat unexpected ways.

3

Robust Computational Vision

Brian G. Schunck[1]

ABSTRACT This paper presents a paradigm for formulating reliable machine vision algorithms using methods from robust statistics. Machine vision is the process of estimating features from images by fitting a model to visual data. Computer graphics programs can produce realistic renderings of artificial scenes, so our understanding of image formation must be quite good. We have good models for visual phenomena, but can do a better job of applying the models to images. Vision computations must be robust to the kinds of errors that occur in visual signals. This paper argues that vision algorithms should be formulated using robust regression methods. The nature of errors in visual signals will be discussed, and a prescription for formulating robust algorithms will be described. To illustrate the concepts, robust methods have been applied to three problems: surface reconstruction, image flow estimation, and dynamic stereo.

3.1 Introduction

There has been considerable progress in the field of computer vision in understanding the physics and mathematics of vision processes. Current understanding of vision enables computer graphics to produce realistic renderings of scenes. Visual processes can be modeled precisely and in many cases the models have been verified experimentally. Unfortunately, progress in using machine vision in applied settings has not met expectations. Apparently, our understanding of vision has not translated into vision algorithms that work on diverse, realistic scenes. If our understanding of vision is sufficient, then the fault must lie in how our knowledge of vision is used in formulating vision algorithms and not in the knowledge itself. The fault must lie in the methods for formulating vision algorithms, rather than in the theory of vision. The lack of success in applying machine vision to practical applications is mirrored by the inability of current vision algorithms to fully account for the performance of natural vision in real life.

Vision is fundamentally a process of discriminating between different elements. Discriminating between elements from two classes is an easy task when the features of elements within a class are nearly identical, the features can be measured accurately and reliably, and the classes are clearly distinct. In vision, there can be as much variation within a class of elements as between elements, and the variations

[1] Artificial Intelligence Laboratory, University of Michigan

in feature measurements can be worse. Consider edge detection as an example. In theory, an edge is the change between regions of nearly constant brightness. But, in practice the regions are not constant, the transition between regions is more complicated than a step discontinuity, the image is noisy, the camera contains defects and the fine-grained texture within regions can produce numerous indications of edges that are difficult to separate from the intended edge. Although the principles of vision may be sound, the practice of vision must survive many hazards.

Applications of vision lead to images that violate the assumptions in vision theory and require solving difficult problems in signal discrimination. The theory of vision must be embodied in computations that are immune to the problems inherent to imagery. There are many ways of formulating vision algorithms, many classes from which the vision computations can be selected. The selection of methods for formulating vision algorithms must be subject to the constraint that the computations be immune to the problems inherent in vision processing. Least squares methods which are popular in many areas of science and engineering are not robust to violations in problem assumptions. There are classes of computations in the field of robust statistics that have been designed to be robust to wide departures from the assumptions. This paper will show how robust methods can be used to formulate vision algorithms that will work in realistic situations.

A necessary condition for an autonomous system to be intelligent is that the system should respond well to unpredicted problems such as violations of the assumptions used to design the system, errors that were not anticipated, and sensor failures. Robust statistics allows calculations to be implemented in such as way that the results are reasonable under a wide variety of unmodeled errors.

This paper is primarily concerned with distributional robustness. The statistical methods presented in this paper were developed to work in cases where few assumptions can be made about the error distribution. Robust statistical methods also provide some tolerance for system failures, such as not properly connecting the camera or not turning on the lights. But system failures may be beyond the capabilities of robust statistical methods to compensate for such errors. However, statistical methods do provide criteria for detecting errors, and these capabilities should be more widely used in vision systems.

Vision processing is the execution of a vision algorithm. Vision algorithms have been developed that are robust to small-scale variations in parameters due to noise. Edge detection provides an example: the edge detection algorithm developed by Canny (1986) uses a Gaussian filter to smooth the input image to the required degree. As another example, regularization (Horn & Schunck, 1981; Poggio, Torre & Koch, 1985) uses a smoothing process to solve ill-posed problems. This paper argues that robustness in the sense of small variations in parameters is not sufficient for vision. Methods adapted from physics and mathematics for analyzing vision problems may not be sufficient for implementing vision algorithms. Statistical methods show how theoretical insights can be cast into reliable systems for vision processing.

Examples of the difficulties in solving vision problems in real domains will be presented in Section 3.2. The concept of immunity to outliers and the problems with common approaches to formulating vision algorithms will be explained in Section 3.3. Robust methods for formulating vision algorithms will be explained in Section 3.4. Examples of the use of robust methods for solving vision problems are included in Section 3.5.

3.2 Vision Problems

The hypothesis of this paper is that computational vision is the process of fitting models of visual phenomena to unreliable data. The data are unreliable because they are contaminated by outliers. A fundamental problem in formulating vision algorithms is to develop calculations that can fit visual models to data and be immune to outliers.

Outliers are data points with large errors that do not fit the error distribution assumed when the theory was developed. Many signal processing algorithms assume that errors have a Gaussian distribution; in this context, outliers are large errors that are not drawn from a Gaussian distribution. Gaussian noise is a relatively well-behaved form of noise: the Gaussian distribution does not have broad tails which means that most errors are small. Calculations that perform optimally with moderate amounts of Gaussian noise may respond badly to outliers.

Another source of outliers in vision is the existence of multiple models such as different surfaces or different objects. Robust statistics has developed methods for measuring parameters when the input data are generated by a combination of models. The parameters of the desired model can be extracted from the confusion of multiple effects. Examples of this problem in vision include model matching with multiple objects in the scene (Knoll & Jain, 1986), shape from shading with multiple reflectance models or different regions of albedo, edge detection with different kinds of edges, and surface reconstruction where there are multiple surfaces separated by discontinuities. outliers.

3.3 Vision Methods

The traditional method for combining measurements into an estimate is least squares regression. Least squares is the basis for many procedures that attempt to minimize the cost of errors in the estimate and is the method behind the derivation of the Kalman filter (Jazwinski, 1970; Kwakernaak & Sivan, 1972). Although least squares criteria may be optimal if the noise can be modeled as a Gaussian process, it is not optimal for other noise distributions. If the noise distribution includes outliers, the distribution is heavy-tailed and least squares criteria can lead to very poor estimators (Hoaglin, Mosteller & Tukey, 1983; Rousseeuw & Leroy, 1987). Formulating a vision algorithm using least squares methods implicitly assumes well-behaved, Gaussian errors and assumes that the input data are free of outliers.

A linear, multivariate model of order n is represented by the equation

$$\hat{y}_i = \hat{a}_1 x_{i1} + \hat{a}_2 x_{i2} + \cdots + \hat{a}_n x_{in} \tag{3.1}$$

for the ith data point, where $\hat{a}_j$ is the estimate of model parameter a_j. The residual for each data point (the deviation of the data point from the estimated model) is $r_i = y_i - \hat{y}_i$. In least squares regression, the estimates of the model parameters are given by minimizing the sum of the squares of the residuals:

$$\min_{\hat{a}} \sum_{i=1}^{n} r_i^2. \tag{3.2}$$

The estimates of the model parameters can take on arbitrary values if only one of the data points is an outlier. Several excellent examples of the effect of outliers on least squares regression are provided by Rousseeuw and Leroy (1987). The least squares criterion can be generalized to

$$\min_{\hat{a}} \sum_{i=1}^{n} \rho(r_i). \tag{3.3}$$

The effect of outliers is reduced by bounding $\rho(r)$ for large r. If $\rho(r)$ goes to zero for large r, then the effect of outliers can be eliminated. This is the class of robust methods called M-estimation (Press, Flannery, Teukolsky & Vetterling, 1988).

The physical analogy shown in Fig. 3.1 may clarify this discussion. Imagine that you want to find the center of mass of a set of points in the plane. Attach springs with equal spring constants to the fixed points and to a small object that can move freely. The object will be pulled to the average of the locations of the points. The springs implement a least squares norm through the spring equation for potential energy. This physical analogy corresponds to the derivation of the calculation of an average from the criterion that the sum of the squares of the residuals, the differences between each point and the average, should be minimized. Now, suppose that one of the points can be moved. Call this point a leverage point. It is possible to force the location of the average to be shifted to any arbitrary point by pulling the leverage point far enough away. This illustrates the extreme sensitivity

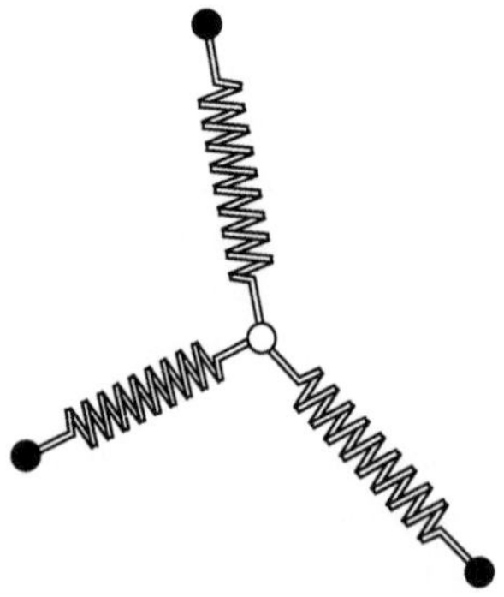

FIGURE 3.1. A physical analogy that illustrates the sensitivity of least squares methods to outliers. Even a single outlier renders a least squares solution useless.

of estimators based on least squares criteria to outliers. Even a single outlier can ruin an estimate. The spring analogy also extends to linear regression with the same conclusions: even a single outlier will distort the regression estimate.

Ideally, one would like to break the spring connected to an outlier so that the estimate remains unharmed. Changing the spring constants so that points that are far away exert little influence on the estimate corresponds to the implementation of robust estimators based on influence functions. Breaking the springs attached to outliers corresponds to resampling schemes where a consistent subset of samples is determined. Resampling plans repeatedly draw random subsets and choose the subset that yields the best estimate. Examples of resampling algorithms include random sample consensus, least median squares regression, and other computer-intensive methods in regression.

Often, the noise and outliers can be modeled as a mixture distribution: a linear combination of a Gaussian distribution to model the noise and a broad-tailed distribution to account for outliers. In this case, it makes sense to formulate an estimator with a norm that resembles a least squares norm for small errors but is insensitive to large errors so that outliers are ignored. This is the influence function approach pioneered by Huber (Huber, 1981; Hampel, Ronchetti, Rousseeuw & Stahel, 1986).

The breakdown point is the smallest percentage of data points that can be incorrect to an arbitrary degree and not cause the estimation algorithm to produce an arbitrarily wrong estimate (Rousseeuw & Leroy, 1987). Let Z be a set of n data points. Suppose that the set Z' is a version of set Z with m points replaced with arbitrary values. Let a regression estimator be denoted by $\hat{a} = T(Z)$. The bias in an estimate due to outliers is given by

$$\text{BIAS} = \sup_{Z'} \parallel T(Z') - T(Z) \parallel . \tag{3.4}$$

The idea behind breakdown point is to consider what happens to the bias as the number of outliers m increases as a percentage of the total number of data points n. Since the data points can be replaced with arbitrary values, for some ratio of m to n the bias can potentially be unbounded. This is the breakdown point. Below the breakdown point, the regression estimator may be able to totally reject outliers or the outliers may have only some small effect on the estimate. Beyond the breakdown point, the outliers can drive the estimator to produce an arbitrary answer in the sense that the answer will depend on the outliers and not on the legitimate data. In other words, the result provided by the estimator is unpredictable. The breakdown point is defined as

$$\epsilon_n^\star = \min\left\{\frac{m}{n} : \text{BIAS}(m; T, Z) \text{ is infinite}\right\}. \tag{3.5}$$

For least squares regression, $\epsilon_n^\star = 1/n$, and in the limit as the number of data points becomes large, $\epsilon_\infty^\star = 0$. In other words, least squares regression has no immunity to outliers; a single outlier can completely ruin the result. For vision applications with the sources of outliers discussed in Section 3.2, the result of applying vision algorithms based on least squares criteria is unpredictable since the calculations are influenced by errors of an extreme and unpredictable nature.

Vision problems are ill-posed in the sense that vision is an inverse problem with insufficient information, relative to the number of unknowns, to provide a well-behaved solution. One approach to solving ill-posed problems is to use regularization which leads to a smoothing process for combining local information (Horn & Schunck, 1981; Poggio, Torre & Koch, 1985). The details of how regularization leads to smoothing operators are explained by Horn (1986). Regularization is a least squares method and the algorithms produced by regularization average data over local neighborhoods. The calculation has the same properties as the spring analogy for computing the location of a cluster of dots. If the observations are related to the parameters of interest by a linear model and if the ill-posed problem is regularized by least squares, then the resulting algorithm is a linear system. The algorithms spawned by regularization are just as sensitive to outliers, including the effects of discontinuities and multiple models, as any calculation created by least squares methods. The smoothing process that is the result of algorithm formulation by regularization is the antithesis of the discrimination capabilities required for rejecting outliers and preserving discontinuities.

3.4 Robust Methods

Vision algorithms must be able to combine data while simultaneously discriminating between data that should be kept distinct, such as outliers and data from other regions. Robust statistics has developed a class of algorithms that are robust to variations in the statistical distribution of errors. In particular, the algorithms are immune to errors from broad-tailed distributions and achieve acceptable results when the data are subject to well-behaved errors, such as from a Gaussian distribution. Robust statistics offers methods for extracting measurements from uncooperative data.

The random sample consensus algorithm, called RANSAC, developed by Fischler and Bolles (1981) handles outliers by computing an estimate from a consistent subset of inconsistent data points. The algorithm randomly selects a set S of the minimum number of points required to fit a model from the set of data points P. The set S is used to compute a model estimate. The consensus set S^* of points in P that are within an error tolerance τ of the model is determined. If the number of points in S^* is greater than some threshold, then the model is recomputed using the data points in S^*; otherwise, a new set S is randomly selected from the data points P and the consensus procedure is repeated. After a predetermined number of iterations, the algorithm either fails or settles for the model obtained from the largest consensus set. The RANSAC algorithm has three parameters that must be chosen by the user: the error tolerance, the minimum acceptable size of the consensus set, and the number of trials.

TABLE 3.1. The least median squares regression algorithm.

Algorithm 1 Assume that there are n data points and p parameters in the linear model.

[1] Choose p points at random from the set of n data points.

[2] Compute the fit of the model to the p points.

[3] Compute the median of the squared residuals.

The fitting procedure is repeated until a fit is found with sufficiently small median of squared residuals or until some predetermined number of resampling steps.

In least median squares regression, the estimates of the model parameters are given by minimizing the median of the squared residuals:

$$\min_{\hat{a}} \operatorname{med}_{i} r_i^2. \tag{3.6}$$

The least median squares algorithm is outlined in Table 3.1. The median has a 50% breakdown point and this property carries over to least median squares regression (Rousseeuw & Leroy, 1987). In other words, even if as many as half of the data points are outliers, the regression estimate is not seriously affected. This situation can occur, for example, in surface reconstruction. As argued by Schunck (1989a), it is realistic to assume that a well-behaved boundary between two surfaces will divide a neighborhood such that at least half of the neighborhood covers the same surface region that contains the center of the neighborhood. What this means is that a local surface reconstruction procedure which is trying to choose the depth value for some point in the image will be faced (in reasonable worst-case circumstances) with the situation where the point is surrounded by a neighborhood where no more than half the points are outliers. In general, the requirements of many vision algorithms match the characteristics of least median squares regression (among other algorithms), but are not satisfied by the breakdown characteristics of least squares methods.

Another example of a robust statistic is given by the trimmed mean. The trimmed mean is an example of an order statistic, which means that the data set is sorted into increasing order and statistics are computed from the order in which data occur in the sorted list. The trimmed mean is an order statistic that is an estimator of the location parameter for a symmetric distribution. Consider the Gaussian distribution

$$G(x; \mu, \sigma) = \frac{1}{\sqrt{2\pi}\sigma} e^{(x-\mu)^2/2\sigma^2} \tag{3.7}$$

as an example. The location parameter μ for the location of the center of the distribution can be estimated by computing the average of a set of samples from the distribution,

$$\hat{\mu} = \frac{1}{n} \sum_{i=1}^{n} x_i \tag{3.8}$$

which minimizes the square norm

$$\frac{1}{n} \sum_{i=1}^{n} (x_i - \mu)^2. \tag{3.9}$$

This estimator works fine if the samples are drawn from a Gaussian distribution, but can be very ineffective for other distributions (Hoaglin et al., 1983). For example, if the samples are drawn from a contaminated distribution represented by mixture of Gaussian distributions,

$$p(x) = (1 - \eta)G(x; \mu_1, \sigma_1) + \eta G(x; \mu_2, \sigma_2), \tag{3.10}$$

where η represents the fraction of points that are outliers, then the average is a very poor estimator of μ_1 even if the fraction of outliers is small.

The trimmed mean is a robust estimator that discards a fixed fraction of points from both ends of the ordered list of data points and computes the estimate of the location parameter by averaging the remaining points. The $p\%$ trimmed mean discards $p\%$ of the points from either end of the ordered list. The median is actually a 50% trimmed mean. Because of the tolerance of the trimmed mean for outliers, the estimator may find use as a smoothing operator in edge detection and surface reconstruction.

3.5 Applications

This section presents three applications of robust methods to problems in vision: surface reconstruction with outliers and discontinuities, estimating the image flow velocity field with multiple moving objects, and combining binocular stereo disparities from multiple views.

3.5.1 Surface reconstruction

Surface reconstruction is a common problem in computer vision (Blake & Zisserman, 1987; Grimson, 1981; Terzopoulos, 1983, 1986a). Surface reconstruction is required for fitting a surface to sparse depth values from binocular stereo and range sensors (Besl & Jain, 1988; Grimson, 1981; Terzopoulos, 1983). Surface reconstruction schemes based on least squares criteria have problems with surface discontinuities and outliers. There have been several attempts to patch the criteria to preserve discontinuities in surfaces (Blake, 1984; Marroquin, Mitter & Poggio, 1987; Terzopoulos, 1986b) with some success but the patches do not address the problem of outliers.

Sinha and Schunck (1990, 1992) published a two-stage algorithm for surface reconstruction. The first stage removes outliers from the depth measurements and interpolates the sparse depth values onto a regular grid. The second stage fits weighted bicubic splines to the output from the second stage. Weighted splines adapt to large step changes in the data and do not smooth across discontinuities (Wang, 1991). It is necessary to remove outliers from the data before fitting splines since the fit is done by least squares approximation. The first stage uses least median squares regression to fit local polynomials to the sparse depth measurements. This gives the first stage a high tolerance for outliers and effectively removes bad depth values from binocular stereo mismatches and glitches in range sensors. The first stage also interpolates sparse depth values onto a regular grid while preserving discontinuities. The tasks of discontinuity preservation, outlier removal, and grid interpolation are accomplished simultaneously.

A straightforward expression for least median squares is difficult to write but the algorithm for implementing it is easy to explain. Assume that planar surface patches are fit to the depth measurements in a local neighborhood about each grid point. The algorithm can be easily extended to fit higher-order surface patches. For each grid point, select the n depth measurements that are closest to the grid point. From this set, try all possible combinations of m data points, where m is the number of data points used to fit the surface patch. For each of the k subsets of the data points in the local neighborhood, $k = \binom{n}{m}$, fit a surface patch to the points in the subset and denote the corresponding parameter vector by a_k. Compare all data points in the local neighborhood with the surface patch by computing the median of the squared residuals:

$$\operatorname*{med}_i \left[\left(z_i - f(x_i, y_i, a_k) \right)^2 \right]. \tag{3.11}$$

After surface patches have been fit to all possible subsets, pick the parameter vector a_k corresponding to the surface patch with the smallest median of squared residuals.

This procedure is computationally expensive since the model fit is repeated $\binom{n}{m}$ times for each local neighborhood; however, each surface fit is independent and the procedure is highly parallelizable. Adjacent neighborhoods share data points and could share intermediate results. In practice, it may be necessary to try only a few of the possible combinations so that the probability of one of the subsets being free of outliers is close to one.

3.5.2 *Image flow*

An image flow algorithm has been developed using least median squares regression (Schunck, 1989b). The only information available at each pixel in the image is the projection d of the local velocity vector onto the gradient direction:

$$d = \rho \cos(\alpha - \beta), \tag{3.12}$$

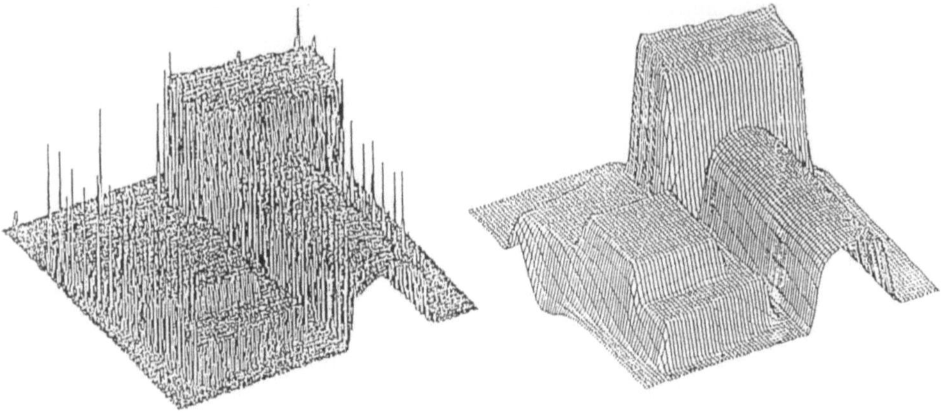

FIGURE 3.2. The results of applying the two stage surface fitting algorithm to a range image from laser radar. Note how the outliers along the boundaries of the surfaces are removed without degrading the sharp transitions between surfaces.

where $\rho(x, y)$ and $\beta(x, y)$ are the speed and direction of motion, respectively, and α is the direction of the image gradient. This is the image flow constraint equation. There are several advantages to this representation (Schunck, 1986). The true velocity vector for the image flow must lie along a line in velocity space, called the constraint line, defined by the image flow constraint equation.

Consider a pair of images from an image sequence. An array of constraint lines is computed from the spatial and temporal derivatives of the frame pair. The least median squares algorithm is applied over all overlapping neighborhoods. Within each neighborhood, the algorithm tries all possible pairs of constraint lines. The intersection of each pair of constraint lines is computed. The median of the square of the residuals is computed. This is the cost assigned to each solution. Each intersection and its cost is stored. After all possible pairs are tried, the intersection corresponding to the minimum cost is used as the estimate for the image flow velocity for the center of the neighborhood.

There are several steps in computing the intersection of the constraint lines and the residuals. The constraint lines are represented in polar form as the distance d of the constraint line from the origin and the angle α of the perpendicular bisector. Let the parameters of the first constraint line be d_1 and α_1 and the parameters of the second constraint line be d_2 and α_2. The position of the intersection in rectangular coordinates is

$$u = \frac{d_1 \sin \alpha_2 - d_2 \sin \alpha_1}{\sin(\alpha_1 - \alpha_2)} \tag{3.13}$$

$$v = \frac{d_2 \cos \alpha_1 - d_1 \cos \alpha_2}{\sin(\alpha_1 - \alpha_2)}, \tag{3.14}$$

where u and v are the x and y components, respectively, of the local image flow velocity vector.

The fit of the model to the constraint lines is the median of the squared residuals:

$$\operatorname*{med}_i r_i^2. \tag{3.15}$$

The residual for each constraint line is the perpendicular distance of the constraint line from the velocity estimate u and v. The residual is given by

$$r = u\cos\alpha + v\sin\alpha - d. \tag{3.16}$$

The position of the intersection of the pair of constraint lines, given by Eqs. 3.13 and 3.14, is a candidate solution. The median of the squared residuals of the constraint lines with respect to the candidate is computed and saved, along with the candidate, as a potential solution. The median of the squared residuals is the median of the square of the perpendicular distance of each constraint line in the neighborhood from the candidate.

The typical neighborhood size used was 5 by 5 pixels. An n by n neighborhood contains n^2 constraint lines. The number of possible pairs of constraint lines in an n by n neighborhood would be

$$\frac{n^2(n^2-1)}{2}. \tag{3.17}$$

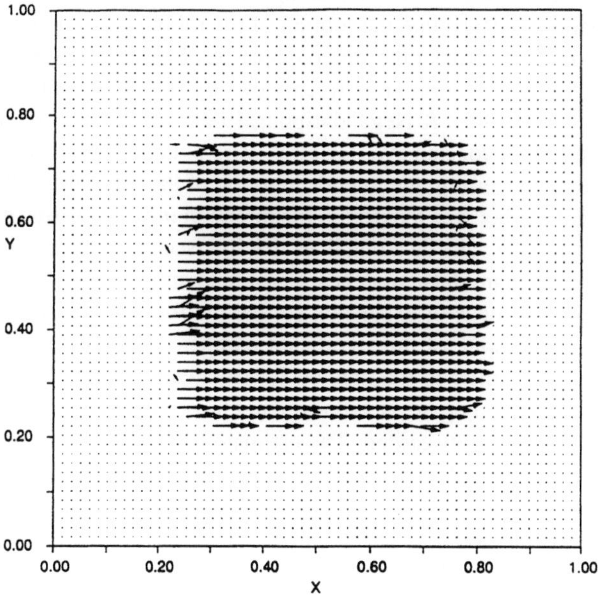

FIGURE 3.3. The results of the robust image flow algorithm based on least median squares regression applied to a synthetic data set that simulates a randomly textured box translating against a randomly textured background.

A 5 by 5 neighborhood would yield 300 pairs. It is not necessary to try all possible pairs if computation time is restricted. Rousseeuw and Leroy (1987, p. 198) provide a table that gives the number of subsamples that must be run to fit a model with p parameters and 95% confidence to data sets with various percentages of outliers. Assume that at most 50% of the constraints in the neighborhood will be outliers. The local estimate of the image flow velocity field requires only two constraint lines. From the table published by Rousseeuw and Leroy, only 11 pairs of constraints would have to be tried to provide a consistent estimate with 95% confidence. Using more pairs would increase the odds of finding a consistent estimate. If fewer than all possible pairs of constraint lines are used, the pairs should be selected so that the constraint lines in each pair have very different orientation. This reduces the problems with ill-conditioning caused by intersecting constraint lines that have nearly the same orientation. A preprogrammed scheme could be used for selecting the constraint line pairs in each neighborhood.

3.5.3 Dynamic stereo

Tirumalai (Tirumalai, 1991; Tirumalai, Schunck & Jain, 1990, 1992) developed a system for assimilating stereo disparities from multiple views that incorporated three robust algorithms. A stereo camera was mounted on a mobile robot. As the robot moved forward, the stereo cameras were pointed approximately in the direction of motion, so the robot was looking where it was going. The system used visual information to determine the motion of the robot between views and used this motion estimate to project the stereo disparities forward in time.

The robot estimated its motion between views by solving the absolute orientation problem (Horn, Hilden & Negahdaripour, 1988) using conjugate points from two successive frames for one eye. Features were obtained from each frame and prior information concerning the intended motion of the robot was used to match features. Since the robot was looking in the direction in which it was moving, the points of the conjugate pairs lie along lines that intersect at the focus of expansion. However, there are many false matches. Least median of squares regression was used to estimate the focus of expansion so that the focus of expansion could be estimated reliably in spite of the significant number of false matches (outliers). The algorithm was similar to the method for calculating image flow outlined in Section 3.5.2 above. Conjugate pairs that did not lie on a line close to the estimated focus of expansion were discarded. The estimate of the focus of expansion was not used in subsequent calculations. The sole purpose of the calculation was to ensure that only correct conjugate pairs were used to solve the absolute orientation problem, since the method for solving the absolute orientation problem was based on least squares error criteria and would not tolerate outliers.

Binocular stereo algorithms are prone to mismatches. The stereo disparities from the initial frames were filtered by using least median squares to fit surface patches to the depth measurements, as in the first stage of the two-stage surface fitting algorithm described in Section 3.5.1 above. For subsequent views, the binocular stereo disparities were projected forward using the estimated motion between views and combined with the disparities computed from the new stereo pair using a robust Kalman filter with a nonlinear gain that discarded new stereo disparities that differed too much from the disparity estimate.

The robust Kalman filter is like an M-estimator (Press et al., 1988). The difference between the new disparity and the current disparity estimate is computed. Note that the current disparity estimate is a dense grid computed with robust surface fitting so there is a current disparity estimate at all points, except in large regions where no disparity information is available. The difference between the new disparity and the current disparity estimate goes through a nonlinear gain $\phi(r)$, which is proportional to r for small r, but tapers to zero for large r. The output of $\phi(r)$ is added to the disparity estimate to compute an updated disparity estimate. Note that for large r, no correction will be added to the disparity estimate and the new disparity (which was probably an outlier) will be ignored.

This system allowed a mobile robot to acquire an accurate disparity map of an unknown scene in spite of errors in binocular stereo matches, incorrect feature matches between views, and inaccurate motion sensors on the mobile platform. Note that this satisfies our necessary condition for an autonomous system, that is, immunity to errors and unknown situations.

3.6 Discussion

Real images contain nasty problems in data analysis of the type addressed by recent work in robust statistics. Images contain outliers (gross errors), multiple models and discontinuities (other forms of outliers). Statisticians have long realized that least squares methods or estimators derived from a least squares optimization criterion are very sensitive to outliers. This paper noted that some methods for developing vision algorithms are essentially least squares methods and will be very sensitive to the kind of outliers that are common in vision problems. This is not a criticism of current viewpoints in vision: the kind of mathematics needed to conceptualize a solution to a vision problem or explain some characteristic of vision processing is different from the kind of mathematics needed to translate an understanding of visual phenomena into a practical vision algorithm. Physics and mathematics have earned their place as tools for developing vision theories; but statistics may provide the tools needed to apply vision models to real applications.

Many of the results developed by vision researchers can be applied to practical applications; but the algorithms will have to be formulated using robust criteria. This paper has described one class of methods, called least median squares regression, for formulating robust regression algorithms, and described the robust Kalman filter based on M-estimation. If vision is the process of fitting models to unreliable data, then robust estimators can be used to fit our understanding of vision processes to the data encountered in realistic scenes. Three applications were presented as a proof of concept.

3.6.1 Comparison with other paradigms

Robust statistics provides a paradigm for handling data that contain extreme errors due to sensor faults and unpredictable situations that violate assumptions. Newell and Simon (1976) argue that intelligence can be modeled as a symbol manipulation system. Later, Horn (1986), Marr (1982), and Poggio (1984) argued that visual perception involves manipulating numerical quantities, not just symbols, and presented numerical paradigms for vision processing. With real world sensory data, either symbols or numbers can be in error. The symbol hypothesis can be patched to handle uncertainty by attaching a measure of certainty to symbols. This approach is exemplified by certainty factors (Shortliffe & Buchanan, 1984), Dempster-Shafer theory (Gordon & Shortliffe, 1984; Shafer, 1976), and adaptations of probability (Cheeseman, 1985). All three of these paradigms, symbols, numbers, and symbols with certainty, are depicted in the diagram in Fig. 3.4. The missing element in the three paradigms summarized above is a way to deal with numerical quantities that are subject to unpredictable errors.

If an agent is to function intelligently in a loosely constrained environment, it must be able to handle sensory information that contains extreme errors and violations in problem assumptions. Tolerance for errors may not be sufficient for intelligence, but robust sensory processing is surely a necessary condition for intelligence. Few would argue that an agent could not be considered intelligent when it is easily confused by a single sensor failure or an obvious violation in assumptions.

Assuming that intelligence at the knowledge level can be modeled by symbol manipulation, the problem of transforming signals from real-world sensors into

	Symbols	Numbers
Certainty	Artificial Intelligence	Mathematical Physics
Uncertainty	Certainty Factors	Robust Statistics

FIGURE 3.4. An illustration of the relationship between four paradigms for processing sensory information. The two paradigms on the left depend on the symbol hypothesis, the other two paradigms work with numerical data; but only one paradigm, robust statistics, deals with real-world numerical measurements that contain errors and violations of model assumptions.

symbols remains to be done. Ideally, real-world signals would be processed into symbols without uncertainty, or at least symbols with an attached measure of certainty that could be propagated by existing methods. Sensory signals which are uncertain, numerical measurements should be transformed into one of the boxes on the left side of the diagram in Fig. 3.4. Since sensory signals are not perfectly certain, the stable of mathematical methods popular in vision today, represented by the upper right box in Fig. 3.4, may not be applicable to real-world processing because methods from mathematical physics assume certain measurements or at least a certain model for errors. If numerical methods have any role in perception, they must follow robust statistical methods so that extreme errors and violations in assumptions are screened out. Numerical methods should be restricted to working on the results of robust regression. The first steps in converting signals to symbols should include robust statistical methods.

Robust statistical methods provide a way to handle numerical measurements that are contaminated by unpredictable errors. If the errors obey a known distribution, then traditional statistics or methods from systems theory can provide information processing methods that may be optimal; but robust methods are required when the nature of the errors is hard to predict. Robust statistics may be the right paradigm for processing uncertain numerical inputs to autonomous systems. The traditional techniques of artificial intelligence work with symbols that are assumed to be correct. A robust estimate is the ideal symbol for artificial intelligence techniques predicated on the symbol hypothesis.

3.6.2 Improving performance

Another element in paradigms for perception is the need for parallelism. The popularity of neural networks is due in part to the natural formulation for parallelism; another reason why neural networks are popular is tolerance for errors. Robust methods provide high tolerance for errors, as well as a sound foundation for algorithm development, and also lead to a parallel implementation. In a resampling plan, such as least median squares regression, the same model is fit to many combinations of data points. Each fit is independent of the others and can be done in parallel. The calculation of the residual between each model fit and the data points can also be done in parallel; in fact, the calculation of each squared residual between a model fit and a single data point can be done in parallel so there is opportunity for massive, fine-grained parallelism. The calculation of the median also leads naturally to massive parallelism.

Besides parallelism, there are many opportunities to improve the run-time performance of robust methods for vision processing. In the surface reconstruction and image flow estimation applications presented in this paper, least median squares regression was applied to overlapping neighborhoods. Much of the calculation performed on data in one neighborhood is repeated in other neighborhoods. There is ample opportunity for improving performance by eliminating duplicate calculations.

3.6.3 Computational resources

The computer intensive methods for regression such as least median squares repeat the same calculation a large number of times. This may seem inefficient, but it is actually a good use for the increased computational capacity of modern computers. Computers are so fast today that computational power is unlimited for all practical purposes in many applications. Often, more computational capacity is available than what is needed for a particular calculation. Excess computational capacity is frequently spent in driving a sophisticated user interface, which is not needed in a fully autonomous system. Algorithms formulated using least median squares regression can use the extra computational capacity to repeat the same calculation on more subsets of the data. The repetition gives the system a very valuable characteristic that would be hard to achieve by other means: immunity to unmodeled errors, violations of assumptions, and sensor failures.

In most cases, the data set has far fewer outliers than assumed for a conservative design. The correct answer will be obtained in the first few iterations. The system can allocate computational power to other calculations that must be done, but any excess capacity can be used later to repeat the calculations over additional subsets of the data to improve reliability. Regression diagnostics can trigger the allocation of more resources to evaluating more samples of the data set for a perception task. Robust methods lead to systems that are very reliable and the computational burden can be managed in an opportunistic fashion.

3.6.4 Further work

There is considerable work left to be done. Least median squares regression and similar statistical techniques assume that at least half of the data points are not outliers. This assumption may fit the needs of vision problems such as edge detection, surface reconstruction, and image flow estimation; but seems too restrictive for many vision problems such as structure from motion. In problems such as structure from motion, a global computation is performed on the points derived from the scene. Multiple objects undergoing different motions can play havoc with structure from motion calculations derived using least squares criteria. Least median squares regression can provide an immediate improvement in tolerance to outliers; but the number of outliers will not always be less that half of the data points. If there is a prominent object that accounts for over half of the data points, then least median squares regression could derive a structure from motion estimate for the prominent object and the process could be repeated for the remaining data points leading to a sequence of robust regression problems. This scheme assumes that each set of data points will include one object accounting for more than half of the data points and this will certainly not generally be true. For example, there could be five moving objects in the scene with each object accounting for around one-fifth of the data points. Robust methods must be used for such applications, but least median squares regression may not be the best method.

Robust methods such as the median perform well with respect to immunity from outliers, but are not optimal in suppressing Gaussian noise. Although outliers are a serious problem in vision and must be addressed in the formulation of vision algorithms, Gaussian noise is also present. Some robust methods based on influence functions may be able to handle outliers and Gaussian noise simultaneously. The two-stage algorithm for surface reconstruction described in Section 3.5.1 uses robust methods in the first stage to remove outliers and weighted least squares methods in the second stage to fit a smooth surface to the cleaned and gridded data points. Further work on systems for handling the combination of well-behaved noise and outliers in vision problems must be done. In the mean time, the combination of resampling methods for the first stage and linear filters for the second stage may solve several difficult problems in bringing vision theory into successful practice.

Acknowledgments: The idea that regularization is essentially a regression came from Larry Maloney during our weekly dinner meetings. I have had many fruitful discussions about robust methods with Ramesh Jain, Greg Wakefield, and my colleagues in the Artificial Intelligence Laboratory at the University of Michigan. Support for this work was provided by a Rackham grant from the University of Michigan, the Air Force Office of Scientific Research under contract DOD-G-AFOSR-89-0277, and grants from Perceptron and Ford.

3.7 References

Besl, P. J. & Jain, R. C. (1988). Segmentation through variable-order surface fitting. *IEEE Transactions on Pattern Analysis and Machine Intelligence, 10*, 167–192.

Blake, A. (1984). Reconstructing a visible surface. In *Proceedings of the National Conference on Artificial Intelligence* (pp. 23–26). Los Altos, California: Morgan Kaufmann.

Blake, A. & Zisserman, A. (1987). *Visual Reconstruction*. Cambridge, Massachusetts: MIT Press.

Canny, J. F. (1986). A computational approach to edge detection. *IEEE Transactions on Pattern Analysis and Machine Intelligence, 8*, 679–698.

Cheeseman, P. (1985). In defense of probability. In *Proceedings of the International Joint Conference on Artificial Intelligence* (pp. 1002–1009). Los Altos, California: Morgan Kaufmann.

Fischler, M. A. & Bolles, R. C. (1981). Random sample consensus: A paradigm for model fitting with applications to image analysis and automated cartography. *Communications of the Association of Computing Machinery, 24*, 381–395.

Gordon, J. & Shortliffe, E. H. (1984). The Dempster-Shafer theory of evidence. In B. G. Buchanan & E. H. Shortliffe (Eds.), *Rule-Based Expert Systems: The MYCIN Experiments of the Stanford Heuristic Programming Project* (pp. 272–292). Reading, Massachusetts: Addison-Wesley.

Grimson, W. E. L. (1981). *From Images to Surfaces: A Computational Study of the Human Early Vision System.* Cambridge, Massachusetts: MIT Press.

Hampel, F. R., Ronchetti, E. M., Rousseeuw, P. J. & Stahel, W. A. (1986). *Robust Statistics: An Approach Based on Influence Functions.* New York: Wiley.

Hoaglin, D. C., Mosteller, F. & Tukey, J. W. (1983). *Understanding Robust and Exploratory Data Analysis.* New York: John Wiley & Sons.

Horn, B. K. P. (1986). *Robot Vision.* New York: McGraw-Hill.

Horn, B. K. P., Hilden, H. M. & Negahdaripour, S. (1988). Closed-form solution of absolute orientation using orthonormal matrices. *Journal of the Optical Society of America A, 5,* 1127–1135.

Horn, B. K. P. & Schunck, B. G. (1981). Determining optical flow. *Artificial Intelligence, 17,* 185–203.

Huber, P. J. (1981). *Robust Statistics.* New York: John Wiley.

Jazwinski, A. H. (1970). *Stochastic Processes and Filtering Theory.* New York: Academic Press.

Knoll, T. F. & Jain, R. C. (1986). Recognizing partially visible objects using feature indexed hypotheses. *IEEE Journal of Robotics and Automation, 2,* 3–13.

Kwakernaak, H. & Sivan, R. (1972). *Linear Optimal Control Systems.* New York: Wiley.

Marr, D. (1982). *Vision: A Computational Investigation into the Human Representation and Processing of Visual Information.* San Francisco: W. H. Freeman and Company.

Marroquin, J., Mitter, S. & Poggio, T. (1987). Probabilistic solution to ill-posed problems in computational vision. *Journal of the American Statistical Association, 82,* 76–89.

Newell, A. & Simon, H. A. (1976). Computer science as empirical inquiry: Symbols and search. *Communications of the Association of Computing Machinery, 19(3),* 113–126.

Poggio, T. (1984). Vision by man and machine. *Scientific American, 250(4),* 105–116.

Poggio, T., Torre, V. & Koch, C. (1985). Computational vision and regularization theory. *Nature, 317,* 314–319.

Press, W. H., Flannery, B. P., Teukolsky, S. A. & Vetterling, W. T. (1988). *Numerical Recipes in C: The Art of Scientific Computing.* Cambridge, England: Cambridge University Press.

Rousseeuw, P. J. & Leroy, A. M. (1987). *Robust Regression and Outlier Detection.* New York: Wiley.

Schunck, B. G. (1986). The image flow constraint equation. *Computer Vision, Graphics and Image Processing, 35,* 20–46.

Schunck, B. G. (1989a). Image flow segmentation and estimation by constraint line clustering. *IEEE Transactions on Pattern Analysis and Machine Intelligence, 11,* 1010–1027.

Schunck, B. G. (1989b). Robust estimation of image flow. In Schenker, P. S. (Ed.), *Sensor Fusion II: Human and Machine Strategies, Proceedings of the SPIE,* Volume 1198 (pp. 116–127).

Shafer, G. (1976). *A Mathematical Theory of Evidence.* Princeton, New Jersey: Princeton University Press.

Shortliffe, E. H. & Buchanan, B. G. (1984). A model of inexact reasoning in medicine. In B. G. Buchanan & E. H. Shortliffe (Eds.), *Rule-Based Expert Systems: The MYCIN*

Experiments of the Stanford Heuristic Programming Project (pp. 233–262). Reading, Massachusetts: Addison-Wesley.

Sinha, S. S. & Schunck, B. G. (1990). A robust method for surface reconstruction. In *IEEE Workshop on Robust Computer Vision* (pp. 183–199).

Sinha, S. S. & Schunck, B. G. (1992). A two stage algorithm for discontinuity-preserving surface reconstruction. *IEEE Transactions on Pattern Analysis and Machine Intelligence, 14*, 36–55.

Terzopoulos, D. (1983). Multilevel computational processes for visual surface reconstruction. *Computer Vision, Graphics and Image Processing, 24*, 52–96.

Terzopoulos, D. (1986a). Image analysis using multigrid relaxation methods. *IEEE Transactions on Pattern Analysis and Machine Intelligence, 8*, 129–139.

Terzopoulos, D. (1986b). Regularization of inverse visual problems involving discontinuities. *IEEE Transactions on Pattern Analysis and Machine Intelligence, 8*, 413–424.

Tirumalai, A. P. (1991). *Constructing Environment Maps with an Active Vision System Through Information Assimiliation*. PhD thesis, University of Michigan, Ann Arbor.

Tirumalai, A. P., Schunck, B. G. & Jain, R. C. (1990). Dynamic stereo with self-calibration. In *International Conference on Computer Vision* (pp. 466–470). Los Alamitos, California: IEEE Computer Society.

Tirumalai, A. P., Schunck, B. G. & Jain, R. C. (1992). Dynamic stereo with self-calibration. *IEEE Transactions on Pattern Analysis and Machine Intelligence, 14*, 1184–1189.

Wang, X. (1991). *Weighted Regularization and Its Applications*. PhD thesis, University of Michigan, Ann Arbor.

4

Eye Movements and the Complexity of Visual Processing

Misha Pavel[1]
Helen A. Cunningham[2]

ABSTRACT We propose the hypothesis that the difficulty of a visual task for human observers can be estimated from its computational complexity, defined with respect to a specific constrained computer architecture. We suggest that eye movements may serve to reduce this constrained complexity for certain visual tasks. We begin by discussing human performance on several examples of perceptual tasks. We then informally introduce analysis of algorithmic complexity and examine the complexity of these tasks for parallel computational networks whose depth is constrained to a small number of levels. Finally, we discuss the ability of the human visual system to perform translation invariant pattern recognition. In accordance with predictions based on the complexity of visual tasks, eye movements can compensate for failures of the translation invariance of the visual system.

4.1 Introduction

A flexible, mobile sensor appears to be an essential component of most biological vision systems. In the human visual system, mobility is achieved by head and eye movements. The degree of importance of a manipulable sensor to a vision system is a critical question both for students of biological vision systems and for designers of artificial vision systems. In humans, eye movements appear to mediate a variety of functions ranging from image stabilization to visual search.

In this chapter, we examine the notion that eye movements mediate a tradeoff between various information processing demands on the visual system. In its most basic form, our hypothesis is that eye movements permit the visual system to convert parallel solutions of certain tasks that would require large amounts of hardware (or "wetware") into sequential algorithms that require considerably less complex signal processing mechanisms, although sacrificing processing speed. We will argue that the usefulness of sequential algorithms will increase with the difficulty or perceptual complexity of visual tasks.

[1]Department of Electrical Engineering and Applied Physics and Department of Computer Science and Engineering, Oregon Graduate Institute

[2]Apple Computer, Inc.

In the first part of this chapter we discuss several examples of perceptual tasks and consider the potential impact of eye movements on human performance. We note that one of the important information processing demands affecting the impact of eye movements is that performance be independent of stimulus location (i.e., translation invariant). We note that eye movements appear generally more useful for those tasks that are more difficult and in which the human visual system is less translation invariant.

The key question in our analysis is how to determine the complexity of any particular visual task. To answer this question, we explore the potential of a formal analysis of algorithmic complexity to allow estimation of the *perceptual complexity* of visual tasks. Although algorithmic complexity theory is most relevant for the analysis of computer algorithms for unbounded problems, we suggest that it can also be used for bounded problems confronting the visual system if appropriate constraints are placed on the computing mechanism. In particular, we examine the complexity of parallel computational networks whose depth is constrained to a small number of levels. Our analyses suggest that the perceptual complexity of a task generally correlates with the difficulty of the problem as measured by human performance on that task.

The final section of this chapter (Section 4.4) presents a detailed discussion of translation invariance. As we noted above, translation invariance is a key requirement of the visual system that may determine the need for eye movements.

4.2 Visual Task Performance

In this section we consider several examples of visual tasks and their perceptual complexity. We also discuss the role that eye movements might have in facilitating performance. In particular, we anticipate that the effects of eye movements on performance will depend on the ability of the visual system to perform various visual tasks equally well at different locations in the field of view. Finding performance to be independent of stimulus location would reflect translation invariance of the visual system with respect to those tasks. The impact of eye movements on task performance, in turn, is probably quite limited for those tasks that can be performed in a translation invariant manner.

4.2.1 Detection

Perhaps the most straightforward visual task is the detection of a luminous target (e.g., luminous disk) on a dark background (Fig. 4.1(A)). In a detection experiment that consists of a sequence of trials, an observer is asked to fixate on a fixation point at the center of the display, indicated by the central cross in Fig. 4.1(A). On some trials, the target is displayed for a brief period of time at a location within

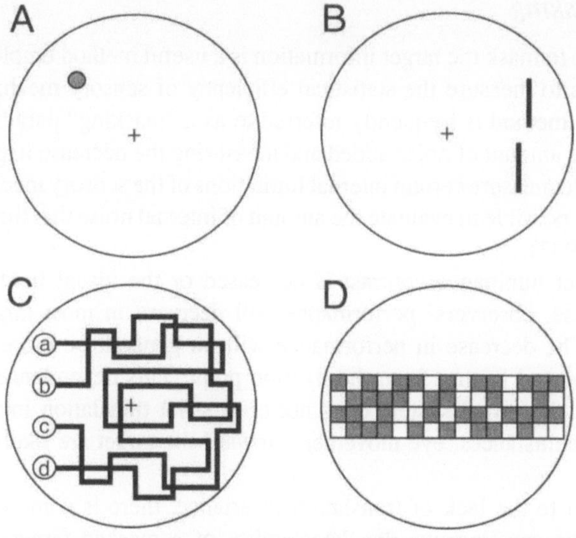

FIGURE 4.1. Examples of various visual tasks. The cross represents the fixation point. (A) Luminous object detected on a dark background. (B) Vernier acuity task. (C) Continuity puzzle. (D) Parity problem.

the display area. On the remaining trials, the target is not displayed at all. On each trial, the observer is asked to indicate whether or not a target was present.

The ability of an observer to detect the target is generally found to depend on the contrast between the background and the target, target luminance, size of the target, and the distance of the target from the fixation point (eccentricity). For luminous, clearly visible targets, the task appears to be very easy and the observers' responses indicate nearly perfect performance. In that situation, the target eccentricity is not a critical variable; the visual system is fairly translation invariant, and we would not expect eye movements to improve performance.

A very similar task involves detecting a bright disk among dark disks (distractors) or a red disk among green ones. If the difference between the distractors and the target is such that the target is easily detected, then the performance is essentially independent of the target eccentricity and even the number of distractors (e.g., Treisman & Gelade, 1980). For these tasks that appear to be fairly translation invariant, eye movements will provide little additional improvement in performance. Observers can perform these tasks without any noticeable effort or focused attention. Because of that, such tasks are sometimes referred to as preattentive (e.g., Bergen & Julesz, 1983).

In addition to the detection of a conspicuous target, the human visual system can also easily perform other tasks in the periphery, such as detection of rapid fluctuations in luminance over time (i.e., flicker), motion at various eccentricities, and so on. Eye movements will provide little help in performing these tasks.

4.2.2 *Masking*

Adding noise to mask the target information is a useful method employed by psychophysicists to measure the statistical efficiency of sensory mechanisms. This experimental method is frequently referred to as a "masking" paradigm. By manipulating the amount of noise added and measuring the decrease in performance, it is possible to measure certain internal limitations of the sensory mechanisms. For example, it is possible to evaluate the amount of internal noise that limits detection (e.g., Pelli, 1983).

If the target luminance contrast is decreased or the visual field is contaminated by noise, observers' performance will decrease in most target detection paradigms. The decrease in performance will, in general, be more pronounced for targets located further from the fixation point. This dependence on location suggests that the visual system does not obey strict translation invariance. Under these circumstances, eye movements toward the target are likely to improve performance.

In addition to the lack of translation invariance, there is some evidence that multiple looks can improve the detectability of a masked target (Levi, Klein & Aitsebaomo, 1985; Rovamo & Virsu, 1979). Under these circumstances, eye movements are likely to facilitate improvements in performance.

It is worthwhile to note that in most detection situations, the amount of signal (contrast increment) that is just detectable is proportional to the amount of noise (e.g., standard deviation of the signal). A similar phenomenon holds for tasks in which observers are asked to detect increments or decrements in contrast. This type of scale invariance in psychophysics is called Weber's Law.

4.2.3 *Localization*

Another important visual task is the localization of objects in the visual field. Whereas absolute localization is relatively poor, the human visual system is capable of making accurate relative location judgments, such as length discrimination (e.g., Burbeck & Hadden, 1993). One way to summarize the empirical results is that the uncertainty in judgments obeys Weber's Law.

An interesting version of the relative location judgment is a task called vernier acuity (Fig. 4.1(B)). Subjects in the vernier acuity task are asked to identify whether the bottom bar is to the left or to the right of the top one. When the vernier stimulus is presented in the fovea, observers can make these judgments extremely accurately. For example, they can discriminate an offset in location down to 6 seconds of arc. As we shall discuss in depth later, performance on the vernier task deteriorates quickly as the stimulus is moved from central vision to the periphery.

The vernier acuity task requires the visual system to perform a more sophisticated task than simple detection. First, it requires the detection of both bars using luminance contrast. Second, it depends on the ability of the visual system to compare locations of two spatially separated objects.

4.2.4 Multidimensional tasks

Spatial relations need not be limited to the relations between two points. One way to increase the requirements on spatial processing is to ask observers to judge spatial contiguity. For example, consider the task of identifying which points are connected in Fig. 4.1(C). There are four starting points on the left of the display and they are pairwise connected. Observers are asked to identify the connected pairs. Although we are not aware of extensive experimental data, our limited observations suggest that eye movements are very useful if not essential for this task.

Each straight line segment in the spatial contiguity task can be interpreted as a dimension of the task. There are ways of increasing the dimensionality of a task that do not depend on spatial location. For example, in the "parity" task (Fig. 4.1(D)) the observer is asked to judge whether the number of dark squares is odd or even. In this task, the position of the squares is irrelevant. The parity task is similar to a counting task. Performance on this task depends on the number of items and the area. For a limited number of items distributed over 2 degrees in the fovea, eye movements are not very helpful (Kowler & Steinman, 1977). As the area and the number of items increase, eye movements appear to be more useful.

In these two examples of multidimensional spatial tasks, eye position might serve as a pointer. One possibility is that the motion of the pointer mediates conversion of a parallel task into a sequential one which is "easier" for humans.

4.2.5 Speed-accuracy tradeoff

Although task difficulty is a central notion of this section, we have not yet defined the relationship between human performance and task difficulty. In fact, an experimenter's choice of a particular empirical performance measure can have critical implications for the assessment of task difficulty.

In psychophysical experiments, the observers' performance is typically characterized by two measures:

1. Accuracy — How accurate are the observers' responses, and

2. Speed — How fast can a task be accomplished.

In most experimental work, researchers have typically focused on one or the other measure. This emphasis might have undesirable consequences because, for most tasks, there is typically a significant tradeoff between speed and accuracy. In particular, faster responses are generally less accurate, and slower responses are typically more accurate. A comprehensive discussion of how these effects are critical for the interpretation of empirical results is beyond the scope of this chapter, but the interested reader can refer to the authoritative work of Sperling and Dosher (1986).

One way to represent the effects of speed-accuracy tradeoff for two different tasks is shown in Fig. 4.2. Each curve in Fig. 4.2, an operating characteristic curve, corresponds to human performance on one task. An operating characteristic curve

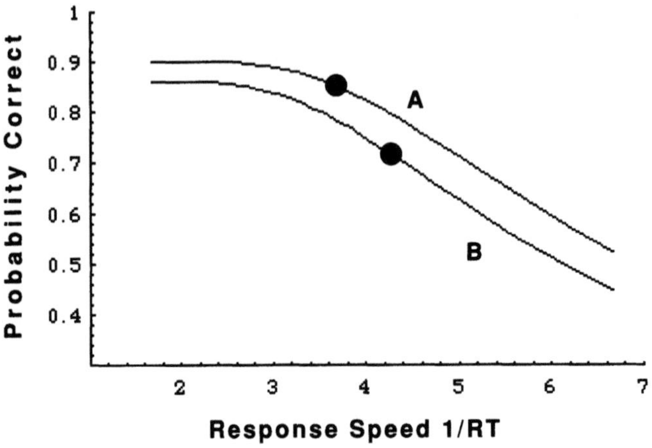

FIGURE 4.2. A tradeoff between speed and accuracy for two different tasks (A and B).

is obtained by repeating the same experiment, but instructing observers to put different emphasis on accuracy or on reaction time. As the observers change their strategies from focusing on accuracy to decreasing their reaction times, they trace out a curve.

Note that if an experimenter performed only a single experiment for each task, he could obtain results indicated by the two black dots. An important implication of this example is that a definition of task difficulty on the basis of reaction time alone, would lead to a conclusion that task A is more difficult than task B. An examination of the operating curves leads to the opposite conclusion. In particular, for any fixed probability of correct responses, task B takes longer to complete than does task A. Thus, task B is actually uniformly more difficult than task A.

4.3 Task Complexity

In this section we consider the notion that the different tasks described in the previous section can be characterized by a single measure of difficulty. The measure that we focus on is the notion of task complexity. We will argue that many tasks that are complicated benefit from using eye movements as a way of converting complex tasks into sequences of simpler ones.

4.3.1 Theory of complexity

One appealing approach to describing task difficulty is the mathematical notion of complexity. A comprehensive introduction to complexity theory is well beyond the scope of this chapter, but we will present certain fundamental concepts that will be useful in our later discussions of the complexity of visual tasks. A clear presentation of some of these concepts can be found in Cover and Thomas (1991).

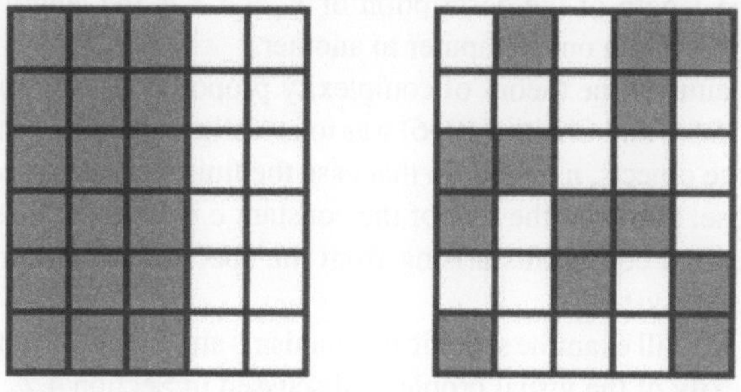

FIGURE 4.3. Example of a "simple" and a "complex" image.

Loosely speaking, the complexity of a pattern is the shortest binary string (interpreted as a computer program) that reproduces the pattern. We introduce the notion of complexity in the following examples.

Consider the two binary $n \times n$ images shown in Fig. 4.3. The left image can be described by specifying the rectangle comprised of dark squares. This description would require on the order of $2 \log n$ bits, that is, two integers describing the corners of the rectangle. In contrast, the image on the right appears to require nearly n bits. According to this analysis, the left image is less complex than the right one. More formally, the complexity K of a string x is the minimum length program p that generates x using computer $\mathcal{U}$:

$$K_{\mathcal{U}}(x) = \min_{p:\, \mathcal{U}(p)=x} l(p). \tag{4.1}$$

The analysis of the example in Fig. 4.3 was based on an assumption that the second image is one of 2^{n^2} possible images. As it turns out, the right image can be described as a repeated sequence $y = \{0101100\}$ that was written horizontally and wrapped around. The sequence y can be thought of as a program that was used to generate the image. The length of the program required to generate the entire image is the length of y, that is, $l(x) = 7 + m$ where m is the length of a small program specifying how y is used to generate the image. This type of efficient representation, however, can be used for a relatively small proportion of possible images. If all 2^{n^2} images are equally likely there is no savings in the length of representation.

This example illustrates that the length of the description depends on the computer, data representation, and possible data. If $\mathcal{U}$ and $\mathcal{A}$ are two computers sufficiently powerful (i.e., universal) computers, then the complexities of x relative to $\mathcal{U}$ and $\mathcal{A}$ differ by a constant independent of the length of x:

$$K_{\mathcal{A}}(x) \le K_{\mathcal{U}}(x) + c. \tag{4.2}$$

The constant c represents the length of a set of instructions that programs computer $\mathcal{A}$ to behave as computer $\mathcal{U}$.

Within this framework, the complexity of string x of n elements can be written as

$$K(x) = c + l(x), \tag{4.3}$$

where $l(x)$ the length of the description of x, and c is the length of a program needed to convert from one computer to another.

One of the aims of the theory of complexity proposed by Kolmogorov (1965), Solomonoff (1964) and Chaitin (1966) was to investigate the algorithmic complexity of very large objects, $n \to \infty$. In that case the finite constant c can be ignored. For our purpose, however, the size of the constant c is likely to be significant because it represents constraints arising from the specific mechanisms underlying the human visual system.

Therefore, we will examine specific mechanisms and the effect of the constraints on the complexity of the visual problems discussed in Section 4.2.

Before we proceed to discuss specific architectures, we note that there is a close connection between complexity theory and information theory. It should be apparent from the examples in Fig. 4.3 that the description of an object depends on the number of other objects that must be distinguished. For example, the number of bits required to describe an integer in a computer depends on the largest possible integer N for the particular machine, and is equal to $l(x) = \log N$. This number of bits is required to distinguish N different integers.

A very closely related notion is the entropy of random variables in information theory. The entropy of a random variable X that takes on values from a set $\mathcal{X}$ is defined by

$$H(X) = -\sum_{x \in \mathcal{X}} p(x) \log(p(x)), \tag{4.4}$$

where $p(x)$ is the probability that X takes on the value x. If each value of an integer-valued X is equally likely, then the entropy of X is equal to $\frac{1}{N} log N$. In general, the entropy of a sequence of N independent random variables $X_i \in \mathcal{X}$, sampled from a finite alphabet $\mathcal{X}$, is approximately equal to the expected value of Kolmogorov complexity divided by N (Cover & Thomas, 1991), that is,

$$\frac{1}{N} E\{K\left[X_1, X_2, ..., X_N | N\right]\} \approx H(X). \tag{4.5}$$

This fact can be useful in relating the complexity of visual tasks to the accuracy of human performance.

4.3.2 Capacity of constrained parallel machines

As we noted above, for finite problems, the length of the computer-specific program c might be an important component of the overall complexity of a problem. Thus, for most visual tasks, the visual system architecture and stimulus representation system will significantly affect the task complexity (difficulty). This effect of architecture on problem complexity is commonly used to infer aspects of the structure of the visual system by measuring the difficulty of different visual problems.

Our analysis here is based on an alternative approach. We start by assuming a two-layer parallel structure for the system architecture and examine its implications for the complexity of visual tasks. In many laboratory visual tasks, the input image and observers' responses are binary. Therefore, we first assume an architecture of a Boolean machine based on Boolean algebra or logic rules (i.e., conjunctions, disjunctions and negations). Subsequently, we consider a more general extension of the Boolean machine based on neural networks.

Any Boolean function can be represented in Disjunctive Normal Form (DNF). DNF consists of disjunctions (*OR*) of binary variables combined by conjunctions (*AND*). An example of a DNF representation for three variables has the form

$$y(x_1, x_2, x_3) = (x_1 \wedge x_2 \wedge x_3) \vee (\bar{x_1} \wedge x_2 \wedge x_3) \vee \ldots, \tag{4.6}$$

where the x_i are Boolean variables, the bar represents negation, and $\wedge$ and $\vee$ represent conjunction and disjunction, respectively. For example, an exclusive OR (*XOR*) can be written as

$$y(x_1, x_2) = (\bar{x_1} \wedge x_2) \vee (x_1 \wedge \bar{x_2}). \tag{4.7}$$

The building elements (basis functions) for Boolean networks are single-valued Boolean functions which can in turn be implemented by switching circuits consisting of *AND* and *OR* gates, and negations. DNF is then represented by two layers of gates, the first layer consists of *AND* gates and the second layer is a single *OR* gate. The complexity of these machines (i.e., the length of a "program") is defined as the number of gates and connections.

A DNF Boolean machine with an unlimited number of gates and connections represents a parallel machine that can compute all Boolean functions. Although DNF is universal, there are several reasons why it is not the most desirable way to implement computations in practice. First, DNF is typically the most complex way of representing a function. There are many functions that can be computed by combinations of considerably fewer gates than prescribed by the complexity of DNF. Second, a DNF representation must in general be extended to more than two layers to accommodate constraints on fan-in, fan-out and connectivity.

Because switching circuits are fundamental components of digital computers, much effort has been devoted to evaluating complexity and finding ways to simplify implementation of Boolean functions (for example, see Wegener, 1987). Many techniques have been developed to take advantage of particular properties of Boolean functions to be minimized. The results of these efforts suggest that major simplifications are achievable for functions that have certain properties, such as symmetry or monotonicity and for those functions that are not completely specified, that is, those that include many "Don't Cares".

We can turn to the problem of estimating the complexity of the visual tasks illustrated in Fig. 4.1 if we assume that all pixels are binary (black or white). We also assume that the number of processing layers is limited to two to minimize processing time.

The simple detection requires a single *OR* combination of all binary pixels. That computation can be accomplished with $O(n)$ complexity (this notation means that complexity is "on the order of n"). The vernier acuity task (Fig. 4.1(B)) requires, in general, a comparison of all pairs of pixels; thus, the complexity of the vernier task is $O(n^2)$. Both of these computations can be accomplished within two or three layers of gates. In contrast, the complexity for a parallel computation of connectivity and parity require exponential complexity for a two-layer circuit. Therefore, this type of computation would be limited to a relatively small field of view, and a sequential type of algorithm may be more desirable. Before we consider a sequential approach, we must ascertain that these conclusions were not the result of limitations due to a binary representation and computation. Therefore, we will examine a more general approach based on adaptive (neural) networks.

To extend the above results to continuous inputs, we consider a class of machines based on a network of units that compute a linear combination of the inputs. The basis functions computed by the units are monotone nonlinear transformations of weighted linear combinations of their inputs. We call these networks adaptive because the parameters of the network can be adjusted in response to the performance of the network on a given task.

If the nonlinear transformation is a step function and the inputs are binary, the adaptive networks reduce to the Boolean machine described above. Thus, any Boolean function can be computed by a machine composed of two layers of these units.

The universality of adaptive networks was first described by Kolmogorov (1957), who demonstrated that with an appropriate choice of the nonlinear transformation, it is possible to approximate any continuous function with arbitrary accuracy using three layers of units[3].

For the purpose of analysis of visual tasks, adaptive networks must behave as classifiers. This is a natural function for an adaptive network. In fact, for continuous inputs, the linear sum followed by a step-function (threshold) behaves as a classifier. In particular, these linear threshold units (LTUs) compute linear discriminant functions. A two-dimensional example is shown in Fig. 4.4. Each input dimension represents the amount of the corresponding feature present in the input. Objects to be classified are represented as points in the feature space. The classification is represented by a surface or surfaces that separate the feature space into subsets corresponding to object categories. For example, a linear threshold unit can classify its input space into two regions separated by a hyperplane.

The separating surfaces are called decision surfaces and are generated by discriminant functions characterizing these surfaces. Fig. 4.4 illustrates two simple two-dimensional feature spaces with two classes of objects to be distinguished.

The complexity of a task is again specified by the number of operations which correspond, in the case of adaptive networks, to the number of units and connections. In the case of a classifier, there is an alternative way to specify complexity, that is, in terms of the complexity of the discriminant surfaces required to perform the categorization task.

[3]There are other bases (units) to model computation. For example, another useful basis involves radial basis functions. The overall complexity arguments will generalize to these bases.

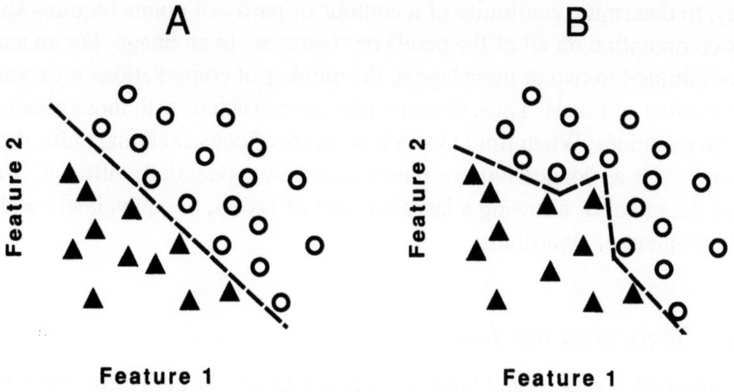

FIGURE 4.4. Classification in a two-dimensional feature space. (A) A linearly separable problem. (B) A problem that can't be solved by a linear combination of features.

Minsky and Papert (1969) provided a useful analysis of various visual tasks. For example, the simplest class of problems that can be solved using a single linear threshold unit is called linearly separable. They classified tasks in terms of the highest order of a predicate[4] that is required to compute a correct response. In a simplified interpretation of their results, the order of a predicate represents the number of pixels that need to be considered simultaneously to perform the task.

The order of predicate is often directly related to the complexity of the machine. For example, according to Minsky and Papert's analysis, linearly separable problems correspond to first-order predicates. That means that each pixel's contribution to a decision is independent of other pixels and the decision can be performed by summing of the contribution of each pixel. Thus, linearly separable tasks can be accomplished by a single layer LTU and their complexity is typically $O(n)$.

We can now turn back to the analysis of the visual tasks presented in Section 4.2. The detection of the presence of a luminous point on a dark background is a first-order predicate, and can be accomplished by a single sum over the image. The sum will be independent of the target location and, therefore, this algorithm is translation invariant.

A different situation arises for the complexity of masking tasks. In a typical masking task, noise is added to each pixel and we cannot assume that the noise variance and the signal values are known exactly. An optimal way to perform a target detection is based on the comparison of each pixel to a number of its nearest neighbors. To perform this detection in a translation invariant manner increases the complexity of the detection task by a factor determined by the size of the neighborhood used for the comparison.

For visual tasks that require assessment of a length or distance, the computer requires at least two pixels. For example, vernier acuity is a second-order predicate.

---

[4]A predicate $\mathcal{P}$ is a binary-valued function of n variables. The order of a predicate is the smallest number of variables $k \leq n$ that can be used to compute $\mathcal{P}$.

Finally, to determine continuity of a contour or parity of points requires simultaneous computation on all of the pixels or "features" in an image. For an adaptive network limited to two or three layers, the number of computations is exponential in the number of pixels. Thus, these results are consistent with those obtained for Boolean machines. When limited to a few layers of computational units, there are problems such as connectivity or parity that are exponentially difficult. We now discuss the effect of allowing a large number of layers, or equivalently and more simply, sequential algorithms.

4.3.3 Sequential machines

With only two layers, visual tasks such as continuity or parity are exponentially complex. For example, a machine that determines parity and is restricted to two layers (e.g., disjunctive normal form or DNF) would require exponentially many gates and connections (Fig. 4.5).

Because of the particular properties of the parity function, it is possible to find more efficient combinatorial circuits by permitting more layers of gates. For example, the parity function can be computed with the order of $O\,(5(n-1))$ gates arranged in $n-1$ layers. The resulting circuit is equivalent to a sequential application of a very simple function such that each level of gates corresponds to a time step for that sequential machine. The only complication of a sequential machine is that it must have memory of the output of its prior output or state.

In particular, the parity $y(n)$ after seeing n inputs can be expressed as a function of the parity after seeing only $n-1$ inputs and the current input x_n,

$$y(n) \;=\; \bar{y}(n-1) \wedge x_n \;\vee y(n-1) \wedge \bar{x_n}. \tag{4.8}$$

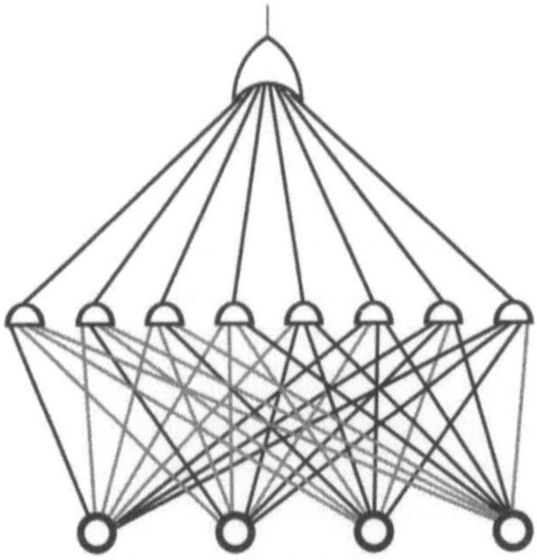

FIGURE 4.5. Disjunctive Normal Form of a combinatorial circuit to determine the parity of the input. The light connections represent negated inputs, the first layer are *AND* gates and the second layer is an *OR* gate.

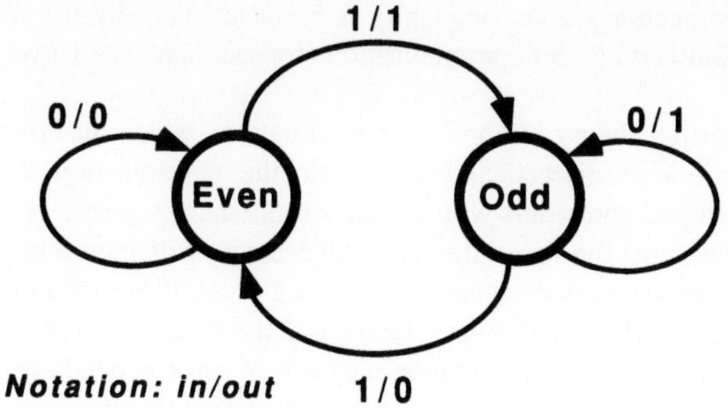

FIGURE 4.6. Sequential machine for the computation of parity for arbitrarily large inputs in the form of a simple logic circuit implementing the state transition diagram.

In this simple case the state (memory) is the output y. A corresponding two-state sequential machine (Fig. 4.6) can determine the parity of an arbitrarily large input in time proportional to the size of the input.

The sequential machine in this example was much simpler than even the simplest corresponding parallel machine. More important, the sequential machine can perform the task (determine parity) for input of any size without prior knowledge of the size.

In general, for some of the computationally "difficult" tasks for parallel machines with a limited number of layers, it is possible to construct sequential machines that can perform the task in polynomial time. We hypothesize that when the complexity of a visual task exceeds the capacity of the available parallel mechanisms, the visual system attempts to convert the parallel task into a sequential one. Although this conversion can be performed by eye movements, it is possible that this tradeoff is also mediated by covert attentional processes. In either case, the sequential approach, which is typically more flexible and requires a less complex mechanism, generally takes longer than a parallel one.

4.3.4 Theoretical speed-accuracy tradeoff

The discussion thus far has been based on the assumption that the results of the computation are always correct. Under this assumption, the complexity of a task is given by the minimum number of steps required to arrive at the correct answer. For sequential algorithms the number of required computations, and the thus the time required to complete a task, can be used as a direct measure of task complexity.

We noted in Section 4.2 that an assessment of human task difficulty must include both accuracy and reaction time because observers have a choice of strategies trading off speed and accuracy. The same type of tradeoff confronts an active sensor.

To interpret accuracy data within the complexity framework we must relate the probability of errors to task complexity. One possible way to bridge the gap between accuracy and complexity is based on rate distortion theory used in information coding and communication (for example, see Cover & Thomas, 1991).

Rate distortion theory is applicable in situations where an errorless code is either impossible or impractical. For example, the representation of arbitrary real numbers in digital computers with a finite word length is generally contaminated by errors. Obviously the size of the error will decrease with increasing length of the binary representation. A designer must make a tradeoff between the complexity of the description (length of binary words) and the resulting error.

We must note that the quantitative measure of error depends on a somewhat arbitrarily selected distortion measure. The theoretical analysis is useful to the extent that the distortion measure, such as Hamming distance or squared error, is relevant to the task.

Loosely speaking, the relationship between the expected distortion D and the length of the optimally selected description is called the *rate distortion function* $R(D)$. If the code is optimized with respect to the selected distortion measure, it is possible to show that the rate distortion function is equal to the mutual information between the coded and the original representations (Cover & Thomas, 1991).

The exact shape of the rate distortion function depends on the distribution of the random variables, as well as on the selected measure of distortion. In situations in which the underlying distribution is Gaussian and the distortion measure D is squared error, the rate distortion function is given by

$$R(D) = \frac{1}{2}\log\frac{\sigma_s^2}{D}, \tag{4.9}$$

where R is in bits, σ_s^2 is the variance of the underlying distribution, and $0 < D \leq \sigma_s^2$.

We entertain the hypothesis that the rate distortion function may account for a portion of the observed speed-accuracy tradeoff. Suppose that the visual system refines its estimate of a normally distributed stimulus parameter over time by a binary search procedure. Then the response time would be proportional to the rate, that is, $T = \alpha R$, where α is a positive constant. In that case, the variance in the stimulus representation after T seconds is

$$\sigma^2 = \sigma_s^2 2^{-2\frac{T}{\alpha}} + \sigma_i^2, \tag{4.10}$$

where σ_i^2 is intrinsic visual system noise. The representation variance is the variability that determines stimulus discriminability and the response error rate.

Although it is not very likely the rate distortion function completely accounts for the empirically observed speed-accuracy tradeoff, it may be a useful way to analyze peripheral coding and processing of visual stimuli.

4.4 Translation Invariance

It is probably not surprising that continuity and parity tasks over a large visual field benefit from eye movements. In this section, however, we argue that even much less complex tasks, such as vernier acuity, can benefit from a sequential approach that compensates for the lack of translation invariance of the visual system.

The capability of translation invariant spatial pattern recognition is of great importance to mechanical and biological visual systems. A translation invariant system is capable of recognizing patterns independently of their position in the visual field, as well as determining their position. The cost of translation invariant pattern recognition is reflected by an increase in complexity because any analysis must be performed at all locations in the visual field. For example, for the vernier acuity task described in Section 4.2 the complexity is $O(n^2)$. The complexity due to translation invariance might be quite high if the number of different patterns at any location increases. It would be useful to determine to what extent the human visual system obeys translation invariance.

Before we proceed any further, we must define translation invariance in terms of observable measures from behavioral experiments. A definition based on the ability to identify an object in the central and peripheral visual fields is not quite sufficient. Such a definition depends on the ensemble of patterns used to test the invariance. For example, consider a letter identification task where observers are asked to identify a large capital letter "A" as different from the letter "B". The differences between the letters are so great that the task is performed perfectly regardless of considerable distortions to the images. A more sensitive measurement (i.e., a stricter definition) is required to determine whether the visual system is translation invariant.

A more strict definition of translation invariance can be based on the probability of discriminating between two similar objects. Let's denote the probability of discriminating between stimuli, defined by their local spatial parameters a and b located at eccentricity r, by $Pr\{a, b; r\}$. The local parameters might represent dimensions or relative positions of object features. For example, the horizontal displacement of the lines could represent the stimulus in the vernier task. It is important that the values of the parameters a and b are chosen such that the discrimination probability is neither zero nor one. Such a choice of stimuli assures that they are not too different (always discriminable) nor too similar (indiscriminable). If a visual system is strictly translation invariant then

$$Pr\{a, b; r\} = Pr\{a, b; 0\}. \tag{4.11}$$

That is, the probability of correctly distinguishing between a and b should be independent of the location in the visual field.

Given this definition, it is obvious that the human visual system is not strictly translation invariant. In addition to the everyday experience that peripheral vision is not as acute as central vision, the lack of translation invariance is suggested by the nonuniform distribution of receptors in the retina. Most of the receptors are located in the central area of the retina called the fovea. The density of the receptors decreases rapidly with distance from the fovea.

Despite this severe violation of our strict definition of translation invariance, it is possible that the visual system is essentially translation invariant except for the nonuniform peripheral representation. This nonuniformity could be achieved by a spatial transformation, such as dilation, whose parameters depend on eccentricity. In practice, such a transformation could be achieved by reducing the density of peripheral sampling and the corresponding representation of the retinal image. There appears to be some evidence of this type of transformation in physiological data.

If the lack of translation invariance can be fully accounted for by the peripheral representation, as some investigators have proposed, it should be possible to compensate for the loss of spatial sensitivity due to translation to the periphery by a change in scale of the objects. Thus, two objects can be equally discriminable in the periphery as they are in the fovea provided that they are appropriately enlarged. Mathematically, we restore translation invariance by enlarging each object by the same factor $m(r)$:

$$Pr\,\{m(r)a,\ m(r)b;\ r\} \;=\; Pr\,\{a,\ b;\ 0\}\,. \tag{4.12}$$

An example of restoration of translation invariance is shown in Fig. 4.7.

The scaling function $m(r)$ represents the derivative of the visual angle at eccentricity r with respect to the corresponding extent in the internal representation. Following the work of Anstis (1974) there are many studies concerning the performance of spatial discrimination in periphery (Levi et al., 1985). The results from various studies with different stimuli s are in general agreement that the scaling functions m_s are linear functions of eccentricity (Fig. 4.8).

The difficulty of the task is expressed in terms of just noticeable differences (JNDs) which represent estimates of the physical offset, δ, that is required for 75% correct responses (the solid line). The scaling function for JNDs can be expressed

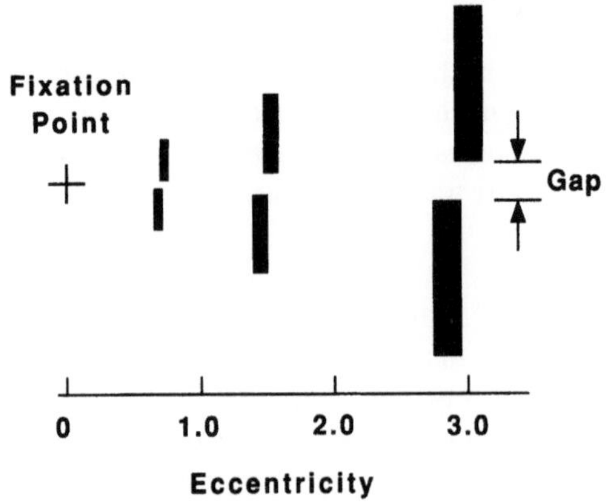

FIGURE 4.7. Vernier acuity task scaled for eccentricity to restore performance

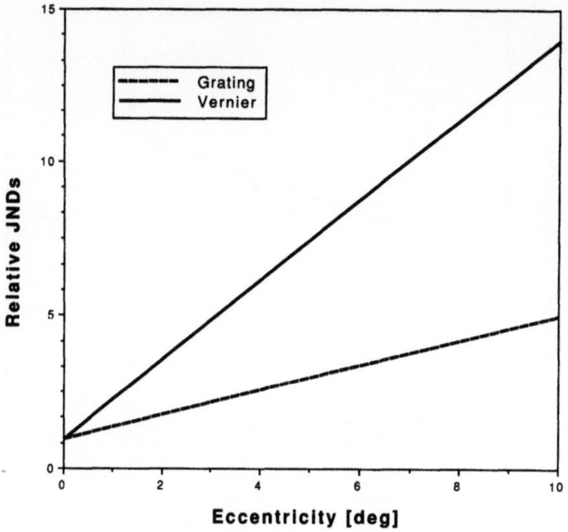

FIGURE 4.8. Plot of Just Noticeable Differences (JND) for a vernier acuity task (solid line) and grating detection (dashed line) as a function of eccentricity. JND is defined as the physical offset required for 75% correct discrimination on the vernier acuity task, and the contrast for 75% detection of sinusoidal gratings. Both functions were normalized to be equal to unity at the center of the visual field (fovea).

as a linear function of eccentricity

$$m_s(r) = 1 + \frac{r}{r_s}, \tag{4.13}$$

where r_s is a positive constant. The linearity of the function m, however, is neither necessary nor sufficient for the existence of translation invariance. That is, the scaling function m could be any positive function of eccentricity.

Recently, more careful investigations of the validity of translation invariance have been undertaken. To investigate whether stimulus scaling can compensate for a translation to the periphery requires that all spatial dimensions of the stimulus be scaled equally. Cunningham and Pavel (1986) have examined the effect of the gap in the vernier acuity target (Fig. 4.1(B)). The resulting performance as a function of gap size for stimuli presented at different eccentricities is shown in Fig. 4.9. The JND is a linear function of gap size at all eccentricities. The fact that these lines are parallel can be used to prove that the visual system is translation invariant after appropriate scaling of all dimensions of the stimuli by the function m.

Whereas these results are encouraging in terms of restoring translation invariance, there is considerable evidence that the scaling function m depends on the specific task. We hypothesize that the scaling function m depends on the task complexity. In particular, the simpler, first-order predicate spatial tasks seem to require scaling functions $m(r)$ with lower slope than do the more complex tasks. In the case of a linear scaling function, the constant r_s appears to be two to four times smaller

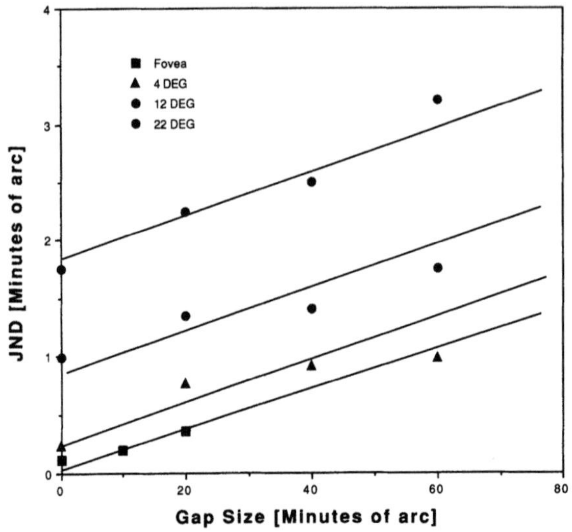

FIGURE 4.9. Plot of Just Noticeable Differences (JND) from a vernier acuity task as a function of the gap between the vertical bars. The parameter is the eccentricity of the stimulus (nasal retina).

for the first-order predicates than it is for higher-order predicates. For example, performance on detection of the presence of a sinusoidal grating in the periphery deteriorates with eccentricity at a slower rate than does the performance of vernier acuity. In general, the performance on tasks requiring second- and higher-order predicates deteriorates more rapidly with eccentricity.

To summarize, the human visual system can recognize patterns in the periphery in a similar manner as in central vision, but with lower spatial acuity. This is because a uniform increase in the size of objects is equivalent to a corresponding reduction in spatial frequencies. Thus, the visual system can be thought of as performing the preliminary, low acuity analysis in the periphery, and a high acuity analysis in central vision. The peripheral analysis provides low acuity and approximate information on object location. This preliminary peripheral analysis is followed by a precise pattern analysis in central vision after the image of the object is centered using eye movements. Thus, a parallel pattern recognition task, requiring a large visual field and high precision, is converted to a serial process consisting of locating, centering, and then analyzing individual patterns.

4.5 Conclusion

Our objective in this chapter was to argue that active vision, as represented by eye movements, can be used to mediate a tradeoff between performance and cost requirements in vision systems. The cost is in the complexity of hardware (wetware), the time, and the accuracy required to perform various tasks.

In the conclusion we would like to reiterate the three main points of our discussion.

1. We argued that computational complexity of a particular task can be used to assess the difficulty of such a task for the human visual system. Although the full power of the Kolmogorov-Chaitin complexity theory may not be applicable directly, the general approach of analyzing the complexity of individual objects was shown to be potentially useful in predicting human performance.

2. Eye movements can be used to mediate a tradeoff between the complexity of a fixed, but fast pattern recognition machine and that of a sequential, slow but flexible pattern recognition system. For complex visual tasks, a parallel solution would require too much parallel hardware that would have to be continuously adapted to each task. The visual system appears to economize on the number of parallel computations by taking advantage of the fact that sequential algorithms are actually preferable for some tasks. For those tasks, eye movements (or a moving camera) can provide a simple and effective means for converting computationally complex parallel tasks into serial ones.

3. The human visual system is, strictly speaking, not translation invariant. However, human performance becomes translation invariant after a particular scaling (dilation) transformation of the input. The resulting representation provides a convenient tradeoff between requirements for high accuracy (high complexity) capabilities, and the size of the visual field that can be monitored in parallel.

We hope that our discussion will motivate more rigorous analyses of the performance of the human visual system. The results of such analyses would improve our understanding of the human visual system and at the same time provide engineers with new directions for designing artificial vision systems.

Acknowledgments: This work was supported by NASA grants NCC 2-269 to Stanford University, and NAG 2-931 to Oregon Graduate Institute.

4.6 References

Anstis, S. M. (1974). A chart demonstrating variations in acuity with retinal position. *Vision Research, 14*, 589–592.

Bergen, J. R. & Julesz, B. (1983). Rapid discrimination of visual patterns. *IEEE Transactions on Systems, Man, and Cybernetics, SMC-13*, 857–863.

Burbeck, C. A. & Hadden, S. (1993). Scaled position integration areas: accounting for weber's law for separation. *Journal of the Optical Society of America A, 10*, 5–15.

Chaitin, G. J. (1966). On the length of programs for computing binary sequences. *Journal of the Association for Computing Machinery, 13*, 547–569.

Cover, T. M. & Thomas, J. A. (1991). *Elements of Information Theory*. New York: Wiley.

Cunningham, H. A. & Pavel, M. (1986). Judgements of position in near and far peripheral fields. *Investigative Ophthalmology & Visual Science, Supplement, 27*, 95.

Kolmogorov, A. N. (1957). On the representation of continuous functions of many variables by superposition of continuous functions of one variable and addition. *Doklady Akademii Nauk USSR, 114*, 953–956.

Kolmogorov, A. N. (1965). Three approaches to quantitative definition of information. *Problems in Information Transmission, 1*, 4–7.

Kowler, E. & Steinman, R. M. (1977). The role of small saccades in counting. *Vision Research, 17*, 141–146.

Levi, D. M., Klein, S. A. & Aitsebaomo, A. P. (1985). Vernier acuity, crowding and cortical magnification. *Vision Research, 25*, 963–977.

Minsky, M. & Papert, S. (1969). *Perceptrons: An Introduction to Computational Geometry*. Cambridge, Massachusetts: MIT Press.

Pelli, D. G. (1983). The spatiotemporal spectrum of the equivalent noise of human vision. *Investigative Ophthalmology & Visual Science, Supplement, 24*, 46.

Rovamo, J. & Virsu, V. (1979). An estimation and application of human magnification factor. *Experimental Brain Research, 37*, 495–510.

Solomonoff, R. J. (1964). A formal theory of inductive inference. *Information and Control, 4*, 224–254.

Sperling, G. & Dosher, B. (1986). Strategy and optimization in human information processing. In K. Boff, L. Kaufman & J. Thomas (Eds.), *Handbook of Perception and Human Performance*, Volume 1 chapter 2. New York: Wiley.

Treisman, A. M. & Gelade, G. (1980). A feature-integration theory of attention. *Cognitive Psychology, 12*, 97–136.

Wegener, I. (1987). *The Complexity of Boolean Functions*. New York: Wiley.

5

Exploratory Vision: Some Implications for Retinal Sampling and Reconstruction

Laurence T. Maloney[1]

ABSTRACT As we move about a room, inspecting its contents, we ordinarily do not confuse our own change of viewpoint with changes in the content of the scene. As we move, the image of a single object or location may fall successively on parts of the retina with markedly different optical qualities and photoreceptor densities. Yet, we typically manage not to confuse the particular characteristics of a retinal region with the appearance of an object imaged on it. An object fixated and then viewed in periphery does not seem to move or change, nor does an object first viewed peripherally and then fixated although, once fixated, we are likely able to answer questions about the detailed appearance of the fixated object that we could not answer when it was viewed peripherally. This *transformational constancy* is all the more remarkable if we examine the initial visual information, the pattern of excitation of photoreceptors in each retinal region. This chapter analyzes the retina as a sampling array in motion, discussing the consequences of motion for reconstruction, aliasing, and visual representation.

5.1 Introduction

Discussions of exploratory vision typically concern strategies for exploring the environment, algorithms for piecing together information derived from multiple views, and specifications of visual representations that would be suitable for a visual system that moves within the represented scenes.[2] This chapter is concerned with the consequences of movement for the earliest stages of human vision, beginning with the encoding of the pattern of light on the retina (*the retinal image*) as a pattern of photoreceptor excitations.

The retinal photoreceptor array, viewed as a sampling grid, differs from the familiar square or hexagonal lattice of sampling theory (Zayed, 1993) in three

[1]Department of Psychology and Center for Neural Science, New York University

[2]For example, Schölkopf and Mallot (1994) develop algorithms that permit a computational vision system to learn the layout of a maze form two-dimensional views of locations in the maze.

respects. First, the distribution of the photoreceptors ("the sampling elements") is markedly inhomogeneous (Hirsch & Miller, 1987; Østerberg, 1935). Second, the receptors in any small retinal region outside of the fovea are not positioned on a precisely regular grid (Hirsch & Miller, 1987). Third, the spatial density of receptors in some parts of the retina may be lower than the density needed to unambiguously encode the light image within that region (Artal, Navarro, Brainard, Galvin & Williams, 1992; Navarro, Artal & Williams, 1993).

When the eye moves, information concerning any given object in the scene may fall on regions of the retina with different densities of photoreceptors, or regions of similar density whose photoreceptors do not form a regular array. The peculiar characteristics of each retinal region are, as a consequence, reflected in the information it records about the retinal image and the scene. It is is not obvious how to decide whether a given object or location in the scene, sampled successively by two retinal regions, has changed in some way. Nor is it obvious how to combine information from different retinal regions that successively view a single object or location.

These two problems, *change detection* and *combination* of information across eye movements, are the organizing themes for the remainder of the chapter. In later sections, we will return to these problems and discuss what kinds of representations are appropriate for an "exploratory sampling system" that embodies a solution to them.

The first sections of this chapter develop a simple mathematical framework for retinal sampling and *linear* reconstruction (and try to eliminate some of the folklore surrounding the "sampling theorem"). The following sections discuss some of the implications of movement for linear and nonlinear reconstruction, analyze the conditions under which reconstruction could benefit from eye movements, and discuss the kinds of representations appropriate for sampling and reconstruction across time.

5.2 From Scene to Sensor to Code

Fig. 5.1 schematizes the physical processes that reduce the full description of the scene to the instantaneous visual information available to the a single eye or camera: (1) *projection* of the scene onto the retina (ignoring the blurring introduced by the optics of the visual system), (2) *blurring* of the retinal image due to the optics of the visual system, and (3) *sampling* of the image by a finite number of photoreceptors. Each process loses information about the scene. The end product, the *sampling code*, is the instantaneous information available about the scene. This section defines, in detail, the terms *ideal image*, *retinal image*, and *sampling code*, making clear what assumptions are being made to reduce the initial stages of vision to the simple diagram in Fig. 5.1. Here, we only consider the case of monocular vision.

The instantaneous excitation of a biological photoreceptor is, in effect, a discrete sample of the light intensity in a small region of the retina, weighted by the

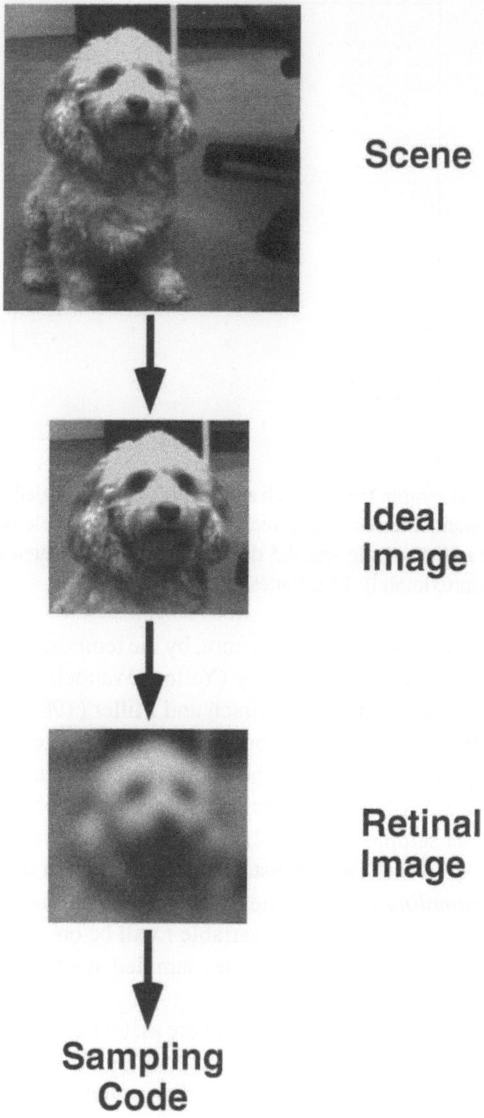

FIGURE 5.1. Successive transformations of the information available to a biological vision system. The scene is *projected* by visual optics. If the the eye were optically perfect, the result would be the *ideal retinal image*. The optics significantly blur this *ideal image* and the result is the actual pattern of light on the retina, the *retinal image*. This image is sampled by photoreceptors. The combined instantaneous excitation of all of the photoreceptors is termed the *sampling code*.

FIGURE 5.2. A retinal sample from Hirsch and Miller (1987) rescaled to lie within the unit square. Each point marks the center of a receptor. This is the sample numbered 20 in their Table 1. The center of the sample was 4.5 degrees from the estimated center of the retina, and the sample is approximately 13 minutes of arc on each side.

spatial profile of the photoreceptor's aperture, by the temporal response of the photoreceptor, and by its spectral sensitivity (Yellott, Wandell & Cornsweet, 1984). Fig. 5.2 shows a retinal sample from Hirsch and Miller (1987). Hirsch and Miller recorded the locations of photoreceptors in 25 square regions in a single human retina. The figure plots the centers of the photoreceptors found in a single region.

Suppose there are N photoreceptors in a given region of the retina ($N = 60$ for the Hirsch-Miller sample in Fig. 5.2) numbered, for convenience, from 1 to N. Let $\rho_i(t), i = 1, \cdots, N$ denote the instantaneous excitation of the ith receptor at time t. Then the *sampling code* for the given region is defined to be the vector[3] $\rho(t) = [\rho_1(t), \cdots, \rho_N(t)]^T$. The time variable t will be omitted in the sequel. We will be concerned only with sampling codes sampled at either one or two instants of time.

In the analyses in this chapter, all receptors are assumed to have the same spectral sensitivity $R(\lambda)$, where λ denotes the wavelength of light. Let $L^{\text{ideal}}(\lambda, x, y)$ denote the mean intensity of light of wavelength λ that would be incident on the retina at location (x, y) if the optics of the eye were perfect. The blurring of any small region of the retinal image introduced by the optical media in the eye can then be approximated by a *space-variant linear operator*,

$$L(\lambda, x, y) = \iint E(\lambda, x, y, u, v) L^{\text{ideal}}(\lambda, u, v)\, du\, dv, \tag{5.1}$$

where the kernel $E(\lambda, x, y, u, v)$ characterizes the effect of the optics on the retinal image for wavelength λ at location (x,y). If we confine attention to a region of the

[3]The superscript T denotes matrix or vector "transpose".

retina across which the effect of the optical media is approximately constant, then we could replace Eq. 5.1 by a *convolution*,

$$L(\lambda, x, y) = \iint E(\lambda, x-u, y-v) L^{\text{ideal}}(\lambda, u, v)\, du\, dv. \tag{5.2}$$

In this chapter, we consider the change detection problem for patches of retina that are sufficiently far apart in the eye that the degree of optical blurring differs. The more general Eq. 5.1 will therefore be used.

We will ignore complexities introduced by the chromatic aberration of the eye (Simonet & Campbell, 1990; Thibos, 1987; Thibos, Bradley, Still, Zhang & Howarth, 1990) and the Stiles-Crawford effects (Wyszecki & Stiles, 1982) and suppress the argument λ in $E(\lambda, x, y, u, v)$. With these simplifications, Eq. 5.1 becomes

$$L(\lambda, x, y) = \iint E(x, y, u, v) L^{\text{ideal}}(\lambda, u, v)\, du\, dv. \tag{5.3}$$

We model the effect of the aperture of a photoreceptor at location (x, y) as a space-invariant linear operator with kernel $A(x, y, u, v)$ ("the blurring due to the aperture"). We assumed above that all receptors have the same spectral sensitivity $R(\lambda)$.

The output of a receptor positioned at (x, y) is assumed to be

$$\rho(x, y) = \iiint A(x, y, w, z) L(\lambda, w, z)\, dw\, dz\, R(\lambda)\, d\lambda. \tag{5.4}$$

Substituting Eq. 5.3 in Eq. 5.4, we have

$$\rho(x, y) = \iiiint\!\!\int A(x, y, w, z) E(w, z, u, v) \\ L^{\text{ideal}}(\lambda, u, v)\, du\, dv\, dw\, dz\, R(\lambda)\, d\lambda, \tag{5.5}$$

and, rearranging the order of integration,

$$\rho(x, y) = \iint \left[\int L^{\text{ideal}}(\lambda, u, v) R(\lambda)\, d\lambda\right] \\ \left[\iint E(w, z, u, v) A(x, y, w, z)\, dw\, dz\right] du\, dv. \tag{5.6}$$

We simplify this expression as follows. Let

$$\rho^{\text{ideal}}(x, y) = \int L^{\text{ideal}}(\lambda, x, y) R(\lambda)\, d\lambda \tag{5.7}$$

(the first expression in brackets in Eq. 5.6). Define

$$F(x, y, u, v) = \iint E(w, z, u, v) A(x, y, w, z)\, dw\, dz. \tag{5.8}$$

(the second expression in brackets in Eq. 5.6). We then rewrite Eq. 5.6 as

$$\rho(x, y) = \iint \rho^{\text{ideal}}(u, v) F(x, y, u, v)\, du\, dv. \tag{5.9}$$

The function $F(x, y, u, v)$ is the space-variant linear filter that results from combining the original optical filter kernel function $E(x, y, u, v)$ and the aperture blurring kernel function $A(x, y, u, v)$. By combining the kernel functions in this way, we can regard the photoreceptors as punctate sampling devices as suggested by Yellott (1982). Of course, what we are now calling the retinal image differs from the mean physical intensity of light on the retina.

Definitions: The function $\rho^{\text{ideal}}(x, y)$ is termed the *ideal image*. It is the mean intensity of light that would reach the retina without blurring, weighted by the spectral sensitivity of the photoreceptors. The function $\rho^{\text{ideal}}(x, y)$ is also the excitation of a punctate receptor centered on location (x, y), if there were no blurring by optical media or aperture. The function $\rho(x, y)$ is the excitation of a punctate receptor centered on location (x,y) after blurring.

The receptors in the Hirsch-Miller sample shown in Fig. 5.2 resemble a distorted hexagonal lattice. We will, however, make no special assumptions concerning the spatial organization (or lack of spatial organization) of the receptors in what follows.

If we denote the coordinates of the ith receptor in the Hirsch-Miller sample by $(x_i, y_i), i = 1, \cdots, N$, then the *sampling code* is just the vector $\rho = [\rho_1, \cdots \rho_N]^T$ where $\rho_i = \rho(x_i, y_i)$.

Terminology: In this chapter we will analyze the sampling and reconstruction properties of small retinal regions. They will be termed *patches* or *retinal patches*. The term *sampling code* will refer to the sampling code of the patch under discussion, and the terms *retinal image* and *ideal image* will refer to the parts of each image that fall within the patch. The physical processes in the environment that give rise to the retinal image are complex. It would be incorrect, in several respects, to refer to "the object that a patch is sampling". Most obviously, there need be no object or surface along a given line of sight, or there may be several (transparency). In addition, the retinal image contains information about the illumination of a surface along a given line of sight as well as about the surface itself. Keeping these factors in mind, I will use phrases such as "the same location (or direction or object) in the scene is sampled in succession by two retinal patches" simply to describe the geometric relation between retina and environment.

5.3 Linear Reconstruction and the Sampling Theorem

Biological visual systems are unreasonably successful at turning the initial visual input, the sampling code, into useful information about objects, scene illumination, surface properties, and the like. Their successes may be termed "unreasonable" in two respects. First, it is generally accepted that there is insufficient information available in the sample vector to identify the contents of the scene outside: visual perception, based on a single glance at a scene, is an *ill-posed* problem (Marr, 1982; Poggio & Torre, 1984; Poggio, Torre & Koch, 1985). Second, human performance in simple hyperacuity tasks indicates that we can, under ideal conditions, judge retinal separations that are small compared to the spacings and diameters of the receptors (Westheimer & McKee, 1975). This level of performance suggests that some estimates of relative position made in constructing a representation of the scene have an accuracy comparable to the spacings between and diameters of the receptors.

For now, we are concerned with the instantaneous contents of the scene and the instantaneous sampling code. We will consider to what extent and under what circumstances it is possible to "reverse" the last arrow in Fig. 5.1 and reconstruct the retinal image $\rho(x, y)$, from the sampling code ρ.

A question that might occur to the reader at this point is "Why study linear reconstruction at all?" It is the case that all of the processes by which the ideal image was reduced to the retinal image and the retinal image, in turn, was reduced to a sampling code, are linear or approximately linear (Yellott et al., 1984). However, the initial projection of the scene onto the retinal is a nonlinear operation, and it is difficult to imagine a reconstruction of the three-dimensional contents of a scene that is linear with respect to the photoreceptor excitations. Indeed, there seems to be no obvious way to define an addition operation on scenes containing objects that occlude one another under projection. If we wish to model how the visual system reconstructs a representation of its environment, linear reconstruction methods would seem to be especially unpromising.

The term "nonlinear" is used in two senses in early vision. In the first sense, it describes nonlinear transformations of the outputs of channels in the visual system (Sperling, 1989). To the extent that these transforms are reversible, they have little effect on the sampling issues treated here, and we may treat such a linear-nonlinear visual system as effectively linear. In the second sense, the term "linear" is used to describe the piecewise smooth mappings of the retina to retinotopic areas later in the visual system. Mallot, von Seelen and Giannakopolous (1990) provide an excellent introduction and analysis of the consequences of space-variant image processing and mapping between cortical areas.

We will return to this point in a later section and argue that an *initial* linear (or effectively linear) reconstruction stage could serve a useful role in visual reconstruction, by addressing the change detection problem, the first of the two thematic problems discussed in Section 5.1. Namely, the resulting representation can be made less dependent on the idiosyncratic layout of receptors in each retinal region, and less sensitive to small eye movements.

The engineering literature on sampling theory concentrates on specific problems where a temporal or spatial signal is sampled at indefinitely many regularly spaced points. Zayed (1993) provides a comprehensive survey of current mathematical results in sampling theory. We are concerned with linear reconstruction of a retinal signal over a finite region given only a finite number of samples at fixed locations, where the layout of the sampling points, their density, and the nature of the signal varies across space. Despite the apparent similarity of this problem and the problems treated by sampling theory, we can (and should) avoid the latter altogether. Instead, we make use of simple results from *interpolation theory* (Davis, 1963) that do not require any special assumptions about the layout of the photoreceptors in the retinal patch or the shape of the retinal patch. The assumptions we make about the class of possible retinal images will be less restrictive than those made in sampling theory. The interpolation theory results are precisely correct for the problem we have set up and, as we will see below, most of what we will be concerned with reduces to simple and well understood linear algebra.

We will model the ideal image $\rho^{\text{ideal}}(x, y)$ as a *linear function subspace* of countably infinite dimension (Apostol, 1969). That is, there are fixed functions $\rho_1(x, y), \cdots, \rho_i(x, y), \cdots$ such that any ideal image can be expressed as

$$\rho(x, y) = \sum_{i=1}^{\infty} \varepsilon_i \rho_i(x, y). \tag{5.10}$$

We assume that the fixed functions $\rho_i(x, y)$ are linearly independent. The fixed functions $\rho_1(x, y), \cdots, \rho_i(x, y), \cdots$ are then termed a *basis* for the linear function space of ideal images.[4] Each is a function from the region of the retina of interest to the real numbers $\mathbb{R}$. Next, we assume that there is a fixed integer P such that any retinal image $\rho(x, y)$ can be expressed as

$$\rho(x, y) = \sum_{i=1}^{P} \varepsilon_i \rho_i(x, y). \tag{5.11}$$

That is, the first P basis elements of the infinite-dimensional space of ideal images form a basis for the P-dimensional subspace of retinal images. With respect to this basis the retinal image in Eq.5.10 can be assigned coordinates

$$\varepsilon = [\varepsilon_1, \cdots, \varepsilon_P]^T. \tag{5.12}$$

In setting up this notation, we only made only one significant empirical assumption: the linear process of filtering which transforms the ideal into the retinal image projects the ideal image into a finite-dimensional linear subspace. For example, a perfect low-pass filter is such a projection, and it is typical to assume that the blurring induced by optical media and receptor apertures effects something like a low-pass filtering of the ideal image. It is useful to assume that the $\rho_i(x, y)$ are arranged in order of the degree to which each is attenuated by the linear optical filter. That is, $\rho_{i+1}(x, y)$ is assumed to be attenuated (by optical filtering) at least as much as $\rho_i(x, y)$. Of course, we continue to assume that the function $\rho_{P+1}(x, y)$ does not pass through the optical filter at all (by whatever criterion we have established). We will make use of this notational assumption in a few places in the following.

---

[4]**Notation:** Note that ρ_i is the excitation of the ith photoreceptor in the sampling code, and that $\rho_i(x, y)$ is the ith basis element, something completely different. Also, ρ denotes the sampling code (a vector) and $\rho(x, y)$, the retinal image, is a function of two variables.

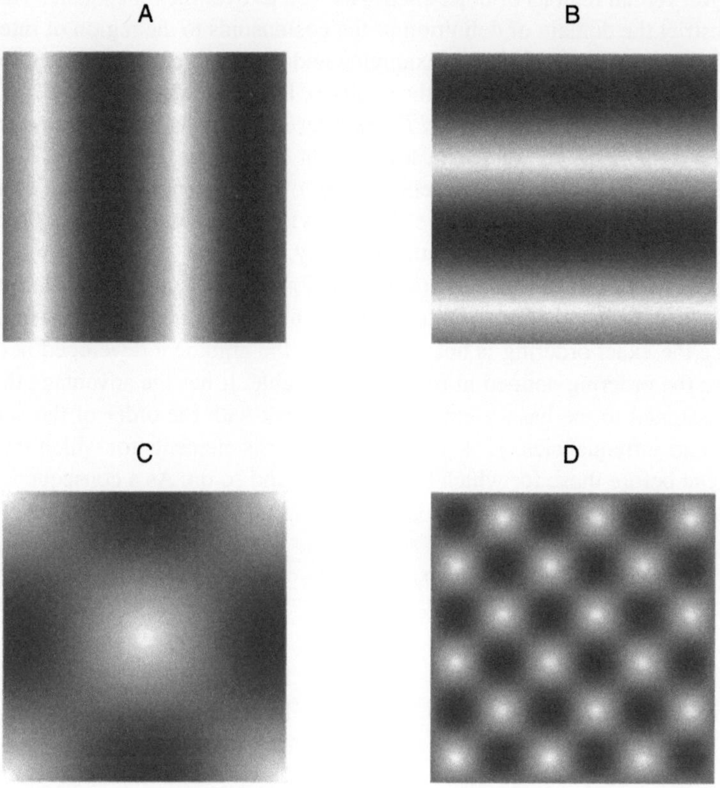

FIGURE 5.3. Four cosinusoids from the Fourier basis: (A) $\sin 2\pi 2x$, (B) $\sin 2\pi 2y$, (C) $\cos 2\pi x \cos 2\pi y$, (D) $\sin 2\pi 3x \sin 2\pi 3y$.

The Fourier Basis. A basis typically employed when the retinal region of interest is square is the set of two-dimensional cosinusoids,[5]

$$\begin{array}{ccc} 1 & \cos 2\pi mx & \sin 2\pi mx \\ & \cos 2\pi ny & \sin 2\pi ny \\ & \cos 2\pi mx \cos 2\pi ny & \sin 2\pi mx \cos 2\pi ny \\ & \cos 2\pi mx \sin 2\pi ny & \sin 2\pi mx \sin 2\pi ny \end{array} \tag{5.13}$$

for $m, n = 1, 2, 3, \cdots$.

Fig. 5.3 shows a few of these cosinusoids. With this choice of basis, Eq. 5.10 is a two-dimensional *Fourier series* and Eq. 5.11 is a *trigonometric polynomial* (Dym & McKean, 1972). With this choice of basis, the possible ideal images include all piecewise continuous functions (Dym & McKean, 1972). This basis can serve as a

[5]The term *cosinusoid* will be used to refer to both sine and cosine functions and the product of any two such functions. It will also include the constant function 1.

basis over retinal regions of other shapes as well as over the unit square. We need only restrict the domain of definition of the cosinusoids to the region of interest.[6]

We will employ this basis in examples and certain analyses below. Nothing about our analyses will depend on the choice of basis (or sampling array). Eq. 5.10 is often referred to as a *generalized Fourier series* or simple a *Fourier series* even if the basis elements are not cosinusoids (Dym & McKean, 1972).

Each cosinusoid has two frequencies, one of the x variable f_x, the other for the y variable f_y. We noted above that we wished to order the basis functions by the degree to which each is attenuated by the linear optical filter. It is not immediately obvious how to do so (e.g., is $\sin 2\pi x \sin 2\pi y$ more or less attenuated than $\sin 2\pi 2x$?), but the ordering could be established by empirical measurement.

Since the exact ordering is not important to the argument developed here, we will use the ordering defined in the following table. It has the advantage that the order assigned to the basis elements is consistent with the order of the *sums* of their x and y frequencies, $f_x + f_y$. That is, all basis elements for which this sum is 8 come before those for which this sum is 9, and so on. As a consequence, the successive indices increase as one goes down and to the right in the table.

	1	$\sin 2\pi y$	$\cos 2\pi y$	$\sin 2\pi 2y$	$\cos 2\pi 2y$	$\sin 2\pi 3y$	$\cdots$
1	1	2	4	7	11	16	
$\sin 2\pi y$	3	5	8	12	17	23	
$\cos 2\pi y$	6	9	13	18	24	31	
$\sin 2\pi 2y$	10	14	19	25	32	40	
$\cos 2\pi 2y$	15	20	26	33	41	.	
$\sin 2\pi 3y$	21	27	34	42	.	.	
$\vdots$							

An alternative numbering scheme is the following. It has the property that, for every nonnegative integer n, the space of images spanned by $\rho_1(x, y), \cdots, \rho_N(x, y)$ where $N = (2n+1)^2$ contains all of the functions whose x and y frequencies are both less than or equal to n.

	1	$\sin 2\pi y$	$\cos 2\pi y$	$\sin 2\pi 2y$	$\cos 2\pi 2y$	$\sin 2\pi 3y$	$\cdots$
1	1	2	5	10	17	26	
$\sin 2\pi y$	3	4	7	12	19	28	
$\cos 2\pi y$	6	8	9	14	21	30	
$\sin 2\pi 2y$	11	13	15	16	23	32	
$\cos 2\pi 2y$	18	20	22	24	25	.	
$\sin 2\pi 3y$	27	29	31	33	.	.	
$\vdots$							

[6]The reader may be aware that the Fourier basis elements are *orthogonal* to each other over the unit square. Over other regions they need not be orthogonal. Since we do not assume that the basis elements are pairwise orthogonal, the analyses apply to nonsquare regions as well.

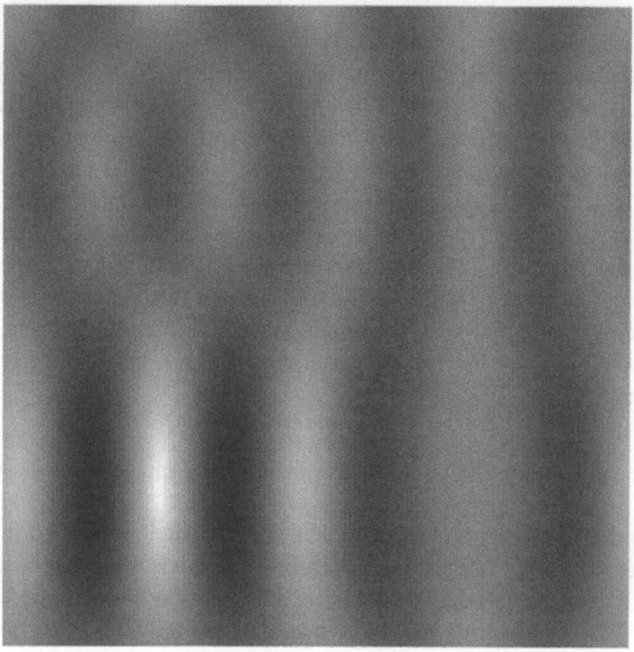

FIGURE 5.4. A retinal image.

Again, the precise choice of ordering is somewhat arbitrary. The analyses below use the first numbering and index individual functions according to it.[7] Fig. 5.4 plots a particular function in the space spanned by the Fourier basis. It is a linear combination of cosinusoids with $m, n \leq 3$.

Recall that we wish to analyze the conditions under which it is possible to reconstruct the retinal image given the sampling code. Let (x_j, y_j), $j = 1, 2, \cdots N$, be the set of points at which we have sample values ρ_j (the receptor coordinates). The possible retinal images fall within a linear subspace with P basis functions, $\rho_j(x, y)$. Let the N by P matrix Ψ have as its ijth entry,

$$\psi_{ij} = \rho_j(x_i, y_i). \tag{5.14}$$

This *interpolation matrix* has as its jth column the sampling code for the basis image $\rho^j(x, y)$.

Then,

$$\rho = \Psi \varepsilon \tag{5.15}$$

expresses the relationship between any retinal image $\rho(x, y)$, with coordinates $\varepsilon = [\varepsilon_1, \cdots, \varepsilon_P]^T$ in the function space, and its retinal sampling code ρ.

---

[7]Note that no finite subset of this Fourier basis (whatever ordering is chosen) contains "all of the cosinusoids below a certain frequency". Any cosinusoid $\cos 2\pi\alpha x$, where α is not an integer, is not in the list of basis elements and is not the weighted sum of any finite number of basis elements.

We are interested in determining the conditions under which it is possible to go from a sampling code ρ back to the retinal image coordinates ε, and it is straightforward to do so. We distinguish three cases: $P > N$, $P = N$, and $P < N$.

When $P > N$ it is not possible to invert Eq. 5.15. We will examine this case in detail in the following section. When $P = N$, the equation is invertible if and only if the matrix Ψ is nonsingular. The matrix Ψ is then said to have the *interpolation property* (Davis, 1963). This simple result is known as the *interpolation theorem.*

Before discussing the third case, let us illustrate the second using the Hirsch-Miller sample. The result above indicates that the 60 photoreceptors are sufficient to reconstruct any 60-dimensional linear subspace of functions across the retinal region, so long as the matrix Ψ determined by the receptor coordinates and the choice of function space has nonzero determinant. Once that is known to be true we can write

$$\varepsilon = \Psi^{-1}\rho, \tag{5.16}$$

allowing us to compute the coordinates of the retinal image from the sampling code.

In practice, the stability and robustness of the computation implicit in Eq. 5.16 can be assessed using standard linear algebraic methods. The matrix Ψ for the Hirsch-Miller sample and the Fourier basis of dimension 60 is readily computed from Eq. 5.14. The resulting matrix is invertible and the ratio of its largest singular value to its smallest (a measure of instability termed *condition number*) is approximately 319 (Ben-Israel & Greville, 1974). The logarithm to the base 10 of the condition number (= 2.5) is, roughly speaking, the number of decimal digits of precision needed in inverting this matrix (Johnson & Riess, 1982, p. 50ff). Put another way, if the computer or visual system which "computes" the inverse of Ψ works using arithmetic operations with fewer than 2.5 digits of accuracy, then Ψ is effectively singular. Thus, the condition number gives an interpretable measure of how close to singular a particular Ψ matrix is and whether a visual system with a given computational accuracy could reliably reconstruct the light image from the given samples.

A large condition number indicates instability, an infinite condition number corresponds to a singular matrix, and a condition number of 1 is optimally stable. Reconstruction of the Fourier basis using a regular array results in an interpolation matrix with condition number 1, and it is interesting that the slight deviations of the receptors from a regular lattice in Fig. 5.2 have cost 2.5 digits of precision.

In the remaining case, $P < N$, there are more samples than there are dimensions in the space of retinal images ("oversampling"). The matrix Ψ is no longer square, but it is still possible to solve the equation $\rho = \Psi\varepsilon$ for ε by multiplying both sides by Ψ^T, the transpose of Ψ. The result is the set of *normal equations*

$$\Psi^T\Psi\varepsilon = \Psi^T\rho, \tag{5.17}$$

where $\Psi^T\Psi$ is a square matrix. Its invertibility is the condition for reconstruction. Eqs. 5.16 and 5.17 specify completely the reconstruction process:

$$\varepsilon = (\Psi^T\Psi)^{-1}\Psi\rho. \tag{5.18}$$

The stability of the reconstruction process is determined by the matrix $(\Psi^T\Psi)^{-1}\Psi$.

When $P < N$, there must be sampling codes that do not correspond to any retinal image and (if sampling is error-free) should "never" occur. When the sampling process is contaminated by error, such nonexistent sampling codes may occur. In this case, Eq. 5.17 can be interpreted as the least-squares solution to the problem of the finding the ε that best accounts for the observed sampling code.

5.4 Linear Reconstruction and Aliasing

In this section we return to the first case, where $P > N$ (the dimension of the linear subspace containing the retinal image is greater than the number of photoreceptors).

The analysis is most easily presented if we divide the possible retinal images into two classes as follows. We consider a region of the retina containing N receptors. The space of possible ideal images is,

$$\rho^{\text{ideal}}(x, y) = \sum_{i=1}^{\infty} \varepsilon_i \rho_i(x, y). \tag{5.19}$$

We assume that the filtering properties of the optics of the eye eliminate all but the first P basis elements from the retinal image, where $P > N$. Let $M = P - N$. The space of retinal images is, therefore, P-dimensional,

$$\rho(x, y) = \sum_{i=1}^{P} \varepsilon_i \rho_i(x, y). \tag{5.20}$$

We can divide this space into the *target space* $\mathcal{T}$ of retinal images,

$$\rho^{\mathcal{T}}(x, y) = \sum_{i=1}^{N} \varepsilon_i \rho_i(x, y), \tag{5.21}$$

and the *distractor space* $\mathcal{D}$ of retinal images,

$$\rho^{\mathcal{D}}(x, y) = \sum_{i=N+1}^{N+M} \varepsilon_i \rho_i(x, y). \tag{5.22}$$

The terms *target* and *distractor* here are merely mnemonic. The target space $\mathcal{T}$ is the space of functions we are attempting to reconstruct from the retinal sample. The space of distractor functions comprises all other functions that pass through the optical filter. The target space and the distractor space are determined by the number of receptors in a patch, its size and shape, the optical quality of the eye over a given patch, and the ordering of the basis functions.

We assume that the receptors in the retinal patch have the interpolation property with respect to the target space. That is, there is an $N \times N$ invertible matrix $\Psi_{\mathcal{T}}$

such that, for any target retinal image

$$\rho^{\mathcal{T}}(x, y) = \sum_{i=1}^{N} \varepsilon_i \rho_i(x, y) \tag{5.23}$$

with samples

$$\rho^{\mathcal{T}} = [\rho^{\mathcal{T}}(x_1, y_1), \cdots, \rho^{\mathcal{T}}(x_N, y_N)]^T \tag{5.24}$$

and

$$\varepsilon^{\mathcal{T}} = [\varepsilon_1, \cdots, \varepsilon_N]^T, \tag{5.25}$$

then $\varepsilon^{\mathcal{T}} = \Psi_{\mathcal{T}}^{-1}\rho^{\mathcal{T}}$. For the distractor space there is an $N \times M$ matrix $\Psi_{\mathcal{D}}$ such that, for any distractor retinal image

$$\rho^{\mathcal{D}}(x, y) = \sum_{i=N+1}^{N+M} \varepsilon_i \rho_i(x, y) \tag{5.26}$$

with sampling code

$$\rho^{\mathcal{D}} = [\rho^{\mathcal{D}}(x_1, y_1), \cdots, \rho^{\mathcal{D}}(x_N, y_N)]^T \tag{5.27}$$

and retinal image coordinates

$$\varepsilon^{\mathcal{T}} = [\varepsilon_{N+1}, \cdots, \varepsilon_{N+M}]^T, \tag{5.28}$$

we have

$$\rho^{\mathcal{D}} = \Psi_{\mathcal{D}}\varepsilon^{\mathcal{D}}. \tag{5.29}$$

$\Psi_{\mathcal{D}}$ need not be invertible (or even square). As noted above, the sampling code ρ is uniquely decomposable into the distractor and target samples codes: $\rho = \rho^{\mathcal{T}} + \rho^{\mathcal{D}}$. The image coordinates ε are the concatenation of the coordinates $\varepsilon^{\mathcal{T}}$ and $\varepsilon^{\mathcal{D}}$.

Suppose that we attempted to estimate $\varepsilon^{\mathcal{T}}$ by inverting $\Psi_{\mathcal{T}}$ and applying it to the sample code ρ:

$$\hat{\varepsilon}^{\mathcal{T}} = \Psi_{\mathcal{T}}^{-1}\rho = \Psi_{\mathcal{T}}^{-1}(\rho_{\mathcal{T}} + \rho_{\mathcal{D}}) = \varepsilon^{\mathcal{T}} + \Psi_{\mathcal{T}}^{-1}\Psi_{\mathcal{D}}\varepsilon^{\mathcal{D}} = \varepsilon^{\mathcal{T}} + A\varepsilon^{\mathcal{D}}, \tag{5.30}$$

where $A = \Psi_{\mathcal{T}}^{-1}\Psi_{\mathcal{D}}$ will be termed the *aliasing matrix*. The aliasing matrix expresses how much our estimate of the target retinal image $\varepsilon_{\mathcal{T}}$ is affected by the presence of the distractor retinal image $\varepsilon_{\mathcal{D}}$.

The aliasing matrix A maps each image in $\mathcal{D}$ to some image in $\mathcal{T}$. Such pairs of images are said to be *aliased*. If M is smaller than N, then some target retinal images will not have distractor aliases. If M is larger than N then each target retinal image will be aliased to infinitely many distractor retinal images. Fig. 5.5 shows pairs of retinal images and their reconstructions from samples taken by the retinal array in Fig. 5.2. The target space $\mathcal{T}$ is spanned by the first 60 elements of the Fourier basis, the distractor space $\mathcal{D}$ by the succeeding 60.

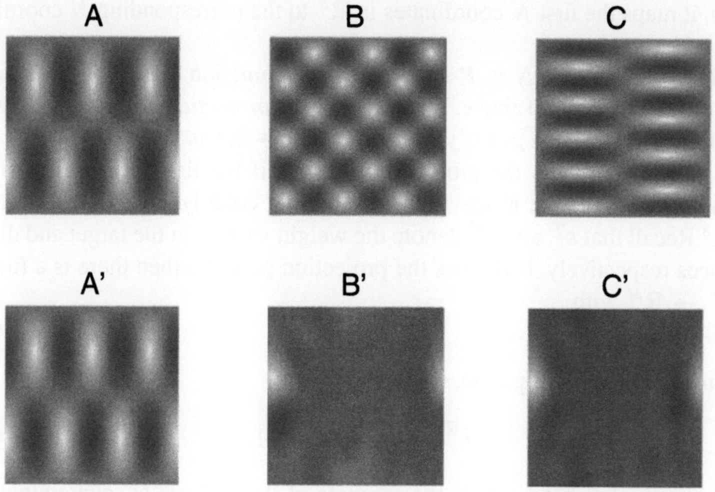

FIGURE 5.5. Three pairs of retinal images and their reconstructions using the Hirsch-Miller sampling array discussed in the text. (A), (A') Basis element 30 is correctly reconstructed. (B) Basis element 61 is not in the target space ($N = 60$) and is replaced by its target alias (B'). (C) Basis element 65 is not in the target space ($N = 60$) and is replaced by its target alias (C').

5.5 Nonlinear Constraints on Possible Images

Blurring the retinal image is one method of reducing the size of the distractor space $\mathcal{D}$ and thereby eliminating aliasing. In this section, we consider a second possible way out of the "aliasing" dilemma that depends not on the visual system, but on constraints on the possible retinal images that can be specified by the environment. The dimension of the target space is determined by the number of receptors in a given retinal area denoted N. The dimension of the distractor space is determined by the number P of linearly independent retinal images that pass through the optical filter. If $N < P$ then there will be aliased pairs of retinal images in $\mathcal{T}$ and $\mathcal{D}$.

Suppose that the possible retinal images in the combined target and distractor spaces include $P > N$ linearly independent images, but do not include all the weighted sums and differences of these P images. Suppose further that there is a continuous function $S : \mathbb{R}^N \to \mathbb{R}^P$ whose image is taken to be the *locus of possible retinal images* Λ in the environment. We assume (for the remainder of this section) that only images drawn from this locus will ever appear on the retinal patch of interest. We need a further assumption concerning the map $S(s_1, \cdots, s_N)$, namely that it has the *projection property*.

Definition: Assume $N \leq P$. The *projection* from $\mathbb{R}^P$ to $\mathbb{R}^N$ is the function $p : \mathbb{R}^P \to \mathbb{R}^N$ defined by

$$p(x_1, \cdots, x_P) = (x_1, \cdots, x_N). \tag{5.31}$$

That is, it maps the first N coordinates in $\mathbb{R}^P$ to the corresponding N coordinates in $\mathbb{R}^N$.

Definition: Assume $N \leq P$ and that $S()$ is a function $S : \mathbb{R}^P \rightarrow \mathbb{R}^N$. Let $p()$ be the projection defined above. Then $S()$ has the *projection property* with respect to the basis $\rho_1(x, y), \cdots \rho_P(x, y)$ if $p \circ S : \mathbb{R}^N \rightarrow \mathbb{R}^N$ is invertible.

The function $S()$ has the projection property if the first N coordinates (corresponding to the target space) of the sampling code $[\rho_1, ...\rho_P]$ determine the others.[8] Recall that $\varepsilon^{\mathcal{T}}$ and $\varepsilon^{\mathcal{D}}$ denote the weight vectors in the target and distractor spaces respectively. If $S()$ has the projection property then there is a function $\mathcal{S} : \mathbb{R}^N \rightarrow \mathbb{R}^M$ with,

$$\varepsilon^{\mathcal{D}} = \mathcal{S}(\varepsilon^{\mathcal{T}}). \tag{5.32}$$

Substituting for $\varepsilon^{\mathcal{D}}$ in Eq. 5.30, we have,

$$\hat{\varepsilon}^{\mathcal{T}} = \varepsilon^{\mathcal{T}} + A\mathcal{S}(\varepsilon^{\mathcal{T}}). \tag{5.33}$$

Recall that the left-hand side is the estimate of the weights $\varepsilon^{\mathcal{T}}$ contaminated by aliasing. A is the fixed aliasing matrix, and the right-hand side is a function of $\varepsilon_{\mathcal{T}}$. Correct recovery of $\varepsilon_{\mathcal{T}}$ is possible precisely when the equation is invertible and it is possible to go from $\hat{\varepsilon}^{\mathcal{T}}$ to $\varepsilon^{\mathcal{T}}$. These two conditions, the projection property and the invertibility of the above equation (*the nonlinear inversion property*), suffice to permit nonlinear reconstruction without aliasing.

An example will clarify the consequences of the projection and nonlinear inversion properties. Fig. 5.6 illustrates how a one parameter nonlinear locus $S(s)$, s a real parameter, can span a two-dimensional linear subspace. The axes are the weights ε_1 and ε_2 applied to the two basis elements that determine the space. The two vectors A and B are in the locus Λ and are linearly independent. They span the entire plane (the entire plane is the smallest linear subspace that can contain them both). Yet only a very small percentage of the patterns in the plane are in the nonlinear locus Λ. Note also that no two points in the locus have the same ε_1 coordinate. The map S(s) has the projection property and, if Eq. 5.33 is invertible for A and $\mathcal{S}()$, nonlinear reconstruction is possible. We assume, for this example, that it is.

Suppose now that a researcher who did not know that all patterns fall within the locus were to investigate the visual system outlined in Fig. 5.6. S/he determines that the linear space of possible images includes both of the images corresponding to A and B and that, therefore, the smallest linear subspace containing the images has (at least) dimension 2. S/he then determines that the visual system only samples the image once (measuring ε_1) rather than twice. The researcher concludes that the visual system is undersampling ($P = 2, N = 1$) and yet the expected effects of the undersampling ("aliasing") are not apparent. Since we know about the constraint embodied in the locus, we know that the single measurement ε_1 does determine the

[8]It would be plausible to generalize the projection property to the case where some N coordinates out of the P determine the others. This is an unneeded complexity as we can simply renumber the basis so that the N coordinates that determine the remainder are the first N.

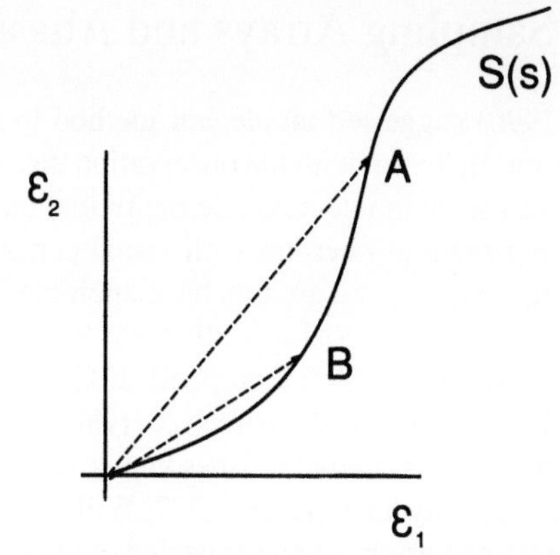

FIGURE 5.6. A nonlinear locus $S(s)$. The locus is one-dimensional, yet the two linearly independent vectors A and B in the locus together span the entire two-dimensional linear subspace. Either coordinate of any point in the locus determines the point. The locus represents the space of all possible images, the plane the linear subspace spanned by these images. Almost all images in the plane are not in the locus of possible images.

image $S(s)$. In this example, then, the solution to the problem of apparent retinal undersampling is very straightforward. A nonlinear reconstruction algorithm that had knowledge of ε_1 and the fixed nonlinear constraint $S(s)$ can determine ε_2. The retina may undersample the *linear* subspace containing the retinal images, but the samples taken are sufficient to determine the retinal image because of the constraint imposed by $S()$. The projection property and the invertibility of Eq. 5.33 guarantee that N samples suffice.

Is it plausible that there are such nonlinear constraints on "naturally occurring" retinal images? Putting aside the problem of defining "naturally occurring", there is some evidence that such constraints are present in the environment and that the human visual system makes use of them. Galvin and Williams (1992) examined whether aliasing affected perception of high-contrast edges presented extrafoveally in normal viewing. They concluded that it did not. They suggest that "The failure to observe aliasing in the laboratory with high contrast edges presented in isolation and in sharp focus on the retina demonstrates the importance of the properties of natural scenes as a protection from aliasing" (p. 2257). and again "the spatial frequency content of the environment is one important reason why we are not troubled by aliasing artifact during normal peripheral viewing" (p. 2258). They note also that blurred edges appear sharp in the periphery and suggest that perhaps a template-matching scheme might account for this phenomenon. The constrain function $S(s_1, \cdots, s_N)$ is very much in the spirit of their suggestion.

Whether such a constraint exists and whether it allows nonlinear reconstruction is a question both about the environment and also about the matrix A, determined by the topography of the retina.

5.6 Irregular Sampling Arrays and Aliasing

Yellott (1982, 1983, 1990) suggested an elegant method to reduce the effect of aliasing in primate vision. He began with the observation that, outside of the fovea, human and rhesus retinas seem to undersample the retinal image, yet the aliasing error term $A\rho^{\mathcal{D}}$ does not seem to interfere with visual performance. Anatomical and psychophysical studies by several authors have attempted to assess the degree of aliasing to be expected in human vision at different locations outside the fovea (Bossomaier, Snyder & Hughes, 1985; Campbell & Green, 1965; Campbell & Gubisch, 1966; Galvin & Williams, 1992; Hirsch & Hylton, 1984; Hirsch & Miller, 1987; Jennings & Charmen, 1981; Snyder, Bossomaier & Hughes, 1986; Snyder, Laughlin & Stavenga, 1977; Snyder & Miller, 1977; Williams & Collier, 1983) and it is not clear to what extent aliasing is to be expected or under what circumstances it occurs in human parafoveal vision. Recent results indicate that aliasing is to be expected in some parts of extra-foveal retina (Artal et al., 1992; Navarro et al., 1993).

Yellott noted that the aliasing matrix A could be altered by perturbations of the positions of receptors in a retinal region, possibly without greatly altering the stability and accuracy of reconstruction of the target signal. Suppose that we could, in effect, choose which distractor and target signals to alias by choosing A. What should we do to reduce the effect of aliasing? If M is much smaller than N, then we could attempt to choose A so as to alias distractor signals to the "least visible" or "least salient" or "least disturbing" of the target signals. An intelligent choice of A would do just that. Of course, we choose A not directly, but by choosing the locations of the receptors in the retinal region under consideration. Changing the location of the receptors alters the matrices $\Psi_{\mathcal{T}}$ and $\Psi_{\mathcal{D}}$, and thereby A.

Yellott (1982, 1983, 1990) examined the aliasing properties of primate retinas with respect to Fourier bases and concluded that the somewhat irregular arrangement of photoreceptors in parafoveal retina (as in Fig. 5.2) mapped cosinusoidal distractor basis elements to target retinal images that were not concentrated in frequency ("broad-band noise"). He suggested that these patterns were less salient in normal vision. Yellott concluded that the irregular positioning of photoreceptors in the retina could serve to reduce the impact of aliased distractor signals in those retinal areas where there are insufficient numbers of receptors to reconstruct the retinal image.

Some issues remain unresolved, notably the question of which (aliased) target patterns are least disruptive when added to target images. A second issue of concern is that the analysis above assumes that M is small compared to N. If M is equal to N, for example, the aliasing matrix A is $N \times N$ and potentially invertible for many arrangements of photoreceptors.[9] In such a case, the matrix A establishes a one-to-one mapping between the space of distractor retinal images and the space

[9] It is invertible for each of the 25 Hirsch-Miller samples (Hirsch & Miller, 1987). The condition number for the distractor interpolation matrix $\Psi_{\mathcal{D}}$ for the sample in Fig. 5.2 and Fourier basis elements 61 through 120 is 315, slightly less than (i.e. better than) the condition number for the target interpolation matrix $\Psi_{\mathcal{T}}$ which was 319.

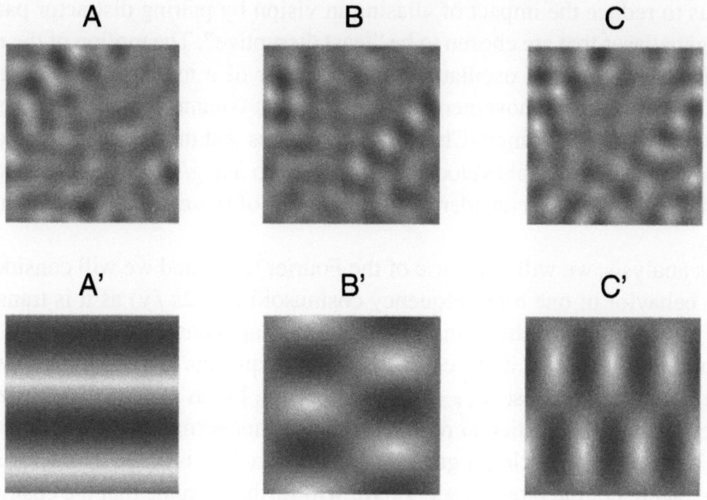

FIGURE 5.7. (A)–(C) Three images drawn from the distractor space whose target aliases (for the Hirsch-Miller retinal sample discussed in the text) are the basis elements 10, 20, and 30, shown as (A')–(C').

of target retinal images. By altering the spacing of the sampling array we can still attempt to banish the spurious images by aliasing them to less salient or intrusive low-pass images. But, so long as A is nonsingular, there will always be distractor retinal images that alias to those target images that are most disruptive. A change in A just permutes the assignment of aliases. Fig. 5.7 shows distractor signals in the Fourier space spanned by basis elements 61 through 120 whose target space aliases are target cosinusoids. With the Fourier basis and the Hirsch-Miller sample, every cosinusoid has such a distractor space alias.

We have not eliminated the distractor signals that alias to "disruptive" target signals. Changing A, so long as it remains nonsingular, just permutes the assignment of aliases, so that some other signal in $\mathcal{D}$ is now aliased to a given signal in $\mathcal{D}$. If the retinal array in a parafoveal region is markedly "undersampling" the signal, then Yellott's approach will not work.

In the previous section, we discussed the possibility that nonlinear constraints on naturally occurring retinal images could reduce the number of samples needed for alias-free reconstruction. A nonlinear constraint on the distractor space that precludes the distractor retinal images in Fig. 5.7(A–C) and their ilk would also serve to reduce aliasing. The "disturbing" target patterns would be aliased to nonoccurring distractor patterns.

5.7 Linear Reconstruction and Movement

So far, we have considered the problem of sampling and reconstruction in a single instant of time, and we have considered a proposal by Yellott that potentially allows us to reduce the impact of aliasing in vision by pairing distractor patterns with target aliases that are chosen to be "least disruptive". The motion of the retina ranges from small, rapid oscillations on the order of a minute of arc or less to large drift and saccadic movements (Steinman, this volume, Chapter 1; Steinman & Levinson, 1990). Steinman (Chapter 1) concludes that the retina is, much of the time, moving at appreciable velocities (1 deg/sec to 5 deg/sec) with respect to the scene. In this section we consider the implications of these retinal movements for aliasing.

In this analysis, we will make use of the Fourier basis, and we will consider the aliasing behavior of one high frequency cosinusoid $cos(2\pi f y)$ as it is translated in the y direction across the retina by a distance that is one-quarter or more of its wavelength $1/f$. That is, if the cosinusoid has frequency $f = 15$ cycles/degree, then the distance of interest is $\frac{1}{4f} = 1$ minute of arc. Let us assume that the density of photoreceptors in a particular retinal patch does not permit reconstruction of the Fourier basis up to 15 cycles/degree, that is, our function is in the distractor space $\mathcal{D}$ of the patch, not the target space $\mathcal{T}$. We will further assume that the cosinusoid is constant in the x direction to simplify the discussion.

We can assume that, at the begin of its traverse, the cosinusoid is in fact a sine, not a cosine (by choice of coordinate system). The initial distractor cosine is aliased to some target pattern $g(x, y)$. After the sine has shifted one-quarter of its wavelength to the left, the pattern is now that of the basis function $\cos 2\pi f x$ with respect to the retinal patch, and this function (also in the space $\mathcal{D}$) is assumed to be aliased to a target pattern $h(x, y)$. With these two pieces of information, $g(x, y)$ and $h(x, y)$, we can now predict the appearance of the original moving cosinusoid for any amount of translation. As a consequence of the simple identity, $\sin(y + a_t) = \alpha_t \sin y + \beta_t \cos y$ where $\alpha_t = \cos a_t$ and $\beta_t = \sin a_t$, and the linearity of the aliasing map A, it will always be mapped to a weighted, time-varying average of the two pattern $g(x, y)$ and $h(x, y)$.[10] Any shift of the original cosinusoid by a distance a_t is equivalent to some weighted mixture of the sine and cosine phases of the cosinusoid. The cosinusoid shifted by a_t is therefore also in $\mathcal{D}$, and is aliased to

$$f_{a_t}(x, y) = \alpha_t g(x, y) + \beta_t h(x, y). \tag{5.34}$$

The weighted, time-varying average of the two pattern $g(x, y)$ and $h(x, y)$ is *completely, physically equivalent* to the moving high-pass cosinusoid at the sampling array.

Suppose that we drift the cosinusoid in the y direction at fixed velocity v so that $a_t = vt$. Then, as the sine wave moves, $h(x, y)$ replaces $g(x, y)$ when the cosinusoid travels one-quarter of its wavelength, going from sine to cosine phase. After another one-quarter wavelength, $-g(x, y)$ replaces $h(x, y)$, and after another one-quarter wavelength $-h(x, y)$ replaces $-g(x, y)$. Yet another quarter wavelength,

[10]The aliases of an arbitrary moving cosinusoid such as $\cos 2\pi mx \sin 2\pi ny$ is the weighted sum of four fixed patterns that are the respective aliases of $\cos 2\pi mx \cos 2\pi ny$, $\cos 2\pi mx \sin 2\pi ny$, $\sin 2\pi mx \cos 2\pi ny$, and $\sin 2\pi mx \sin 2\pi ny$.

and we are back where we began: the sine wave has translated one full period. The sine wave is spatio-temporally aliased to rapidly changing mixtures of two patterns and those same patterns with contrast inverted, something like a commercial on MTV. If v were, for example, 1 degree/second, then the patterns go through a full cycle every 66.7 msec as the cosinusoid drifts. The change from $g(x, y)$ to $h(x, y)$ takes only 16.7 msec. As the frequency f increases, the time decreases in inverse proportion as does the total distance that must be traversed to pass from $g(x, y)$ to $h(x, y)$.

The two patterns $g(x, y)$ and $h(x, y)$ are determined by the aliasing matrix A, that is, by the locations of the receptors within the retinal patch. If the receptors fall on a square lattice, then the aliases of the sine and cosine phases of the cosinusoid will themselves be cosinusoids of a different, lower frequency. Then the weighted mixtures of $g(x, y)$ and $h(x, y)$ will correspond to different phases of a low-frequency cosinusoid. The drifting distractor cosinusoid will alias over time to a drifting target cosinusoid drifting at a different velocity and in a different direction.

When the array is not regular, the relation between $g(x, y)$ and $h(x, y)$ is more complicated. Fig. 5.8 shows the aliases corresponding to a drifting cosinusoid drawn from the distractor space of the Hirsch-Miller sample. Note that the target aliases are not simply shifted copies of one another.

The first conclusion to be drawn from this analysis is that Yellott's proposal could be restated as a statement about *spatio-temporal aliasing*. The layout of the receptor array in a retinal patch determines the spatio-temporal aliases to which drifting distractor images are aliased. It is the visibility of this spatio-temporal aliasing that determines whether the visual impact of distractor images has been reduced.

A second, related conclusion to be drawn is that sampling with photoreceptor arrays where the receptors form no regular pattern may serve to eliminate coherent

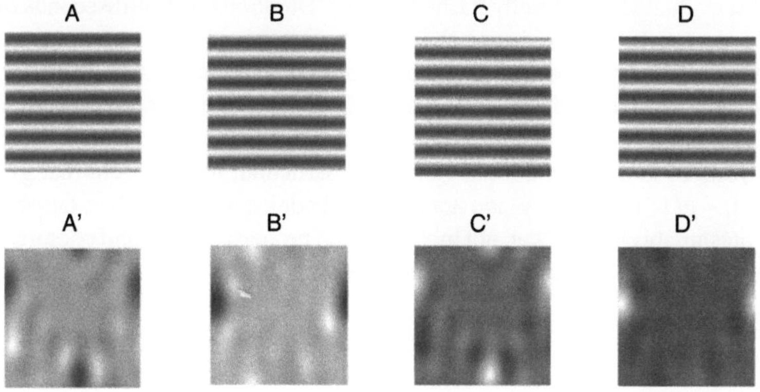

FIGURE 5.8. Low-pass aliases of a distractor drifting grating. (A) $\sin 2\pi 7x$, (B) $\sin 2\pi 7x$ shifted one-quarter of its period down, (C) $\sin 2\pi 7x$ shifted one-half of its period down, (D) $\sin 2\pi 7x$ shifted three-quarters of its period down. The corresponding target aliases, reconstructed using the Hirsch-Miller retinal sample as discussed in the text, are shown in (A')-(D').

motion in aliased signals and turn them into high temporal frequency flicker. If, for example, a vertical sine wave grating $\sin 2\pi f_x x$ of frequency f_x cycles/deg moves horizontally with velocity v deg/sec, then the rate at which the alias moves from each of its quarter-phase patterns ($g(x, y)$, $h(x, y)$, etc.) is $4 f_x v$. If v is 1 deg/sec then any grating with a frequency of above 15 cycles/deg will change quarter phase patterns with temporal frequency 60 Hz or higher, at or above the human temporal frequency cutoff (Woodhouse & Barlow, 1982).

The descriptions by subjects of aliased patterns induced by bypassing the normal optical filtering of the eye are consistent with this view of spatio-temporal aliasing. Williams (1985) describes the appearance of patterns of fine gratings imposed onto the retina by laser interferometry. At frequencies of 90-100 cycles/deg, "Observers describe [the] pattern as resembling a finger print or pattern of zebra stripes. ... The pattern is small and scintillates markedly" (p. 200). The frequency range chosen is well above the reconstruction limit of central foveal vision which is estimated to be about 60 cycles/degree. The alias is as much the scintillation as it is the zebra stripe pattern.

Artal, Derrington, and Colombo (1995) measured the performance of three observers detecting or discriminating the direction of motion of high-contrast gratings in the periphery. They report that "Gratings which can be detected but whose direction of motion cannot be discriminated appear as irregular speckle patterns whose direction of motion varies from trial to trial" (p. 939). Their observations are consistent with the present analysis.

These results raise a methodological issue. Experimental studies of retinal sampling, studies of acuity, or studies of aliasing, that present stimuli for extended periods of time cannot be easily interpreted in terms of the instantaneous sampling theory presented in the preceding sections. "Extended period of time" is here the time the eye needs to move a fraction of the spacing between photoreceptors which may be on the order of a few milliseconds. The sampling characteristics of the retina are best evaluated with briefly flashed stimuli. Durations of 5 milliseconds or less would seem to be a conservative choice.

These results raise theoretical and psychophysical issues as well. What features of the spatiotemporal aliases $g(x, y)$ and $h(x, y)$ control the degree to which they disrupt normal vision? What are the choices of $g(x, y)$, $h(x, y)$ that minimize the effects of aliasing in normal vision? In particular, to the extent that $g(x, y)$, $h(x, y)$, $-g(x, y)$, $-h(x, y)$ are not translated copies of one another, the observer will (presumably) see flicker, not motion. What properties of A (and photoreceptor layout) determine the degree to which these patterns are or are not translates of one another?

5.8 Linear Reconstruction with Multiple Sampling Arrays

So far we have considered the consequences of movement for sampling when we are in effect reconstructing the scene at each instant using only the instantaneous sample information. In successive glances at a scene, we obtain information about some parts of the scene twice. Can we somehow combine these pieces of information so as to get a single combined estimate that is better than either estimate alone?

The idea that information from two glances can be combined to produce enhanced estimation of visual properties, or discrimination, has a long history (See Steinman & Levinson, 1990, for a review). As Packer and Williams (1992) note, there is no firm evidence indicating that multiple, monocular glances at a single scene can result in improved estimation, acuity, or whatever.[11] In this section, the problem of combining multiple glances at a single scene is considered, as well the conditions under which such combinations might improve performance. The analysis suggests that there could be a modest increase in the sampling performance of a given retinal patch under certain conditions, and that peripheral rather than foveal vision might be the place to look for such an increase.

If the retinal image were constrained to be from the target space $\mathcal{T}$, then so long as the assumptions of the interpolation theorem were satisfied, a single glance at the scene is enough to reconstruct the retinal image. But, if the retinal image is drawn from the combined spaces $\mathcal{T}+\mathcal{D}$, then the sampling array cannot accurately reconstruct $\mathcal{T}$ without aliasing the component of the retinal image that lies in $\mathcal{D}$.

Suppose that we could move the sampling array and take a second glance at the scene. Fig. 5.9 shows the Hirsch-Miller sample and the same array shifted by 0.9 minutes of arc in the horizontal direction, and 0.6 minutes of arc in the vertical.[12] The total shift corresponds to an eye movement that is a bit more than 1 minute of arc and is somewhat smaller than the spacing of the receptors in the sample. The combined arrays have $2N = 120$ samples and, if we had the values of all the samples simultaneously available, we could reconstruct a function space of dimension $2N$. That is, so long as $M \leq N$ and the resulting interpolation matrix for the combined samples were nonsingular, we could reconstruct $\mathcal{T}+\mathcal{D}$ without aliasing. This analysis assumes that we know precisely where the eye has moved relative to its original position.

If we can accurately estimate the magnitude of the eye movement, we can reconstruct samples that would otherwise be aliased. How well can we estimate the position of the eye from visual information alone? Suppose that we returned to the space $\mathcal{T}$ only. We can reconstruct any image in $\mathcal{T}$ from either of the two viewpoints, and in the absence of aliasing (and measurement error) we expect to get an accurate estimate of the retinal image. That is, in the case portrayed in Fig. 5.9, we will reconstruct $\rho(x, y)$, the image in $\mathcal{T}$, and in the other case $\rho(x - 0.9, y - 0.6)$, that is, the reconstruction will reflect the eye shift.

Suppose now that the retinal image is drawn solely from $\mathcal{T}$, and we sample it twice, with an unknown eye movement $(\Delta x, \Delta y)$ between the two samples. We

[11] In contrast, the binocular vision experiments of Burr and Ross (1979) indicate that information presented at different times to the two different retinas can be integrated across time to estimate motion trajectories not otherwise visible.

[12] The Hirsch-Miller sample is plotted on the unit square and the precise shifts (in the units in which the samples were plotted) are 0.07 horizontal and 0.04 vertical.

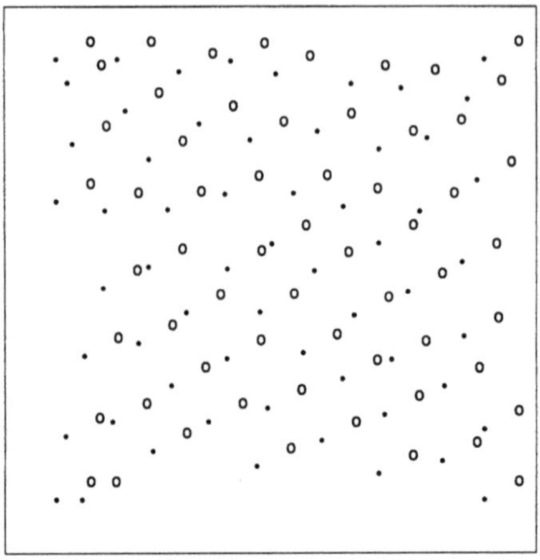

FIGURE 5.9. The Hirsch-Miller array (•) and a translation of the same array (o) by 0.9 minutes of arc in the horizontal direction and 0.6 minutes of arc in the vertical direction.

expect to reconstruct a "before" image $\rho_B(x, y) = \rho(x, y)$, and an "after" image $\rho_A(x, y) = \rho(x + \Delta x, y + \Delta y)$. We could then attempt to estimate Δx and Δy by minimizing,

$$\iint (\rho_B(x, y) - \rho_A(x - \Delta x, y - \Delta y))^2 \, dx \, dy, \tag{5.35}$$

by choice of Δx, Δy. *The accuracy of this estimation process will depend on the contents of the visual field,* the function $\rho(x, y)$. If, for example, $\rho(x, y)$ is constant (a uniform field), then no worthwhile estimate of Δx, Δy is possible.[13]

Therefore, if the retinal image is drawn from the target space $\mathcal{T}$, and is sufficiently nonuniform, then it is possible to estimate the eye movement shift. But then, by assumption, there are no distractor images from $\mathcal{D}$ to reconstruct. When the retinal image is drawn from $\mathcal{T} + \mathcal{D}$, it would seem that we must reconstruct the scene to get the eye shift, but that we need the eye shift to reconstruct the scene!

For human vision, though, there is a straightforward escape route from this vicious circle. The quality of vision in central fovea is plausibly matched to the spacing of receptors in the array. That is, in central fovea, the retinal image is drawn from $\mathcal{T}$ only, not $\mathcal{D}$, and it is plausible that the eye can localize itself. Steinman (1965) measured the ability of human observers to stabilize gaze and found that eye fixation errors had a standard deviation of 2-5 minutes of arc. Since the observed error of the control system stabilizing gaze is greater than the error of the localization signal available to it, this figure is an upper bound for the accuracy of localization of the eye.

---

[13]This analysis ignores the possibility that the retina may cyclorotate $\Delta\theta$ degrees or that small head/body movements could induce an approximately affine transformation of the retinal image. Additional parameters can, of course, be fitted by using the least-squares approach described here.

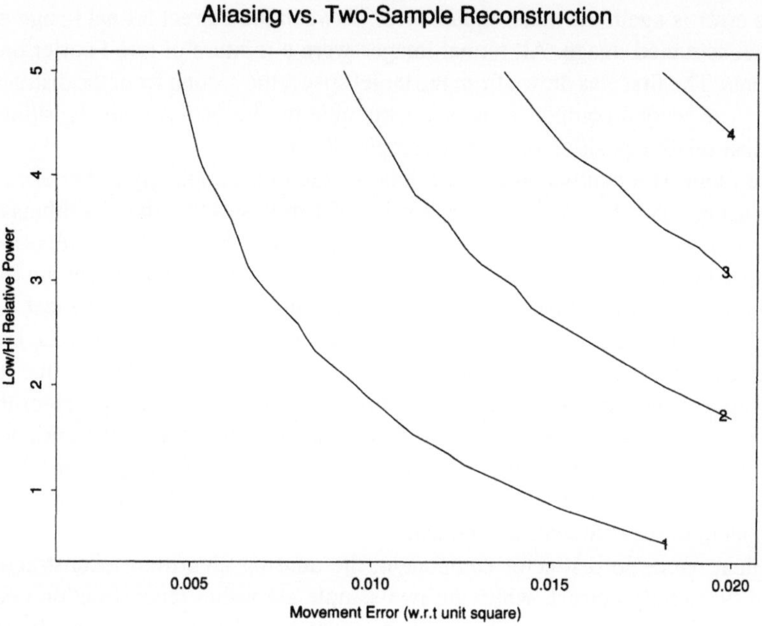

FIGURE 5.10. A comparison of two-sample and aliasing reconstruction. The dependent variable is the ratio of the mean square error of reconstruction using a two-sample algorithm and the mean square error using an aliasing algorithm. This ratio is plotted as a contour map and, when the ratio is less than 1, the the reconstruction by the two-sample algorithm is superior to that of the aliasing algorithm (below and to the left of the lowest contour line). The x-axis is the simulated error in the estimate of the relative positions of the two samples. The y-axis plots that ratio of the weights of the target and distractor components in the retinal image. See text for further explanation.

This same retinal image motion is then available to allow multiple images to be combined in extrafoveal regions where receptor density is lower and there may be significant potential for aliasing. Note also that sampling and reconstruction across multiple views requires that there be sufficient detail in the scene to permit accurate foveal estimates of position.

Fig. 5.10 compares the performance of two algorithms when the retinal array undersamples the space of possible retinal images. The first algorithm (*The Aliasing Algorithm*) reconstructs the retinal image from a single sampling code. Components of the retinal image drawn from the distractor space are transformed into their aliases in the target space. As noted in the discussion of the aliasing matrix above, these terms produce errors in reconstruction. The measure of the error is the mean square error between the correct retinal image and the reconstructed image. The second algorithm (*The Two-Sample Algorithm*) reconstructs the retinal image from two samples, assuming that the two samples were taken at the locations shown in Fig. 5.9. The true relative location of the second sample could differ from the value assumed by the algorithm, introducing error into the reconstruction. The measure of the error is again the mean square error between the correct retinal image and the reconstructed image. All retinal images were a mixture of two Fourier basis elements. The first was drawn from the target space, the second from the distractor space. The second component is reconstructible by the two-sample algorithm if the exact relative position of the two samples is known.

The figure is a contour plot of the ratio of the two-sample algorithm error to the aliasing algorithm error. That is, a value of 1 indicates that the algorithms are performing equally well; a value greater than 1 indicates that the aliasing algorithm is outperforming the two-sample algorithm, and a value less than 1 that the two-sample algorithm is outperforming the aliasing algorithm. Only in the last case would it make sense to use the two-sample algorithm. The x-axis plots the error introduced into the relative position of the two sampling arrays. That is, the true shift between the arrays differs from the shift assumed in the two-sample algorithm by this amount. This error affects only the two-sample algorithm. All scenes were composed of a single component from the target space and a single component from the distractor space. The y-axis plots the ratio of the weight of the target component to the distractor component.

If there were no distractor component, the aliasing algorithm reconstruction would be exactly correct, while the two-sample algorithm reconstruction could have significant error due to errors in the assumed relative position of the two retinal samples. If, on the other hand, there were *only* a distractor component, then the aliasing algorithm would simply convert it to its alias in the target space while the two-sample algorithm would (for small enough relative error) reconstruct it accurately. In brief, we expect a tradeoff between relative position error and the ratio of target to distractor weights. Fig. 5.10 shows such a tradeoff.

The results of this section lead to an hypothesis concerning human vision, that eye movements serve to increase the effective sampling density in parafoveal vision. The key to testing this hypothesis is to perturb the positional signal available for reconstruction. Consider the stimulus in Fig. 5.11. The pattern on the left is fixated and (presumably) provides the positional information (the location of the eye within the scene). The small grating on the right is drawn from the distractor space $\mathcal{D}$ for the retinal patch it covers when the pattern on the left is fixated. The remainder of the visual field is uniform. The two patterns are "jittered" randomly at temporal frequencies and with excursion magnitudes similar to small eye move-

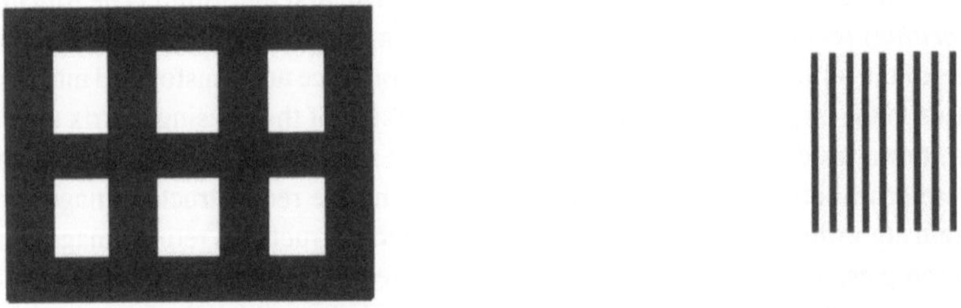

FIGURE 5.11. An experimental stimulus. See text for explanation.

ments. In the first condition, the same motion is imposed on the two stimulus components. In the second condition, the two stimulus components move independently. If the positional signal is, in fact, controlled by the stimulus component on the left, then the positional signal is, in the first condition, the correct signal needed to reconstruct the pattern on the right across time. In the second case, it is not, and accurate reconstruction should not be possible.

5.9 Ideal Arrays

So far, we have examined the conditions under which one of the three arrows in Fig. 5.1 can be reversed, the arrow joining the retinal image and the sampling code. One consequence of a successful (linear) reconstruction of the retinal image is that it is then possible to predict the excitation of an idealized punctate photoreceptor at any location within the retinal patch. We could, for example, predict the excitations of a regular array of photoreceptors across the same retinal patch (an *ideal array* in the terminology of Maloney and Ahumada, 1989), and we could equally well predict the excitation of mechanisms with more complicated receptive fields as well, that is, a receptive field made up of *virtual photoreceptors* that need not be present in the retinal patch. Once we have estimated the retinal image, we can, in effect, simulate the excitations of whatever linear or nonlinear mechanisms we find useful.

Fig. 5.12 shows a regular array containing S linear mechanisms with center-surround receptive fields made up of virtual photoreceptors (not shown). Fig. 5.12(A–B) shows two different arrangements of photoreceptors. So long as both sets of photoreceptors are sufficiently numerous to reconstruct, then the arbitrary receptive fields can be computed from the real photoreceptor excitations (the sampling code). Let the instantaneous excitations of the S mechanisms be denoted by $\sigma = [\sigma_1, \cdots, \sigma_S]^T$. As the mechanisms are linear, and ε denotes the coordinates of the retinal image over a given retinal patch, there is a matrix Σ such that, $\sigma = \Sigma\varepsilon$. As before, the excitations of the true photoreceptors in that patch are related to the coordinates of the retinal image by $\rho = \Psi\varepsilon$. Under conditions where Ψ is invertible, we have $\sigma = \Sigma\Psi^{-1}\rho$, and it is clear that we can bypass the somewhat artificial coordinates ε and pass directly from the excitations of photoreceptors ρ to the ideal array shown in Fig. 5.12. If the mechanisms are a nonlinear function of the excitations of virtual photoreceptors in the retinal patch, then they can also be computed from the sampling code. The only requirements are that the receptive fields (virtual photoreceptors) fall within the retinal patch under consideration and that reconstruction be possible given the real receptors in the patch.

Suppose now that the linear or nonlinear mechanisms are not retinotopic. As the eye moves, the receptive field of each mechanism moves relative to the retina. Then Fig. 5.12 could represent the same mechanisms in correspondence with two different patches of retina with a different layout of photoreceptors. Assume that all the receptive fields have shifted as a whole on the retina to a new location.

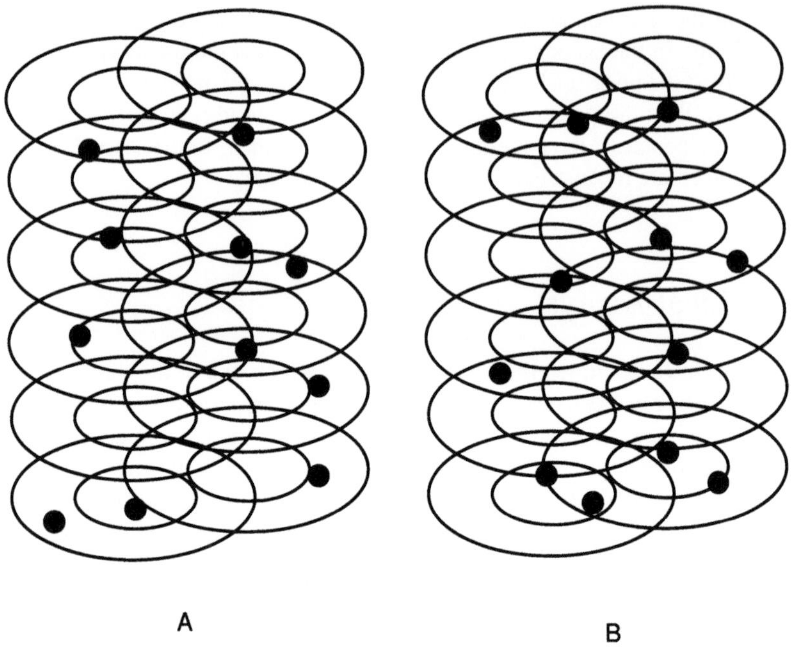

FIGURE 5.12. An ideal array of linear center-surround mechanisms with receptive fields made up of virtual photoreceptors (not shown) and two different distributions of real receptors over it.

Let the sampling code in the new patch (B) be $\tilde{\rho} = [\tilde{\rho}_1, \cdots, \tilde{\rho}_N]$. Then there is a matrix $\tilde{\Psi}$ corresponding to this retinal patch and, if the conditions of the interpolation theorem are satisfied, Ψ is invertible, we have $\sigma = \Sigma\Psi^{-1}\rho$, and when $\tilde{\Psi}$ is invertible, we have $\tilde{\sigma} = \Sigma\tilde{\Psi}^{-1}\tilde{\rho}$. The change from the matrix Ψ^{-1} in patch (A) to the matrix $\tilde{\Psi}^{-1}$ (B) permits the excitations of the retinal array to be independent of the precise patch of retina that it is, at the moment, employing to see the world.

Of course, as soon as the retina moves enough so that regions of the retina with markedly different photoreceptor densities fall onto a single part of the ideal array, then this scheme can break down. For small movements that do not greatly change the density of photoreceptors over each patch of real array, it is possible to compute the excitations of an array of ideal receptors from the available information.

The number of matrices Ψ needed, though, is large. One Ψ is required for each patch of retina and each location on that retina with respect to the ideal array. There is, however, a simple way to break this computation down into two stages. The first stage produces an ideal array that is regular but that moves with the retina (i.e., is *retinotopic*). The second computation goes from this ideal array to a nonmoving representation (see Fig. 5.13). The transition between the retina and corresponding points in the retinotopic ideal array requires a fixed transformation connecting

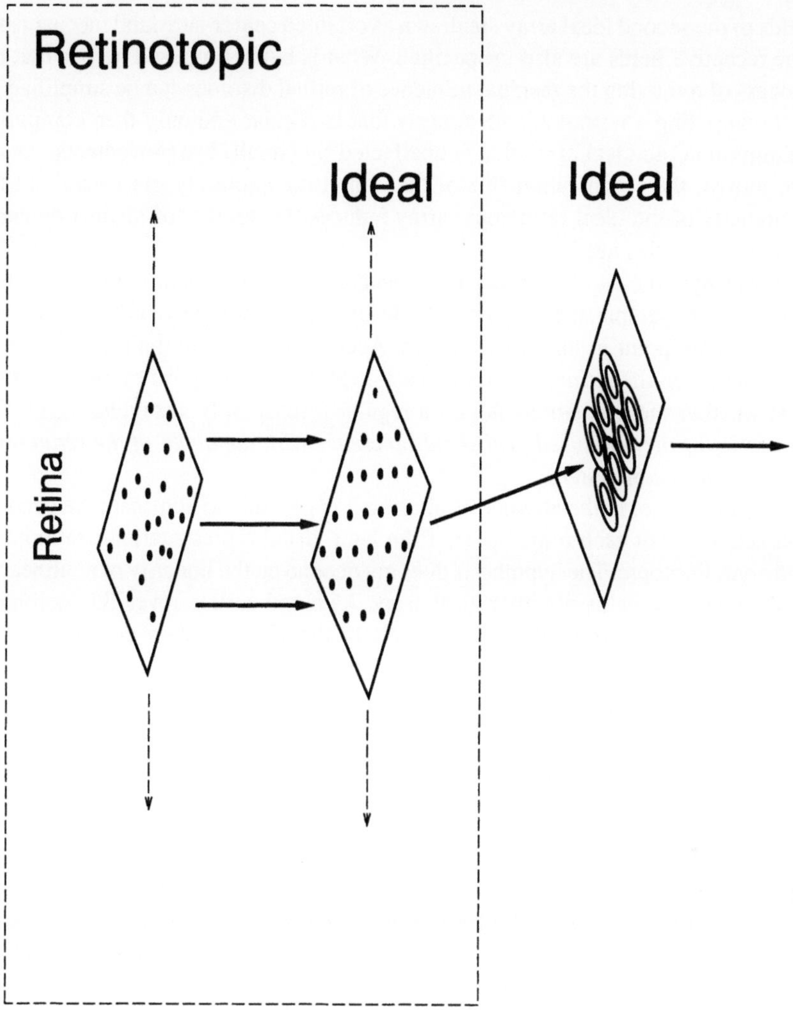

FIGURE 5.13. A proposed two-stage ideal array. The first array is retinotopic and regular, the second is not retinotopic. As the eye moves, the correspondence between the retina and the first ideal array does not change, while the correspondence between the first and second ideal arrays does change.

the two arrays that is unaffected by eye position. The transition from the ideal retinotopic array to the second ideal array requires that the transformation between the two change.

The figure is misleading in one respect. The first ideal array is drawn as a regular hexagonal array of points to emphasize that the transition from the real to the first ideal array eliminates the influence of the idiosyncratic layout of photoreceptors in a given retinal patch. The receptive fields of th first ideal array need not be photoreceptor-like. They are intentionally left unspecified. Similarly, the receptive fields of the second ideal array are drawn as oriented center-surround mechanisms. The receptive fields are also unspecified. What is being proposed here is that the process of removing the residual influence of retinal disorder can be simplified by first computing a retinotopic ideal array that is regular and only then computing a nonretinotopic ideal array that is unaffected by (small) eye movements. As the eye moves, the computation rule of the nonretinotopic array must vary, but the periodicity of the ideal retinotopic array reduces the number of distinct dynamic computation rules needed.

These hypotheses concerning representation can be tested. Consider testing whether, for example, the ideal retinotopic array corresponds to a given retinotopic map at some point in the visual system. Accurately measure the receptive fields of each cell on the retina, and assign each receptive field a position on the retina. Test whether these positions fall on a regular (presumably hexagonal) grid. *The regularity of the measured grid should increase with the depth of the retinotopic layer in the visual system.*

To summarize, in the absence of aliasing it is possible to eliminate the peculiar characteristics of each retinal patch from later visual representations, retinotopic and nonretinotopic. The hypothesis does not depend on the linearity or nonlinearity of the receptive fields of either ideal array. As noted earlier, reversible nonlinear transformations of the output of channels in the visual system do not alter the conclusions drawn here.

5.10 Visual Representation and Transformational Constancy

The end-product of perception is typically taken to be a *representation* of some aspects of the environment. So long as the *visual system* (biological or robotic) is stationary, the representation is simply an enrichment and transformation of the current visual image. When the visual system moves about in the environment and acquires multiple views of the same scene, though, we must consider how to represent the accumulated information. The typical solution is to attempt to build a world-centered coordinate system that comprises all of the information available from all of the many views. The status of the visual system within this representation is that of one object among many.

In human vision, certain implications of such a world-centered representation are termed *perceptual constancies*. Rock (1983) implicitly defines the term *perceptual constancy*: "If perception remains constant and does not mirror proximal stimulus variation, it is correlated with the external object, the distal stimulus" (p. 24). *Size constancy* and *location constancy* require, respectively, that the perceived size and location of objects be independent of viewpoint (Rock, 1983). The conditions under which and the extent to which human vision achieves these constancies is a separate issue. Here, we are concerned only with the natural connection between exploration and constancy.

The notion of perceptual constancies in Rock's sense is of great value, but it misses an important sort of consistency or constancy in a visual representation that is especially germane to exploratory vision systems (and "exploratory sampling systems"). An exploratory visual system will, as it explores, acquire more information about its environment and, presumably, alter its representation of its environment. This enrichment of the representation disallows defining "detail constancy" or "what's-in-back-of-the-opaque-object constancy" since we expect these aspects of the representation to change with exploration.

We can attempt to capture this consistency by defining *transformational constancy*. Suppose that a visual system changes viewpoint in a scene. Let $\phi(S)$ be any assertion concerning the scene S. That is, $\phi(S)$ is any statement such as "there is a one-inch white cube at location (x,y,z)". Suppose that location (x,y,z) is in the visual field from the first viewpoint, that there are no occluding objects between the visual system and (x,y,z), and that the resolution of the visual system is good enough to reliably decide whether there is a one-inch white cube at (x,y,z). Then, from the first viewpoint, the assertion $\phi(S)$ can be correctly judged to be true or false. Let's suppose it is true. There is a one-inch cube at the specified location. Now the visual system moves to the second viewpoint. If location (x,y,z) is still readily visible, then the assertion $\phi(S)$, judged from the second viewpoint, should remain true. Otherwise, the scene has changed, and the cube has vanished.

From any given viewpoint, many assertions about the scene cannot reliably be judged to be true or false. The objects described in the assertion may be too far away, or occluded, or out of the field of view. An assertion, applied to a specific scene, can have one of three possible values: YES, NO, and *insufficient information* (II). A first definition of *transformational constancy* is that, in going from one viewpoint to another, the judged values of predicates may go from II to either TRUE or FALSE, but not from TRUE to FALSE or FALSE to TRUE (unless, of course, the scene has really changed). Assertions that refer to the visual system itself (treated as an object in the scene) present no further difficulties. If the visual system moves, then an assertion such as "the sword of Damocles is hanging over my head" may pass from TRUE to FALSE, reflecting an actual change in the scene induced by movement of the visual system.

A visual system with transformational constancy, as just defined, must know something about its own capabilities. At a minimum, it must know whether it can correctly judge an assertion TRUE or FALSE from a given viewpoint. A slightly more sophisticated visual system could, perhaps, judge whether a given assertion could be judged TRUE or FALSE from any given viewpoint, not just the one it is currently occupying. Such a visual system could predict that a one-inch cube at (x,y,z) would be visible from a specified viewpoint, and could consequently plan its choices of viewpoints to answer certain questions about the scene.

The operation of an exploratory visual system with transformation constancy can be summarized by the following algorithm:

1. $n \leftarrow 0$. Set all assertions to II.

2. Look at the scene and capture View_n

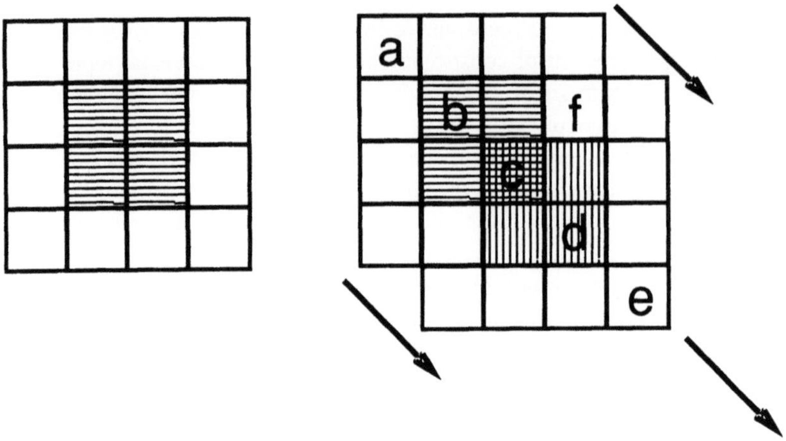

FIGURE 5.14. Left: an idealized retina with high-density patches (shaded) and low-density patches (unshaded). Right: the idealized retina has moved down and to the right. Each patch shown was viewed by either a high- or low-density patch in either the first or second position or in both positions. Region a, for example, was sampled by a low-density patch in the first position, and not sampled in the second. Region c was the sole region to be sampled by high-density patches in both positions. See text for discussion.

3. For each of the assertions $\phi_i(S), i = 1, 2, 3, \cdots$ that we can determine to be either TRUE or FALSE from View$_n$,
   - If the assertion is set to to II, set it to TRUE or FALSE as judged from the current view.
   - If the assertion was assigned TRUE or FALSE in an earlier view, check whether the new and previous assigned values are the same. If they are not, record a CHANGE in the scene.
4. Move within the scene. Increment n. Go to 2.

Figure 5.14 displays a simplified, idealized retina that is divided into small regions containing either a low density of receptors (unshaded) or a high density (shaded). We assume that the optical quality of the eye over each patch is matched to its density. Therefore, there is no aliasing. In the absence of aliasing, the precise locations of receptors within each patch are of no consequence and are not shown. Since the optical quality of a patch is matched to its photoreceptor density, the high-density patches can record more information about the scene. When the retina moves (as shown in the figure) different regions of the scene are viewed by two patches in succession. Consider how we might go about judging whether a change in the scene has occurred during the eye movement. Certain patches (labeled a and e) are seen only in one position and not the other. We cannot judge whether a change occurred in the scene locations corresponding to these patches. Some regions are seen twice by patches of the same density (regions c and f). We need only reconstruct the retinal image by means of each patch and compare the reconstructions to detect any change.

The two regions of interest (in considering transformational constancy) are b and d. Region d is viewed successively by low- and high-density patches. The high-density array (we assume) estimates the first N_h weights of a linear function space. We assume that the low-density array computes only the first $N_l < N_h$ weights. In the language developed above, the arrays can assign values of TRUE, FALSE or II to assertions of the form $\varepsilon_i = \alpha$ where ε_i is the ith weight, and α is any fixed real number. (When we talk about setting a particular such hypothesis, for example $\varepsilon_1 = 2.3$, to be TRUE, we tacitly imply that all other hypotheses $\varepsilon_1 = \alpha$, for any α other than 2.3, are set to be FALSE.)

After region d of the scene is viewed by a low-density patch, there are assertions $\varepsilon_i = \alpha_i$ for $i = 1, \cdots, N_l$, while all assertions of the form $\varepsilon_i = \alpha$ for $i = N_l + 1, \cdots, N_h$, for any value of α, are set to II. When region d is then viewed with the high-density patch, some assertion of the form $\varepsilon_i = \alpha_i$ for $i = 1, \cdots, N_h$ is set to be TRUE. The algorithm above, applied to these assertions concerning the basis weights can detect changes in the weights seen by both patches, but will not see the additional weights available with the high-density patch as a change in the scene. A similar analysis for region b (first high-density, then low) suggests that the algorithm above is at least reasonable.

A visual system has *transformational constancy* if its performance agrees with the algorithm described above. What is perhaps most radical in this view is the idea that claims about the scene that cannot be verified are not "filled in" in any way. An edge in peripheral vision is neither sharp nor blurry, if the information needed to discriminate sharp from blurry is not estimatable in peripheral vision. If the edge is now fixated, it may prove to be sharp or blurry, but in neither case does it trigger a perception of change. Whether the visual system behaves according to the algorithm above is an empirical question, but it is not obvious how to test it without appeal to introspective reports.

We have defined *transformational constancy* without choosing any particular world-centered coordinate system, or a particular kind of scene representation. Instead, we have discussed the sum total of the judgments that the visual system can make about a scene, effectively replacing claims about the nature of the representation by claims about the capabilities of the visual system. There is no intent here to claim that a visual system is in any sense a collection of assertions, only that it is useful to characterize it in such a way. In addition, psychophysical experiments typically reduce to measurements of the accuracy with which a visual system can assess the truth or falsehood of a single assertion. This method of specification lends itself to psychophysical testing.

5.11 Conclusion

This chapter concerns the consequences of sampling across time when the retinal surface is in motion. It is organized around two thematic questions concerning visual information available through small retinal patches (*sampling codes*). The first question is, *How is it possible to decide whether a change has occurred in the world outside by comparison of the sampling codes of two retinal patches that have successively sampled the same object or location in the scene?* We saw that it is sufficient that both patches have the *interpolation property*: each can reconstruct the possible retinal signals that fall on it.

The interpolation property fails precisely when aliasing occurs, and we examined three ways to eliminate or reduce the effects of aliasing: nonlinear environmental constraints on naturally occurring retinal images, the anti-aliasing properties of irregularly spaced retinal sampling arrays, and sampling across time. In the last of the three, we examined conditions that permit two sampling codes from successive glances at a single location in the scene to be combined into a single sampling code and thereby reduce aliasing. At his point the discussion also addressed the second thematic question, *How can the two sample codes be combined to produce a better estimate of the contents of the scene?*

When the interpolation property holds, the sampling properties of a retinal patch are determined by its size, shape, and the number of receptors within it. It is then possible to pass from the particular idiosyncratic layout of photoreceptors to the excitations of any set of *virtual photoreceptors* within the region. The excitation of a cell mechanism with a linear or nonlinear receptive field made out of virtual photoreceptors within that region is also computable. Such *ideal arrays* provide convenient ways to eliminate the peculiar character of a retinal patch from the information about the world it represents.

In the penultimate section of the chapter, I discussed *transformational constancy*. Transformational constancy is a possible property of an exploratory visual system that would permit it to distinguish changes in the world from changes in its knowledge about the world. Transformational constancy is illustrated by considering the problem of combining information from different regions of an idealized moving retina that differ markedly in photoreceptor density.

Acknowledgments: This work was supported by grant F49620-92-J-0187 from the Air Force Office of Scientific Research to Laurence T. Maloney, and by grant EY08266 from the National Eye Institute. Portions of this work were presented at Annual Meetings of the Association for Research in Vision and Ophthalmology, Sarasota, Florida, in May 1988 and May, 1989. I thank Albert J. Ahumada, David H. Brainard, Wilson S. Geisler, Stanley Klein, Michael S. Landy, Walter Makous, David Meyer, John Robson, David R. Williams and John I. Yellott, Jr. for comments on earlier presentations of this material, and Michael Tarr for formatting Fig. 5.1. I especially thank Heinrich Bülthoff and the staff at the Max-Planck-Institut für Biologische Kybernetik for providing a friendly and stimulating environment in which this chapter could be written.

5.12 References

Apostol, T. M. (1969). *Calculus* (2nd Ed.), Volume II. Waltham, Massachusetts: Xerox.

Artal, P., Derrington, A. M. & Colombo, E. (1995). Refraction, aliasing, and the absence of motion reversals in peripheral vision. *Vision Research, 35*, 939–947.

Artal, P., Navarro, R., Brainard, D. H., Galvin, S. J. & Williams, D. R. (1992). Off-axis optical quality of the eye and retinal sampling. *Investigative Ophthalmology and Visual Science (Suppl.), 33*, 1342.

Ben-Israel, A. & Greville, N. E. (1974). *Generalized Inverses; Theory and Applications.* New York: Wiley.

Bossomaier, T. R. J., Snyder, A. W. & Hughes, A. (1985). Irregularity and aliasing: Solution? *Vision Research, 25*, 145–147.

Burr, D. C. & Ross, J. (1979). How does binocular delay give information about depth? *Vision Research, 19*, 523–532.

Campbell, F. W. & Green, D. G. (1965). Optical and retinal factors affecting visual resolution. *Journal of Physiology, 181*, 576–593.

Campbell, F. W. & Gubisch, R. W. (1966). Optical quality of the human eye. *Journal of Physiology, 186*, 558–578.

Davis, P. J. (1963). *Interpolation and Approximation.* New York: Blaisdell.

Dym, H. & McKean, H. P. (1972). *Fourier Series and Integrals.* New York: Academic Press.

Galvin, S. J. & Williams, D. R. (1992). No aliasing at edges in normal viewing. *Vision Research, 32*, 2251–2259.

Hirsch, J. & Hylton, R. (1984). Quality of the primate photoreceptor lattice and the limits of spatial vision. *Vision Research, 24*, 1481–1492.

Hirsch, J. & Miller, W. H. (1987). Does cone positional disorder limit resolution? *Journal of the Optical Society of America A, 4*, 1481–1492.

Jennings, J. A. M. & Charmen, W. N. (1981). Off-axis image quality in the human eye. *Vision Research, 21*, 445–455.

Johnson, L. W. & Riess, R. D. (1982). *Numerical Analysis* (2nd Ed.). Reading, Massachusetts: Addison-Wesley.

Mallot, H. A., von Seelen, W. & Giannakopoulos, F. (1990). Neural mapping and space-variant image processing. *Neural Networks, 3*, 245–263.

Maloney, L. T. & Ahumada, Jr., A. J. (1989). Learning by assertion: A method for calibrating a simple visual system. *Neural Computation, 1*, 387–395.

Marr, D. (1982). *Vision: A Computational Investigation into the Human Representation and Processing of Visual Information.* San Francisco: Freeman.

Navarro, R., Artal, P. & Williams, D. R. (1993). Modulation transfer of the human eye as a function of retinal eccentricity. *Journal of the Optical Society of America A, 10*, 201–212.

Østerberg, G. (1935). Topography of the layer of rods and cones in the human retina. *Acta Ophthalmologica (supplement), 6*, 1–103.

Packer, O. & Williams, D. R. (1992). Blurring by fixational eye movements. *Vision Research, 32*, 1931–1939.

Poggio, T. & Torre, V. (1984). Ill-posed problems and regularization analysis in early vision. In *Image Understanding Workshop* (pp. 257–263). New Orleans, Louisiana.

Poggio, T., Torre, V. & Koch, C. (1985). Computational vision and regularization theory. *Nature, 317*, 314–319.

Rock, I. (1983). *The Logic of Perception*. Cambridge, Massachusetts: MIT Press.

Schölkopf, B. & Mallot, H. A. (1994). View-based cognitive mapping and path planning. Technical Report 7, Max-Planck-Institut für Biologische Kybernetik, Tübingen, Germany.

Simonet, P. & Campbell, M. C. W. (1990). The optical transverse chromatic aberration on the fovea of the human eye. *Vision Research, 30*, 187–206.

Snyder, A. W., Bossomaier, T. R. J. & Hughes, A. (1986). Optical image quality of the cone mosaic. *Science, 231*, 499–501.

Snyder, A. W., Laughlin, S. B. & Stavenga, D. G. (1977). Information capacity of eyes. *Vision Research, 17*, 1163–1175.

Snyder, A. W. & Miller, W. H. (1977). Photoreceptor diameter and spacing for highest resolving power. *Journal of the Optical Society of America, 67*, 696–698.

Sperling, G. (1989). Three stages and two systems of visual processing. *Spatial Vision, 4*, 183–207.

Steinman, R. M. (1965). Effect of target size, luminance, and color on monocular fixation. *Journal of the Optical Society of America, 35*, 1158–1165.

Steinman, R. M. & Levinson, J. Z. (1990). The role of eye movement in the detection of contrast and spatial detail. In E. Kowler (Ed.), *Eye Movements and their Role in Visual and Cognitive Processes* (pp. 115–212). Amsterdam: Elsevier.

Thibos, L. N. (1987). Calculation of the influence of lateral chromatic aberration on image quality across the visual field. *Journal of the Optical Society of America A, 4*, 1673–1680.

Thibos, L. N., Bradley, A., Still, D. L., Zhang, X. & Howarth, P. A. (1990). Theory and measurement of ocular chromatic aberration. *Vision Research, 30*, 33–49.

Westheimer, G. & McKee, S. P. (1975). Visual acuity in the presence of retinal-image motion. *Journal of the Optical Society of America, 65*, 847–850.

Williams, D. R. (1985). Aliasing in human foveal vision. *Vision Research, 25*, 195–205.

Williams, D. R. & Collier, R. J. (1983). Consequences of spatial sampling by a human photoreceptor mosaic. *Science, 221*, 385–387.

Woodhouse, J. M. & Barlow, H. B. (1982). Spatial and temporal resolution and analysis. In H. B. Barlow & J. Mollon (Eds.), *The Senses* (pp. 133–164). Cambridge, England: Cambridge University Press.

Wyszecki, G. & Stiles, W. S. (1982). *Color Science: Concepts and Methods, Quantitative Data and Formulae*. New York: Wiley.

Yellott, Jr., J. I. (1982). Spectral analysis of spatial sampling by photoreceptors: Topological disorder prevents aliasing. *Vision Research, 22*, 1205–1210.

Yellott, Jr., J. I. (1983). Spectral consequences of photoreceptor sampling in the Rhesus monkey. *Science, 221*, 385–387.

Yellott, Jr., J. I. (1990). The photoreceptor mosaic as an image sampling device. In *Advances in Photoreception* (pp. 117–133). Washington, DC: National Academy Press.

Yellott, Jr., J. I., Wandell, B. A. & Cornsweet, T. N. (1984). The beginnings of visual perception: The retinal image and its initial encoding. In *Handbook of Physiology: The Nervous System* (pp. 257–316). New York: Easton.

Zayed, A. I. (1993). *Advances in Shannon's Sampling Theory*. Boca Raton, Florida: CRC Press.

6

Calibration of a Visual System with Receptor Drop-out

Albert J. Ahumada Jr.[1]
Kathleen Turano[2]

ABSTRACT Maloney and Ahumada (1989) have proposed a network learning algorithm that allows the visual system to compensate for irregularities in the positions of its photoreceptors. Weights in the network are adjusted by a process tending to make the internal image representation translation-invariant. We report on the behavior of this translation-invariance algorithm calibrating a visual system that has lost receptors. To attain robust performance in the presence of aliasing noise, the learning adjustment was limited to the receptive field of output units whose receptors were lost. With this modification the translation-invariance learning algorithm provides a physiologically plausible model for solving the recalibration problem posed by retinal degeneration.

6.1 Introduction

During the course of the degenerative disease retinitis pigmentosa (RP), patients experience progressive visual field loss, raised luminance and contrast thresholds, and night blindness. Visual field loss typically begins in the midperiphery as a ring scotoma and spreads both centrally and peripherally, resulting in severely contracted visual fields (Massof & Finkelstein, 1987). Ultrastructural studies of the RP eye indicate that even in the early stages of the disease there is a diffuse loss of photoreceptors in all regions of the RP-affected eye and the remaining photoreceptors are enlarged (Flannery, Farber, Bird & Bok, 1989; Szamier, Berson, Klein & Meyers, 1979).

6.1.1 Retinal degeneration and bisection judgments

Turano (1991) studied the perceptual effects of retinal cone loss caused by RP. She investigated spatial position judgments in RP patients using a bisection task. Some patients with RP exhibited spatial position distortions (i.e., constant errors or biases) ranging from 2 to 5 standard deviations beyond the normal range. Other

[1]NASA Ames Research Center
[2]Wilmer Institute, Johns Hopkins University

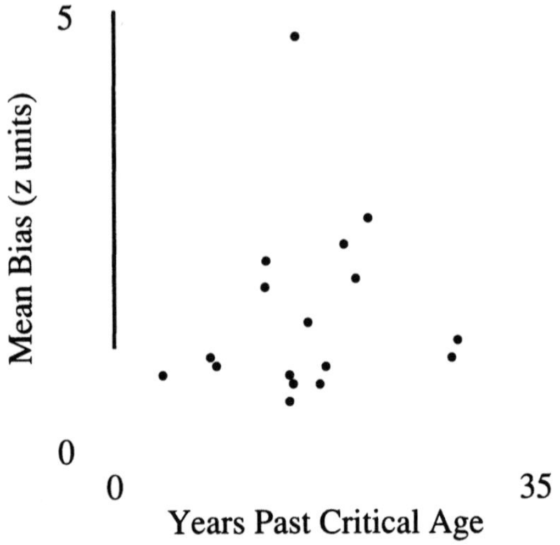

FIGURE 6.1. Mean absolute distortion (bias) of RP patients in units of normal subjects' standard deviation as a function of the estimated number of years of disease progression based on visual field size (Massof & Finkelstein, 1987). The data are from Turano (1991).

RP subjects were able to judge the relative position with the same accuracy as normal subjects. She was somewhat surprised by the lack of correlation between the magnitude of the position distortions and increased pathology (disease progression) as indexed by visual field loss. As illustrated in Fig. 6.1, the correlation was 0.124, indicating that increased pathology is a surprisingly small factor. One possible explanation for the lack of correlation is that RP patients learn to compensate for the loss-induced distortions, especially when the rate of disease progression is slow. Our work on learning theories for geometric calibration (Maloney & Ahumada, 1989) led us to try to see whether these theories, originally designed to calibrate the cone positions of a developing visual system, could be used to explain recalibration of a degenerating one.

6.1.2 Cone position calibration models

Maloney and Ahumada (1989) proposed a network learning algorithm as a solution to the problem of how the visual system knows the positions of the photoreceptors (Ahumada, 1992; Ahumada & Mulligan, 1990). The network transforms the sampled image values to new values from which an internal image is interpolated. The weights of the network are adjusted by a learning algorithm whose goal is to make the internal image translation-invariant, that is, look the same except for a translation when the eye position is changed. The weight adjustment rule can be regarded as a modification to the Widrow and Hoff (1960; Widrow & Stearns, 1985) adaptive linear weight adjustment procedure or delta rule. The delta rule is an error-correcting feedback rule and the error is computed as the difference between the output computed by the network and the desired output of the network. The Maloney and Ahumada rule does not assume that the correct desired output is available. The correct desired output of the delta rule is replaced by a translated and resampled version of the current internal image in a different retinal position. Here, we report on the behavior of this translation-invariance (TI) algorithm calibrating a visual system in which the receptor array experiences random drop-outs. We hoped that the model behavior might provide some insight into the recalibration problem faced by people suffering retinal degeneration and that we would also learn about the suitability of the algorithm for recalibrating remote sensor systems suffering such loss.

6.2 The Learning Algorithms

6.2.1 The visual system model

To keep calculations manageable, we will simulate a simplified model of the visual system. We assume the input receptors are in a regular rectangular array. We also assume an internal array of corresponding units. The input receptors are connected to the internal units by a linear weighting network. Initially, the weights to corresponding units will be unity and the weights to others zero; the network will just copy the receptor inputs to the internal units. An internal image is interpolated from the internal units using a discrete equivalent of sinc interpolation functions, pulses filtered by a rectangular filter in the spatial frequency domain. Initially, if an input image has no spatial frequency components above the Nyquist frequency of the sampling array, the image will be correctly reconstructed. The initial state is illustrated on the left side of Fig. 6.2. The input function illustrated is a pulse centered over the central unit. The pulse has been low-pass filtered at the Nyquist

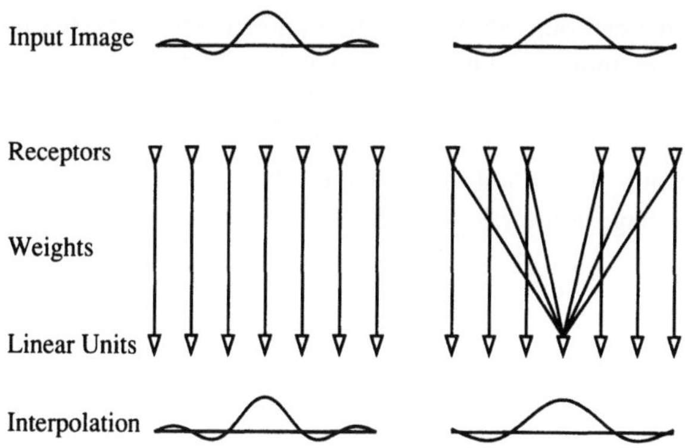

FIGURE 6.2. (Left) An input image is sampled by receptors, copied to the internal units and reconstructed from a linear combination of interpolation functions. (Right) Collateral weights added to the network to compensate for a missing receptor.

frequency of the sampling array. It has zero contrast at the other receptors, so the interpolated image is the interpolation function from the central unit alone. The correctness of the interpolation illustrates that the interpolation functions are filtered pulses.

We simulate the loss of a receptor by removing it and its connections from the network. If we did not change the network, the output image would also have this zero. Connections from the remaining receptors to the unit corresponding to the missing receptor allow them to fill in an estimate of what the missing receptor would have seen, as illustrated on the right side of Fig. 6.2. This network cannot reproduce the input of the left side, which does not now activate any receptors, but it can reproduce correctly stimuli which are restricted to the Nyquist frequency of the less dense sampling array. The image on the right is the same pulse with enough of the highest spatial frequencies removed so that the pulse is sub-Nyquist for the less dense array. It is now reconstructed from a weighted sum of interpolation functions having the shape of the one on the left. Learning rules will allow the development of appropriate weights.

Let s_i be the value of the input image at the ith sample point (receptor position), and let r_j be the value of the internal representation at the jth position. We compute the internal values as a linear combination of the remaining receptors. Let s_i' represent the values of the remaining receptors: $s_i' = s_i$, if receptor i is present; $s_i' = 0$, if receptor i is missing. Then

$$r_j = \sum_i s_i' w_{i,j}, \tag{6.1}$$

where $w_{i,j}$ is the network weight from receptor i to the internal unit at position j.

6.2.2 *The delta rule*

The delta rule was developed to compute weights from a sequence of inputs and corresponding desired outputs. From a rich enough set of images for which the missing points are known, compensating weights can be found by the delta rule. This rule is also called the Widrow-Hoff rule, least mean square learning, outer product learning and back propagation (Stone, 1986). The delta rule is error correcting. The error e_j for output unit j is the difference between the output of the unit r_j and the image at the jth position s_j,

$$e_j = r_j - s_j. \tag{6.2}$$

The weight from receptor i to unit j, $w_{i,j}$ is adjusted by subtracting a fraction λ of the error multiplied by the output of receptor i

$$w_{i,j} \leftarrow w_{i,j} - \lambda s_i' e_j. \tag{6.3}$$

The solid curve in Fig. 6.3 shows the error decreasing over trials in an example run of the delta rule learning weights to fill in for one receptor missing from a 7

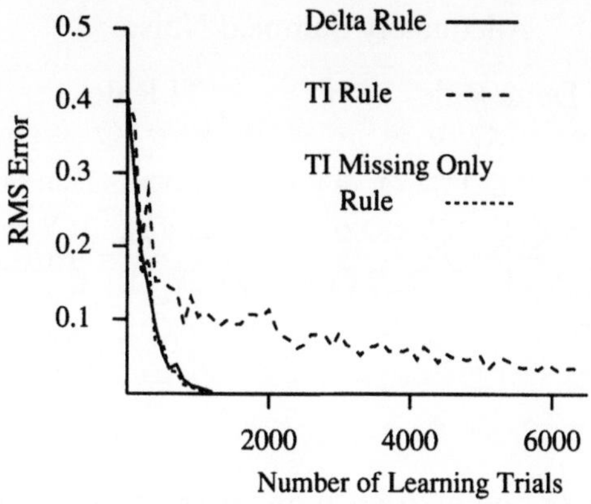

FIGURE 6.3. Root-mean-square (RMS) difference between the weighted network outputs and the actual image samples as a function of trials for the learning rules described in the text. Weights are being adjusted to fill in for one receptor missing from a 7 x 7 array. Images are white noise, low-pass filtered so that the sampling is still adequate for reconstruction.

by 7 array. The images used in the simulation were noises synthesized from Fourier components with equal expected amplitudes. The amplitudes of the sine and cosine components were independent, identically-distributed Gaussian random variables with zero mean. The images included the 25 sine and cosine components having horizontal or vertical spatial frequencies as high as 2 cycles per image. The image resolution was 16 (4 by 4) pixels per receptor. Images were selected for a trial by translating the position of the noise image to a new position selected at random except that it had to move at least one receptor spacing in each dimension (x and y). To simplify calculations, wrap-around motion was simulated. After each block of 100 trials a new noise image was computed. The learning rate λ was set to the value 0.5 divided by the sum of the squared s_i' values. The weight matrix was initialized to be the identity matrix. The left side of Fig. 6.4 shows the weights that were learned for the output whose input is now missing. The weights to the other units are not affected by the delta rule because their error is always zero.

In general, the delta rule can be shown to provide weights that minimize the expected squared error if the learning rate is slowly brought toward zero. In cases, as above, in which a perfect (zero expected error) solution exists, the rule causes the weights to converge towards a solution for a range of fixed learning rates (Widrow & Hoff, 1960; Widrow & Stearns, 1985). Unfortunately, although this rule uses biologically plausible computations, is well behaved, and well understood, the information that it needs, the stimulus input to the missing receptor, is not known.

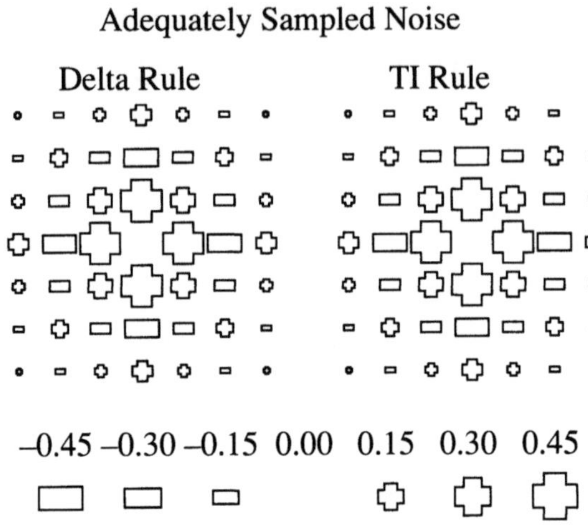

FIGURE 6.4. Weights computed by the simulations giving the learning curves of Fig. 6.3 for the delta rule and the TI rule.

6.2.3 *The TI rule*

The TI rule was developed for the case that the input image is not known. It bases its adjustments on the output of the weighting network. The output is computed for an input image. The eye is moved a known amount and the output is computed again. This second output plays the role of r_i in the delta rule. The output from the previous image is then translated to compensate for the eye movement and then interpolated and sampled. These samples are used in place of s_j in the above formula for e_j. (Complete formulas for the TI rule are given elsewhere (Ahumada, 1992; Ahumada & Mulligan, 1990; Maloney & Ahumada, 1989).) This feedback does not directly drive the weights toward a correct solution, but it drives them toward a translation-invariant solution, in which the interpolated output translates along with translations in the input. Correctness is obtained by forcing one of the units to be correct. We do this by connecting one output to only one input, making that weight unity, and not changing it. The translation-invariance learning algorithm then tries to copy this fixed receptive field to all the others. Fig. 6.3 also contains the learning curve for the TI rule. Although the TI rule learns much more slowly, at the end of 6400 trials the weights were indistinguishable from those of the delta rule after 1200 trials, as shown in Fig. 6.4. It is easy to understand why the learning is so much slower for the TI rule. For the delta rule, errors only occur at the output whose input is missing, so only the weights in the receptive field of that unit are altered. The TI rule, however, will see a translation error at any position translated to or from the image region near the missing input. Initially it changes all the correct weights and then puts them back as it also finds the weights which fill in for the missing input.

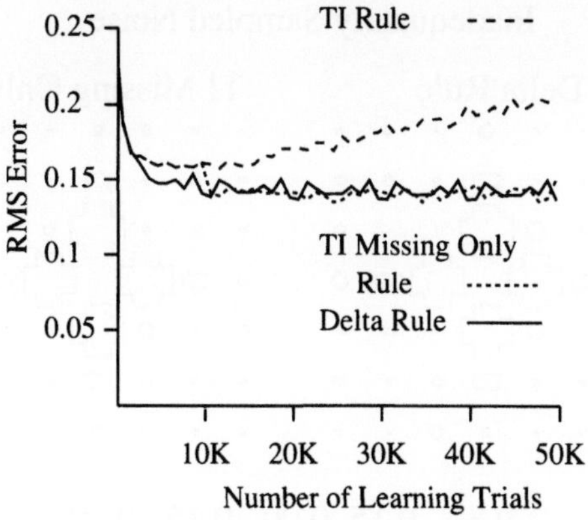

FIGURE 6.5. RMS errors from simulations with images that are "pink" noise, low-pass filtered so that the sampling was adequate for reconstruction before the receptor was lost but not after.

6.2.4 *Inadequate sampling*

In the previous example, the sampling density of the receptors was high enough so that a perfect reconstruction of the band-limited stimulus was possible from the remaining receptors. In this case both rules find weights which provide perfect reconstruction. When the sampling array is no longer capable of reconstructing the stimuli, we encounter a situation where there is no perfect solution. In this case, the delta rule can be shown to find an average least squared error solution, if the learning rate is slowly decreased to zero. The solid and dashed lines in Fig. 6.5 show the learning performance of the two rules for such an example. The parameters were the same as the previous example except that the noise bandwidth extends to the Nyquist frequency of the original 7 by 7 sampling array. Also, since the coefficients now depend on the content of the stimuli, the amplitudes of spatial frequency components of the noise were made inversely proportional to their radial frequency. Finally, the learning rate coefficient was decreased from 0.5 to 0.01, so that the learning itself would not be a large source of error at asymptote. If white noise had been used, rather than this "pink" noise, the missing sample would have been independent of the other samples and no weightings would improve performance. The Gaussian nature of the noise ensures expected squared error optimality of a linear predictor of the missing sample, and thus the delta rule weightings will be nearly optimal.

The figure shows that the delta rule error rate becomes stable, but while the TI rule performs well initially, its error rate later becomes as poor as at the start of learning. A check of the weights during learning revealed that the originally correct weights keep dropping in value. They do not return to their correct value as they do when the number of sampling receptors is adequate for correct reconstruction. The rule is probably hunting for a compromise between the least squared error solution and the only solution with no translation error, the zero weight solution.

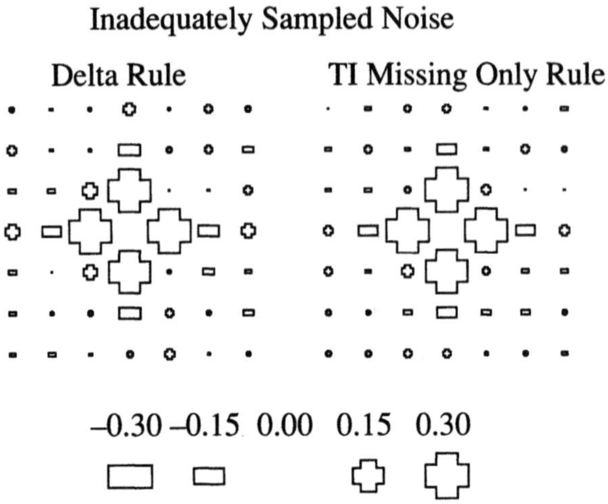

FIGURE 6.6. Weights from the inadequate sampling simulations for the delta rule and the TI rule that only adjusts the weights to the output unit which lost its receptor.

6.2.5 *A new rule*

One possible solution to this problem is to increase the number of receptive fields constrained to be correct. If the loss of a receptor triggers weight learning for weights to its output unit and not to others, the output units for the remaining receptors remain correct. It seems plausible that the physiological consequences of losing a receptor could trigger the learning process. The lack of activity of the internal unit could also initiate the learning process. The dotted line in Fig. 6.3 shows that this rule can learn almost as fast as the delta rule when the sampling is adequate. The dotted line in Fig. 6.5 shows that this rule can learn the same stable solution as the delta rule when the sampling is inadequate. Fig. 6.6 shows the weights learned by the new rule and by the delta rule when the sampling is inadequate.

6.2.6 *A final example*

To illustrate that the above conclusions hold when more than one receptor is missing, we show a final example simulation. The conditions were the same as the previous case of inadequate sampling except that the size of the sampling array was increased to 11 by 11 and 30 per cent of the units were deleted at random. Fig. 6.7 shows that in this situation the TI rule performs worse than it does when only one receptor is missing. The larger number of missing receptors causes a faster erosion of the correct coefficients. The performance of the TI rule applied only to the receptive fields of the missing receptors' output units is also worsened, but by a relatively small amount. Fig. 6.8 shows the weights learned by the new rule and the delta rule in a run where the sampling is inadequate and 30 per cent

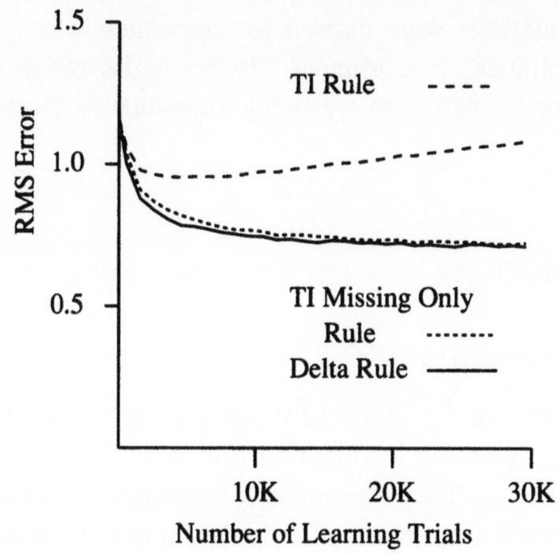

FIGURE 6.7. RMS errors for the three algorithms for the 11 x 11 sampling array with 36 (30 per cent) of the receptors missing.

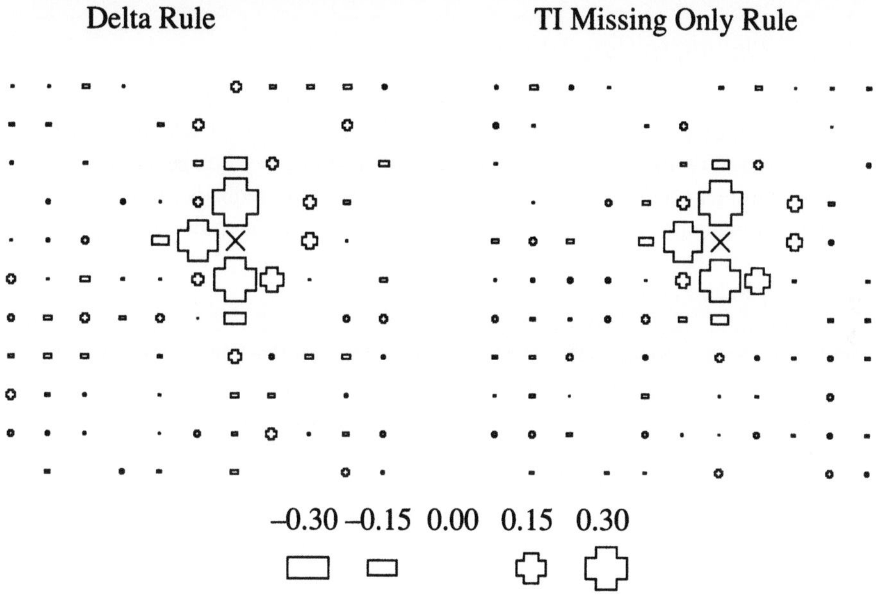

FIGURE 6.8. Weights for the 11 x 11 sampling array computed by the delta rule (left) and the TI rule adjusting only the weights to the output units which lost their receptors (right). The X marks the position of the missing receptor for which these weights are compensating. The other 35 missing receptors have no symbol.

of the units are missing. These weights are after 100,000 learning trials. The weights for the revised TI rule are similar to those of the delta rule after 50,000 trials and, presumably, are converging to a similar asymptote. The learning rates for these simulations were chosen for convenience and were not intended to optimize performance. Performance closer to the range of the adequately sampled case might be achieved by using schedules with decreasing learning rates.

6.3 Discussion

6.3.1 *Known translations*

One apparent weakness of the TI model is that the eye movement is assumed to be known exactly. Ahumada and Mulligan (1990) have shown that this assumption can be relaxed somewhat. They showed that when the TI rule is learning weights to compensate for small amounts of jitter in the input unit positions, the exact eye movement knowledge can be replaced by an estimate based on maximizing the correlation between the two output images. We speculate that the same situation will hold for moderate fractions of missing receptors.

6.3.2 *The interpolated image*

Barlow has proposed an anatomical layer of fine cells in the cortex as the site for the interpolation of a cortical image (Barlow, 1979). The need for such an image has been criticized on the grounds that a homunculus would then be needed to look at it. The TI model needs an interpolated image to obtain its feedback samples unless the eye movements are made in integral multiples of the internal spacing. Other visual processing channels could sample the interpolated image according to their own independent needs.

On the other hand, separate processing channels might carry out their own calibration independently. If a channel can be characterized by a linear impulse response that it computes over space in a translation-invariant manner, the TI rule can be used to calibrate it separately. Ahumada and Tabernero (1994) have shown that separate spatial frequency channels can be calibrated independently with the benefit that associative learning processes can be used instead of a fixed correct output unit to provide a target receptive field. Our TI schemes above are useless if the fixed unit is the one that turns out to have its input missing. In addition, a calibration scheme based on separate bandpass spatial frequency channels should be more resistant to the aliasing problems caused by missing receptors, since bandpass images are just as easy to reconstruct as lowpass images using linear weighting functions.

6.3.3 *Two views from two eyes*

It is possible that the two views used by the TI method might come from separate eyes rather than two views from the same eye over time. We have not tried to implement this interesting idea. We do know that to obtain adequate calibration, the views must differ in more than a single direction (the eyes cannot be assumed to diverge in only the horizontal egocentric direction). Craik (1966) "sunburned" one retina and describes distortions in his undamaged eye apparently induced from adaptations to damage in the other eye, consistent with a model in which inputs from both eyes are mutually calibrated. If there were three eyes, majority rule could prevent miscorrection of an undamaged eye. Similarly, three views over time could prevent the problem of the TI rule messing up the good connections and make unnecessary the restriction of the learning to the region of the missing receptors.

6.3.4 Partial damage

It seems unlikely that any destructive process causing diffuse degeneration of receptors would be completely all or none. Cells would be expected to suffer partial damage, which could be modeled as some combination of position distortion, gain change, and noise level increase. We have seen the TI model work for position and gain changes, but as yet have no information about its performance in the case of noisy receptors.

6.4 Conclusions

We have investigated the ability of Maloney's TI rule to recalibrate a receptor array after receptor loss. We confirmed the expected result that when the remaining array is adequate to reconstruct the stimuli, the TI rule can find a compensating transformation. However, when there is no perfect solution, the TI rule initially improves the situation by computing weights that fill in, but eventually erodes the correct values of weights from receptors to their corresponding outputs. This problem is easily removed by allowing the TI rule only to adjust the weights in the receptive fields of units that have lost receptors. This modification of the TI rule provides a model for how recalibration for missing receptors might take place.

Acknowledgments: This work was presented to the 13th European Conference on Visual Perception, Paris, France, in September, 1990 (Ahumada & Turano, 1990), and parts of it were presented to the SPIE Meeting in San Jose, California, in February, 1991 (Ahumada & Mulligan, 1991). This work was supported by NASA RTOP 506-71-51 and NIH/NEI Grant EY07839. Helpful comments were provided by R. L. Gregory, M. Shiffrar and M. S. Landy.

6.5 References

Ahumada, Jr., A. J. (1992). Learning receptor positions. In M. S. Landy & J. A. Movshon (Eds.), *Computational Models of Visual Processing* (pp. 23–34). Cambridge, Massachusetts: MIT Press.

Ahumada, Jr., A. J. & Mulligan, J. B. (1990). Learning receptor positions from imperfectly known motions. In Rogowitz, B. & Allebach, J. (Eds.), *Human Vision, Visual Processing, and Digital Display, Proceedings of the SPIE*, Volume 1249 (pp. 124–134).

Ahumada, Jr., A. J. & Mulligan, J. B. (1991). Network compensation for missing sensors. In Rogowitz, B., Brill, M. H. & Allebach, J. (Eds.), *Human Vision, Visual Processing, and Digital Display, Proceedings of the SPIE*, Volume 1453 (pp. 134–146).

Ahumada, Jr., A. J. & Tabernero, A. (1994). Anti-hebbian learning and cortical receptive field calibration. *Investigative Ophthalmology and Visual Science, 35(4, ARVO Suppl.)*, 1257 (Abstract).

Ahumada, Jr., A. J. & Turano, K. (1990). Calibration of a visual system with progressive receptor drop-out. *Perception, 19*, 337 (Abstract).

Barlow, H. B. (1979). Reconstructing the visual image in space and time. *Nature, 279*, 189–190.

Craik, K. J. W. (1966). *The nature of psychology*. Cambridge, UK: Cambridge University Press.

Flannery, J. G., Farber, D. B., Bird, A. C. & Bok, D. (1989). Degenerative changes in a retina affected with autosomal dominant retinitis pigmentosa. *Investigative Ophthalmology & Visual Science, 30*, 191–211.

Maloney, L. T. & Ahumada, Jr., A. J. (1989). Learning by Assertion: Two methods for calibrating a linear visual system. *Neural Computation, 1*, 392–401.

Massof, R. W. & Finkelstein, D. (1987). A two-stage hypothesis for the natural course of retinitis pigmentosa. In E. Zrenner, H. Krastel & H. Goebel (Eds.), *Advances in the Biosciences: Research in Retinitis Pigmentosa* (pp. 29–58). New York: Pergamon Press.

Stone, G. O. (1986). An analysis of the delta rule and the learning of statistical associations. In D. E. Rumelhart & J. L. McClelland (Eds.), *Parallel Distributed Processing, Vol. 1* (pp. 444–459). Cambridge, Massachusetts: MIT Press.

Szamier, R. B., Berson, E. L., Klein, R. & Meyers, S. (1979). Sex-linked retinitis pigmentosa: Ultrastructure of photoreceptors and pigment epithelium. *Investigative Ophthalmology and Visual Science, 30*, 191–211.

Turano, K. (1991). Bisection judgments in patients with retinitis pigmentosa. *Clinical Vision Science, 6*, 119–130.

Widrow, A. B. & Hoff, M. E. (1960). Adaptive switching circuits. In *WESCON Convention Record, Part 4* (pp. 96–104).

Widrow, A. B. & Stearns, S. D. (1985). *Adaptive signal processing*. Englewood Cliffs, New Jersey: Prentice-Hall.

7

Peripheral Visual Field, Fixation and Direction of Heading

Inigo Thomas[1]
Eero Simoncelli[1]
Ruzena Bajcsy[1]

ABSTRACT Although moving human observers actively fixate points in the world with their eyes, computer vision algorithms designed for the estimation of structure-from-motion or egomotion typically do not make use of this constraint. In this paper, we investigate the computational advantage of fixation. The main contribution of this work is to specify precisely the form of the optical flow field for a fixating observer moving in a rigid world. In particular, we show that the use of a hemispherical (retinal) imaging surface combined with the active process of fixation generates an optical flow field of a particularly simple form. A further contribution is the finding that the sign of retinal flow at the retinal periphery can be used to predict collisions.

7.1 Introduction

Introspection reveals that when a human observer moves, the eye continually fixates on targets in the world. Although fixation is a common process in human vision, the potential computational advantage of this behavior has not been established. The main contributions of this paper involve (1) formalizing retinal flow for a fixating observer and thereby (2) isolating simple patterns of flow that are useful in determining an observer's direction of heading. In particular we explore the role of the retinal periphery in predicting collision. We expect fixation to be especially fruitful in light of recent advances in computer vision for constructing active head/eye systems (Pahlavan, Uhlin & Eklundh, 1993).

This work is part of ongoing research on active vision pursued at the General Robotics and Active Sensory Perception (GRASP) Laboratory. Active vision, a paradigm introduced by Bajcsy (1985, 1988), provides constraints that allow certain ill-posed problems in computer vision to be converted into well-posed ones. These constraints arise due to simplifying assumptions about the world and its visual appearance that are satisfied when the camera moves actively. For example, Krotkov (1989) describes the active vergence of a stereo camera pair to facilitate

[1]Department of Computer Science, University of Pennsylvania

figure-ground segmentation. Hager (1988) fused constraints from multiple sensory sources (vision and touch). Bajcsy and Maver (1993) used information derived from shadows to actively move the camera to a more informative position.

In the tradition of active vision research, the present article discusses the constraints on optical flow fields that are imposed by the process of active fixation. The crucial difference between the present work and traditional optical flow analyses (cf. Thomas, 1993, for a review) is *fixation*. In the computer vision literature, a previous analysis of optical flow field under fixation has been provided by Raviv and Herman (1990). As in this paper, they determine the loci of zero flow. They also describe the temporal evolution of these loci of zero flow. In contrast, our development is done in a different coordinate system using a different representation of optical flow, we analyze the pattern of flow across the entire retina, and we use the *sign* of a particular projection of flow vectors to determine direction of heading.

The notion of *sign of flow* has been previously exploited in work by Fermüller (1993), who analyzes components of the flow field (so-called *normal* flow) in terms of sign (see Fermüller & Aloimonos, this volume, Chapter 9). Due to the assumption of normal flow she arrives at very different patterns from ours. The simple four-way distinction which we isolate at the periphery cannot be observed using normal flow, which gives rise to areas bounded by conic sections. Furthermore, the search over the solution space that is required by her algorithm may make it unrealistic for implementation on a mobile robot. In Section 7.7, we demonstrate an implementation of our closed-form solution being used to guide a robot to a target.

In this work we make the following assumptions: (1) the observer moves with respect to the world and fixates on a target; (2) the world is rigid, with no independently moving elements; and (3) the possible rotation axes of the eye lie on a plane orthogonal to the direction of gaze. Assumption (1) is a behavioral one that is observed in humans and animals. Assumptions (2) and (3) will allow us to write the retinal flow field in a particularly simple form; (3) also corresponds closely to Listing's Plane for the human eye.

We first define retinal flow for a 2-D universe and then extend it to the full 3-D case; the flow in 2-D provides one component of the flow in 3-D. The retinal flow in 3-D is then decomposed into longitudinal and latitudinal flow, and we then show that the longitudinal component depends only on the translational direction of heading and not the rotation of the eye. We show that longitudinal flow, especially at the periphery, can be analyzed to determine the direction of heading. Finally, we test the performance of this model in experiments involving a simulated environment as well as a real world setting and a mobile robot platform.

7.2 Retinal Flow in a Rigid 2-D Universe

For ease of exposition, we first consider a reduced case of a 2-D universe in which we define the flow on the retina for any given point in the universe; as the observer moves, the flow determines how each point projected on the retina moves.

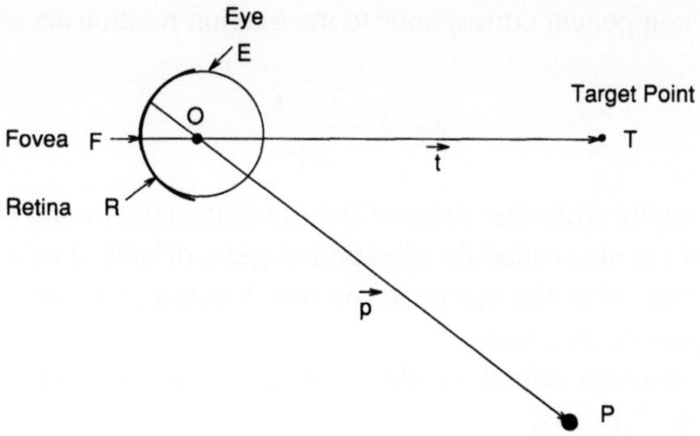

FIGURE 7.1. Model of the eye in a 2-D universe.

7.2.1 Calculating retinal flow

In a 2-D planar universe, the eye of the observer corresponds to a circle (E in Fig. 7.1), and the retina corresponds to a semicircle (R).[2] As the observer moves, the center of the eye (O in Fig. 7.1) translates on the 2-D plane. In addition, the eye may also rotate about its center (O). A combination of these two types of motion is sufficient to capture all possible rigid movements of the eye in this 2-D universe.

When the observer fixates on a target point (such as a corner of an object) this point, by definition, remains projected at the center of the retina, that is, on the fovea (F in Fig. 7.1). To maintain fixation while moving, the observer has to rotate the eye about its center (O). Although the target point on which the observer fixates (T) is stationary at the fovea, the remainder of the retinal image (e.g. P) can be expected to change; this change will be precisely defined below.

The *instantaneous* movement of the retinal image will be referred to as *retinal flow*, and it will be defined in the present formalization in terms of angular coordinates. We parameterize the retinal flow of a point P as an angular velocity, that is, the temporal derivative of an angle between two rays: (1) the direction of gaze (ray OT in Fig. 7.1) and (2) the ray from the point in the world to the center of the eye (OP in Fig. 7.1).

The retinal flow may be decomposed into two components: one due to observer *translation* and the other due to the fixating *rotation*. The first component is written as

$$\omega_1 = \frac{\vec{\mathbf{v}} \times \vec{\mathbf{p}}}{|\vec{\mathbf{p}}|^2}, \tag{7.1}$$

where $\vec{\mathbf{v}}$ is the translational velocity of the center of the eye, $\vec{\mathbf{p}}$ is the vector from the center of the eye to an arbitrary point in the world P, and $\times$ indicates a standard vector cross-product.

[2]The eye and the retina considered in this paper will only correspond to the human eye in terms of optics and not in terms of the actual physical structure.

The second component corresponds to the fixation rotation about the center of the eye:

$$\omega_2 = -\frac{\vec{\mathbf{v}} \times \vec{\mathbf{t}}}{|\vec{\mathbf{t}}|^2}, \tag{7.2}$$

where $\vec{\mathbf{t}}$ is the vector from the center of the eye to the target point (T in Fig. 7.1); the direction of $\vec{\mathbf{t}}$ is also called the direction of gaze, or optical axis. The negative sign indicates that when the eye rotates in one direction, the points on the retina move in the opposite direction.

Finally, the resultant retinal angular velocity of a point is the sum of the two angular velocities[3], that is,

$$\begin{aligned} \omega &= \omega_1 + \omega_2 \\ &= \frac{\vec{\mathbf{v}} \times \vec{\mathbf{p}}}{|\vec{\mathbf{p}}|^2} - \frac{\vec{\mathbf{v}} \times \vec{\mathbf{t}}}{|\vec{\mathbf{t}}|^2}. \end{aligned} \tag{7.3}$$

Given the observer translation velocity $\vec{\mathbf{v}}$, the vector representing gaze $\vec{\mathbf{t}}$, and the vector from the eye to any point in the world $\vec{\mathbf{p}}$, Eq. 7.3 defines the retinal flow of that point in a 2-D universe.

7.2.2 *Level sets of retinal flow*

In this section we will consider the points in the 2-D universe that give rise to the same value of retinal flow. Let us first isolate those points in the 2-D universe that correspond to zero retinal flow. The projections of such points on the retina come to rest (for an instant) while the observer moves and fixates. To find such points, we set Eq. 7.3 to zero and solve for $\vec{\mathbf{p}}$:

$$\frac{\vec{\mathbf{v}} \times \vec{\mathbf{p}}}{|\vec{\mathbf{p}}|^2} - \frac{\vec{\mathbf{v}} \times \vec{\mathbf{t}}}{|\vec{\mathbf{t}}|^2} = 0. \tag{7.4}$$

Factoring the magnitude of $\vec{\mathbf{v}}$ from both terms results in

$$\frac{\hat{\mathbf{v}} \times \vec{\mathbf{p}}}{|\vec{\mathbf{p}}|^2} - \frac{\hat{\mathbf{v}} \times \vec{\mathbf{t}}}{|\vec{\mathbf{t}}|^2} = 0, \tag{7.5}$$

where $\hat{\mathbf{v}}$ is the unit vector in the direction of $\vec{\mathbf{v}}$.

Although the points in the 2-D universe satisfying the above equation lie on a simple curve, reducing the solution to a recognizable form requires further vector algebraic manipulations. Introduce unit vectors $\hat{\mathbf{u}}$ perpendicular to $\hat{\mathbf{v}}$ and $\hat{\mathbf{z}}$ perpendicular to the plane containing $\hat{\mathbf{v}}$ and $\vec{\mathbf{t}}$; then $\hat{\mathbf{z}} \cdot (\hat{\mathbf{v}} \times \vec{\mathbf{t}}) = \hat{\mathbf{u}} \cdot \vec{\mathbf{t}}$. We can now rewrite

---

[3] In general, the sum of two rotation vectors does *not* produce a rotation vector corresponding to the composition of rotations. The angular velocities may be added here because they are *instantaneous* measurements.

equation 7.5 in terms of $\hat{\mathbf{u}}$:

$$\frac{\hat{\mathbf{u}} \cdot \vec{\mathbf{p}}}{\vec{\mathbf{p}} \cdot \vec{\mathbf{p}}} - \frac{\hat{\mathbf{u}} \cdot \vec{\mathbf{t}}}{\vec{\mathbf{t}} \cdot \vec{\mathbf{t}}} = 0. \tag{7.6}$$

This can be rewritten as

$$\vec{\mathbf{p}} \cdot \vec{\mathbf{p}} - \hat{\mathbf{u}} \cdot \vec{\mathbf{p}} \left(\frac{\vec{\mathbf{t}} \cdot \vec{\mathbf{t}}}{\hat{\mathbf{u}} \cdot \vec{\mathbf{t}}} \right) = 0. \tag{7.7}$$

and then, by completing the square, as

$$\left\| \vec{\mathbf{p}} - \frac{1}{2} \left(\frac{\vec{\mathbf{t}} \cdot \vec{\mathbf{t}}}{\hat{\mathbf{u}} \cdot \vec{\mathbf{t}}} \right) \hat{\mathbf{u}} \right\|^2 = \left[\frac{1}{2} \left(\frac{\vec{\mathbf{t}} \cdot \vec{\mathbf{t}}}{\hat{\mathbf{u}} \cdot \vec{\mathbf{t}}} \right) \right]^2 . \tag{7.8}$$

For $r = \frac{1}{2} \left(\frac{\vec{\mathbf{t}} \cdot \vec{\mathbf{t}}}{\hat{\mathbf{u}} \cdot \vec{\mathbf{t}}} \right)$ and $\vec{\mathbf{c}} = r\, \hat{\mathbf{u}}$, we obtain the familiar equation of a circle:

$$(\vec{\mathbf{p}} - \vec{\mathbf{c}})^2 = r^2. \tag{7.9}$$

The vector $\vec{\mathbf{c}}$ is the center of this circle (with respect to the center of the eye). Recall that $\hat{\mathbf{u}}$ is a vector perpendicular to the velocity of the observer $\vec{\mathbf{v}}$; this means that the center of the circle lies in a direction perpendicular to the direction of movement. This circle, corresponding to zero flow in the retina, is depicted by the solid line in Fig. 7.2. The circle passes through the target point of fixation and through the center of the eye. All points on the circle, including these two points, behave in the same way: momentarily, they are stationary. Furthermore, the points *within* this circle all move in the same direction on the retina, whereas the points *outside* of this circle move in the opposite direction.[4]

A similar analysis can be performed for any other value of retinal flow besides the zero flow. For each such value, the result corresponds to a circle of points in the 2-D universe; the radius of the circle varies depending on the particular value chosen. Sample circles which correspond to points with equal retinal flow are shown as dotted curves in Fig. 7.2. Note that the centers of all such circles lie on a straight line.

An interesting boundary case involves fixating straight ahead. In this case the direction of fixation coincides with the direction of observer movement. The resulting circle of zero flow has an infinite radius (i.e., it is a *line*), as illustrated in Fig. 7.3. In this situation, points lying to one side of this line move in one direction, while points on the remaining half plane move in the opposite direction. This result fits intuition: when looking and moving straight ahead, points on the left half of the visual field move leftward, and points on the right half move rightward.

---

[4]The retinal flow associated with each point in the world forms a vector field or dynamical system (Abraham & Shaw, 1993) with a *separatrix* corresponding to the circle of zero flow which separates the two regions of opposite flow. This suggests an interesting connection between the present analysis and the dynamics of well known systems.

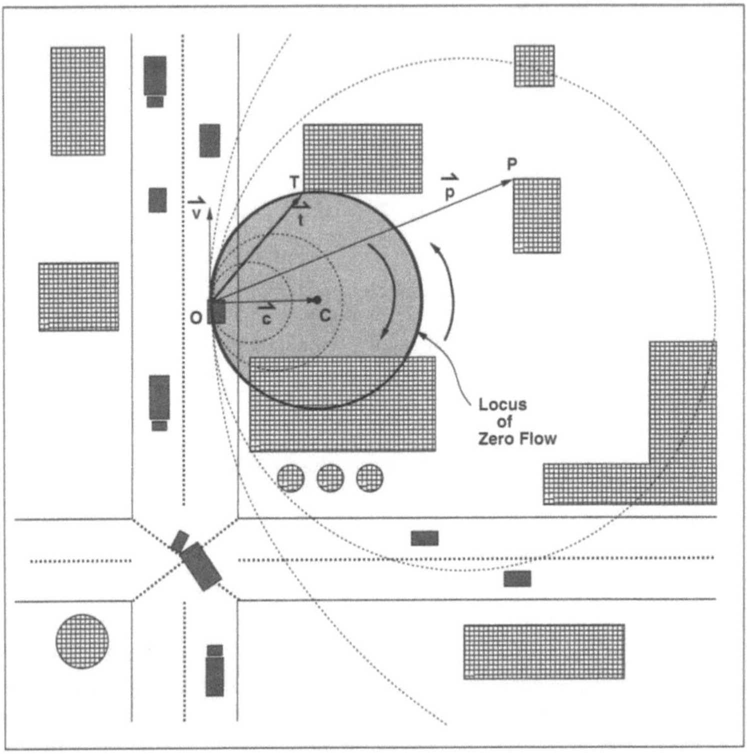

FIGURE 7.2. The level sets of retinal flow in a 2-D universe. This scene depicts a traffic intersection, where the observer is moving along the road with translational velocity $\vec{v}$, and fixates on a corner of a building T. O is the center of the eye of the observer, and P is an example point in the universe, the retinal velocity of which is being calculated. The points with zero retinal flow lie on the solid circle with center C. Points in the shaded area have flow in the clockwise direction while points in the unshaded area have flow in the counter-clockwise direction.

7.3 Retinal Flow in a Rigid 3-D Universe

The case of moving and fixating in a 3-D universe is clearly more complicated than the 2-D case. However, the 3-D case can be elegantly decomposed into two modules: one involving the retinal flow just as in the 2-D case, and the other involving a new component.

7.3.1 Calculating retinal flow

In a 3-D universe, the eye corresponds to a sphere (rather than a circle) with a center O (Fig. 7.4) and the retina is a hemisphere (rather than a semicircle). As in

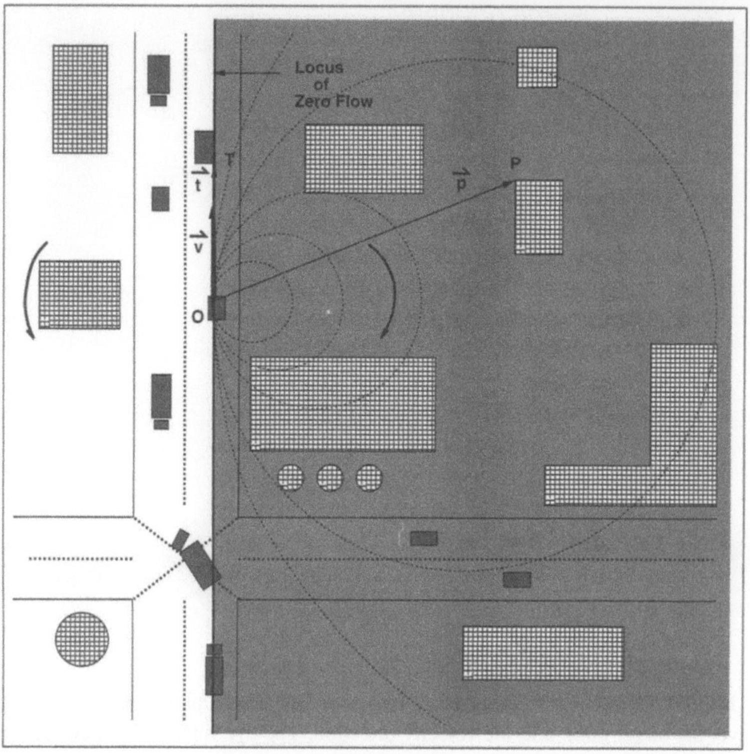

FIGURE 7.3. The special case of moving and fixating in the same direction. See Fig. 7.2 for explanation of the symbols.

the 2-D case, the eye rotates to fixate on a target. However, the 3-D rotation that accomplishes fixation is no longer unique.[5]

To make the problem manageable, we constrain the way in which the eye can rotate in order to fixate. The constraint we impose is that the axis about which the eye rotates is always perpendicular to the direction of gaze; that is, the possible rotation axes lie in a plane. Although this is an arbitrary constraint, the physiology of the eye suggests that a similar constraint operates in humans (involving the so-called Listing's Plane, Yarbus, 1967). Furthermore, this particular formulation of the constraint allows us to decompose the retinal flow into two components.

To represent the two components of retinal flow in a 3-D universe, we impose a grid of longitudes and latitudes on the hemispherical retina. These longitudes and latitudes are comparable to the standard grid used to specify coordinates on the earth. In the present analysis, we wish to fix this grid on the retina in a such a way

[5]For any fixating rotation, an additional instantaneous rotation about the optical axis can be added without loss of fixation. The family of such fixation rotation axes are obtained by varying the amount of this additional rotation.

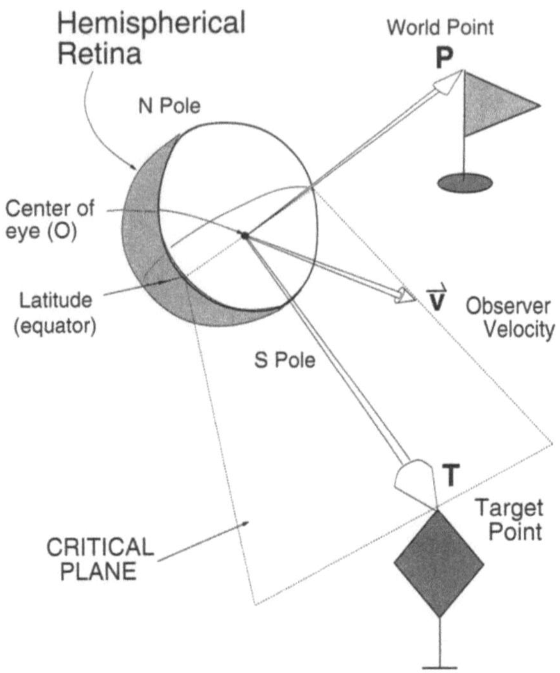

FIGURE 7.4. Model of the eye in a 3-D universe. The critical plane contains the center of the eye, the target point and the velocity direction. The eye rotates about an axis passing through the north and south poles to fixate.

that the center of the eye, the target point (that is being fixated), and the direction of movement all fall on a plane containing the equator. This plane will be referred to as the *critical plane* (Fig. 7.4). The other latitudes are semicircles on the retina lying on planes parallel to the critical plane. The longitudes are semicircles[6] on the retina starting at the north pole and ending at the south pole. As is standard, the north pole, center of the globe (eye) and the south pole lie on an axis perpendicular to the equatorial (critical) plane. Note that this line is the axis about which the eye rotates in order to fixate the chosen target.

The retinal flow corresponding to points lying on the critical plane is identical to the retinal flow of points in the 2-D universe, and was given by Eq. 7.3. As before we will use angular velocity to represent retinal flow. However, in the 3-D case angular velocity is no longer unique: any component of angular velocity in the direction of $\vec{\mathbf{p}}$ is not observable on the retina. To achieve uniqueness of representation, we constrain angular velocity of a point $\vec{\mathbf{p}}$ to be perpendicular to $\vec{\mathbf{p}}$:

$$\vec{\omega} \;=\; \frac{\vec{\mathbf{v}} \times \vec{\mathbf{p}}}{|\vec{\mathbf{p}}|^2} + \left(\left(\frac{\vec{\mathbf{v}} \times \vec{\mathbf{t}}}{|\vec{\mathbf{t}}|^2} \right) \times \hat{\mathbf{p}} \right) \times \hat{\mathbf{p}}. \tag{7.10}$$

[6]These semicircles are half of the so-called great circles.

$\vec{\omega}$ is the retinal flow corresponding to a 3-D world point P, represented as an *angular velocity*. This equation is a generalization of the 2-D result given in Eq. 7.3. The double cross-products with $\hat{\mathbf{p}}$ in Eq. 7.10 ensure that $\vec{\omega}$ (for any given world point) is always perpendicular to $\hat{\mathbf{p}}$, the unit vector in the direction of that world point; however, the double cross-product changes the sign of the term.

7.3.2 Points with zero flow in the 3-D universe

The points with zero flow are interesting boundary cases that enhance an intuitive understanding of how flow is related to positions in the world. The points that are momentarily stationary on the retina are those that have neither latitudinal nor longitudinal flow.

To determine the points in the world that have no retinal flow, we rewrite Eq. 7.10 as

$$\vec{\omega} = \left[\frac{\vec{\mathbf{v}}}{|\vec{\mathbf{p}}|} + \frac{(\vec{\mathbf{v}} \times \vec{\mathbf{t}}) \times \hat{\mathbf{p}}}{|\vec{\mathbf{t}}|^2}\right] \times \hat{\mathbf{p}}. \tag{7.11}$$

When $\vec{\mathbf{v}}$ coincides with $\vec{\mathbf{t}}$, $\vec{\mathbf{v}} \times \vec{\mathbf{t}} = 0$ and there is no rotation due to fixation. In such a special case, all infinitely far away points have zero flow, as well as points along the line of sight. In what follows we concentrate on the more typical case with rotation due to fixation (i.e., $\vec{\mathbf{v}} \times \vec{\mathbf{t}} \neq 0$).

Assuming that $\vec{\mathbf{v}} \times \vec{\mathbf{t}} \neq 0$, the points with zero flow are those where $\vec{\omega}$ is identical to zero, that is, when $\left[\frac{\vec{\mathbf{v}}}{|\vec{\mathbf{p}}|} + \frac{(\vec{\mathbf{v}} \times \vec{\mathbf{t}}) \times \hat{\mathbf{p}}}{|\vec{\mathbf{t}}|^2}\right]$ is either (a) equal to zero or (b) parallel to $\hat{\mathbf{p}}$; we shall denote this term by $\vec{\mathbf{m}}(\vec{\mathbf{p}})$.

7.3.2.1 Case (a): $\vec{\mathbf{m}}(\vec{\mathbf{p}}) = 0$

From this condition it follows that

$$\frac{\vec{\mathbf{v}} \times \vec{\mathbf{t}}}{|\vec{\mathbf{t}}|^2} \times \hat{\mathbf{p}} = -\frac{\vec{\mathbf{v}}}{|\vec{\mathbf{p}}|}. \tag{7.12}$$

The above vector equation holds only when the vectors on both sides of the equation have the same direction and magnitude. One solution for $\vec{\mathbf{p}}$ is either of a pair of points infinitely far away in the direction of the north and south pole (i.e., in the direction of $\vec{\mathbf{v}} \times \vec{\mathbf{t}}$).

An additional set of solutions may be derived as follows. The left-hand side gives rise to vectors orthogonal to $\vec{\mathbf{p}}$. If these vectors are to be in the direction of $-\vec{\mathbf{v}}$, then $\vec{\mathbf{p}}$ must lie on the plane perpendicular to $\vec{\mathbf{v}}$ (labeled H in Fig. 7.5). Furthermore, for any given direction of $\vec{\mathbf{p}}$ in this plane, there is one particular magnitude that will satisfy the magnitude component of the above equation. This magnitude is such that

$$\frac{|(\vec{\mathbf{v}} \times \vec{\mathbf{t}}) \times \vec{\mathbf{p}}|}{|\vec{\mathbf{t}}|^2} = |\vec{\mathbf{v}}|. \tag{7.13}$$

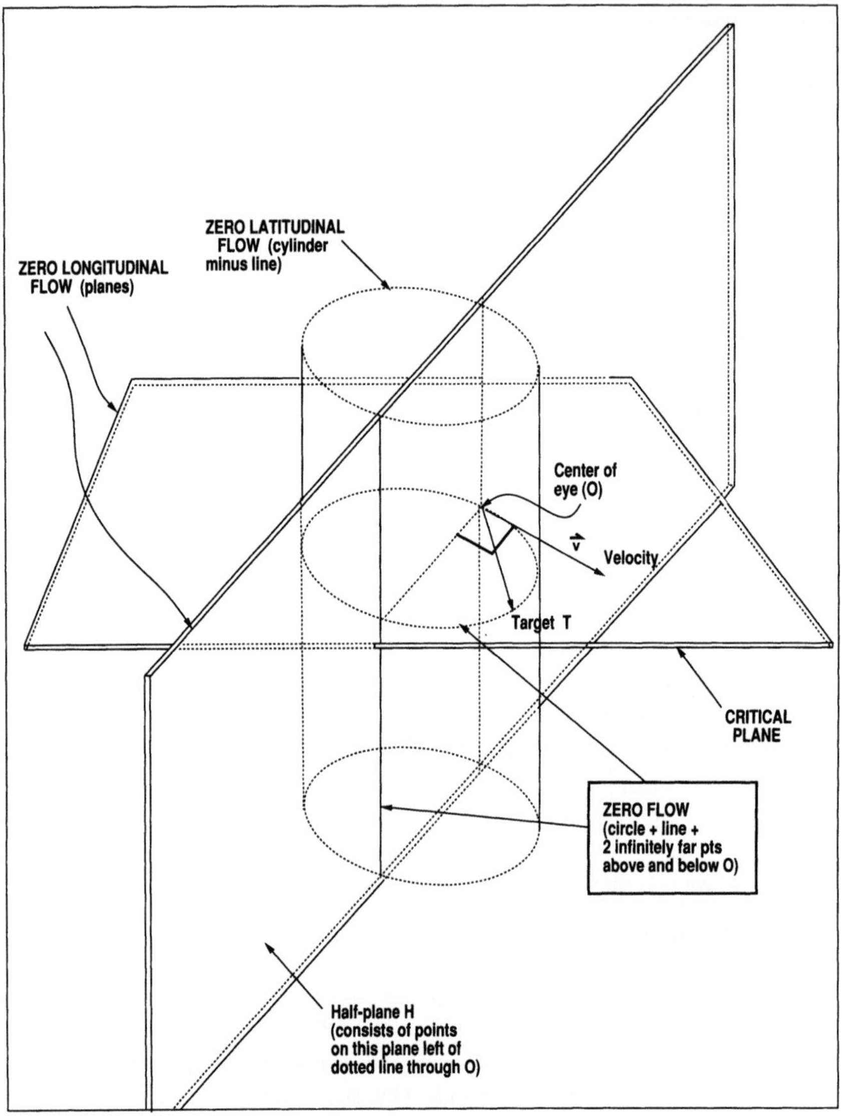

FIGURE 7.5. Points with zero flow

By simple vector algebra we get

$$\hat{v}_p \cdot \vec{\mathbf{p}} = \frac{|\vec{\mathbf{t}} \cdot \vec{\mathbf{t}}|}{|\hat{\mathbf{v}} \times \vec{\mathbf{t}}|}, \tag{7.14}$$

where $\hat{v}_p$ is a unit vector perpendicular to $\vec{\mathbf{v}}$ and $\vec{\mathbf{v}} \times \vec{\mathbf{t}}$. The magnitudes of $\vec{\mathbf{p}}$ that satisfy Eq. 7.14 correspond to a line parallel to $\vec{\mathbf{v}} \times \vec{\mathbf{t}}$ at a distance $\frac{|\vec{\mathbf{t}} \cdot \vec{\mathbf{t}}|}{|\hat{\mathbf{v}} \times \vec{\mathbf{t}}|}$ (on the plane H), as shown in Fig. 7.5.

7.3.2.2 Case (b): $\vec{\mathbf{m}}(\vec{\mathbf{p}})$ parallel to $\hat{\mathbf{p}}$

The second term of $\vec{\mathbf{m}}(\vec{\mathbf{p}})$ always corresponds to vectors on the plane perpendicular to $\vec{\mathbf{v}} \times \vec{\mathbf{t}}$; i.e., these vectors lie on the critical plane spanned by $\vec{\mathbf{v}}$ and $\vec{\mathbf{t}}$ (Fig. 7.4). By definition, the first term of $\vec{\mathbf{m}}(\vec{\mathbf{p}})$ also lies on the critical plane, and hence the sum of the two terms of $\vec{\mathbf{m}}(\vec{\mathbf{p}})$ lies on the critical plane. If this sum vector $\vec{\mathbf{m}}(\vec{\mathbf{p}})$ is to be parallel to $\hat{\mathbf{p}}$, it is necessary that $\hat{\mathbf{p}}$ lie on the critical plane. Thus, the situation reduces to the 2-D case already discussed in Section 7.2.2. As in the 2-D case, the points with zero flow lie on a circle (defined by Eq. 7.8).

In conclusion, apart from the special case of $\hat{\mathbf{v}} = \hat{\mathbf{t}}$, the points with zero retinal flow are:

1. the two infinitely far away points in the direction of the north and south poles,
2. the points on the line parallel to the rotation axis (Eq. 7.14), and
3. the circle on the critical plane.[7]

7.4 Latitudinal and Longitudinal Flow

7.4.1 Calculating latitudinal and longitudinal flow

In general, the retinal flow may be decomposed into two components: latitudinal flow and longitudinal flow. It is crucial for our purposes later in the chapter to isolate the longitudinal component of the flow in Eq. 7.10. The second term of this equation is already in the correct form: since $\frac{\vec{\mathbf{v}}\times\vec{\mathbf{t}}}{|\vec{\mathbf{t}}|^2}$ is a vector perpendicular to the critical plane, the resulting flow is entirely latitudinal.

To perform the desired decomposition of the first term of Eq. 7.10, we decompose the observer's velocity $\vec{\mathbf{v}}$ into two components: (1) $\vec{\mathbf{v}}'$, which is the projection of $\vec{\mathbf{v}}$ onto a plane containing P, the north pole and the center of the eye (this plane will be referred to as the *longitudinal plane*) and (2) $\vec{\mathbf{v}}''$, which is perpendicular to the longitudinal plane, such that:

$$\vec{\mathbf{v}} = \vec{\mathbf{v}}' + \vec{\mathbf{v}}''. \tag{7.15}$$

Since $\vec{\mathbf{v}}'$ lies in the longitudinal plane, it generates flow only within that plane (i.e., longitudinal flow). On the other hand, $\vec{\mathbf{v}}''$ is orthogonal to the longitudinal plane, and thus produces flow out of that plane (i.e., latitudinal flow).

Thus, $\vec{\omega}$ can be decomposed into $\vec{\omega}_x$, the latitudinal flow, and $\vec{\omega}_y$, the longitudinal flow, as follows:

$$\vec{\omega}_x = \frac{\vec{\mathbf{v}}'' \times \vec{\mathbf{p}}}{|\vec{\mathbf{p}}|^2} + \frac{\vec{\mathbf{v}} \times \vec{\mathbf{t}}}{|\vec{\mathbf{t}}|^2} \times \hat{\mathbf{p}} \times \hat{\mathbf{p}}, \tag{7.16}$$

[7] Under a different formulation, Raviv and Herman (1990) identify the line and the circle as involving zero flow. However, our decomposition in the remainder of this paper is unique and distinct from that of Raviv and Herman. The level sets of our latitudinal flow do not correspond to any of their level sets.

$$\vec{\omega}_y = \frac{\vec{\mathbf{v}}' \times \vec{\mathbf{p}}}{|\vec{\mathbf{p}}|^2}. \tag{7.17}$$

As Eq. 7.17 shows, the longitudinal flow of a point P depends only on the movement of the observer. The direct relationship between the observer motion and longitudinal flow will be exploited in the latter half of this paper. This relationship is a result of the particular choice of the latitude and longitude in this formalization.

7.4.2 Points with zero longitudinal flow in the 3-D universe

Although we have already obtained expressions for the world points that produce zero flow, we can gain additional intuition by considering those points that generate either zero latitudinal flow or zero longitudinal flow.

In the case of longitudinal flow, all points with zero flow lie on either of two planes (Fig. 7.5). One such plane is the critical plane. All points on the critical plane project onto a single latitude, the equator. Any translation within or rotation perpendicular to the critical plane will not induce the points to change latitudes; that is, the points remain on the equator regardless of observer motion or fixation. Since the points remain on the same latitude, they have zero longitudinal flow.[8] The second plane with zero longitudinal flow is perpendicular to the direction of the observer's velocity $\vec{\mathbf{v}}$ and passes through the center of the eye. For points on this plane, the modified velocity ($\vec{\mathbf{v}}'$ in Eq. 7.17) is zero, resulting in no longitudinal flow.

7.4.3 Points with zero latitudinal flow in the 3-D universe

In the case of latitudinal flow, all points with zero flow lie on a cylinder but not all points on the cylinder have zero flow. The points with *nonzero* flow on this cylinder form an open interval line passing through the north and south poles (see Fig. 7.5). This can be shown by first writing an explicit expression for $\vec{\mathbf{v}}''$:

$$\vec{\mathbf{v}}'' = \frac{[\vec{\mathbf{v}} \cdot ([\vec{\mathbf{v}} \times \vec{\mathbf{t}}] \times \hat{\mathbf{p}})](\vec{\mathbf{v}} \times \vec{\mathbf{t}}) \times \hat{\mathbf{p}}}{|(\vec{\mathbf{v}} \times \vec{\mathbf{t}}) \times \hat{\mathbf{p}}|^2}. \tag{7.18}$$

After substituting this expression for $\vec{\mathbf{v}}''$ into Eq. 7.16 and performing some simple algebraic manipulations, we arrive at the following constraint on the world points $\vec{\mathbf{p}}$ with zero latitudinal flow:

$$\frac{\vec{\mathbf{v}} \cdot [(\vec{\mathbf{v}} \times \vec{\mathbf{t}}) \times \vec{\mathbf{p}}]}{|(\vec{\mathbf{v}} \times \vec{\mathbf{t}}) \times \vec{\mathbf{p}}|^2} + \frac{1}{|\vec{\mathbf{t}}|^2} = 0. \tag{7.19}$$

Note in the above equation that only the component of $\vec{\mathbf{p}}$ which is orthogonal to $\vec{\mathbf{v}} \times \vec{\mathbf{t}}$ matters. Hence the circle of zero flow on the critical plane extends to a cylinder of points above and below the critical plane.

[8]For points on the critical plane, note that $\vec{\mathbf{v}}'$ and $\vec{\mathbf{p}}$ are in the same direction, leading to a zero cross-product term in Eq. 7.17.

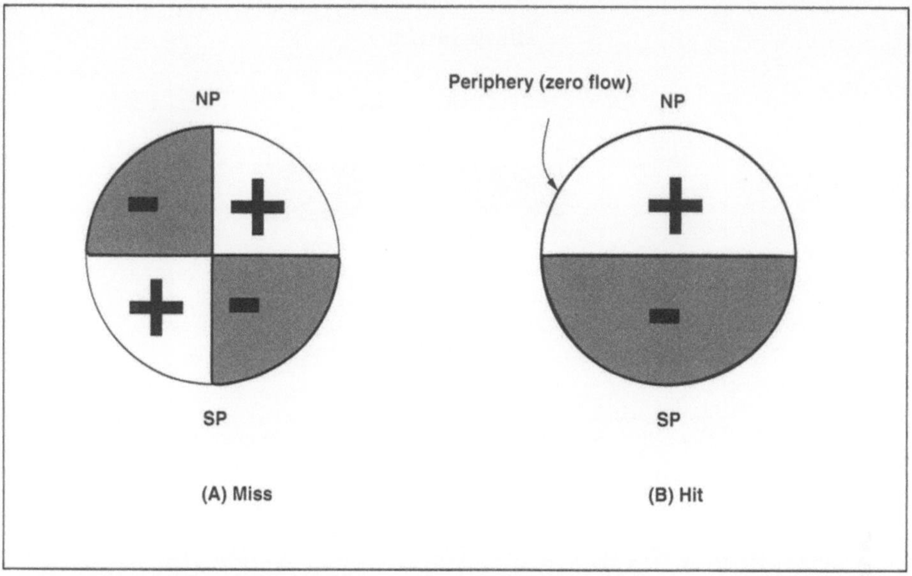

FIGURE 7.6. Sign of longitudinal flow on the hemispherical retina, projected (for illustration purposes) onto the plane perpendicular to the viewing direction.

7.5 A Systematic Pattern at the Periphery

We will show that when a target is fixated, the retinal *periphery* has a unique invariant property: it is the only longitude that is a constant across all possible rotations of the eye. This makes the retinal periphery an interesting location to use for certain visual tasks. We define the periphery to be that longitude (or great circle) lying in a plane perpendicular to the direction of gaze.

Let us consider a situation where the moving observer has to decide whether he/she is heading towards the fixated target or not. In the former case, the observer will *hit* the target if he/she *continues* in the current direction of motion, whereas in the latter case the observer will *miss* the fixated target.[9] Such an ability to predict *hit* and *miss* situations (assuming that the current direction of movement is maintained, and assuming that the target does not move) should be useful in navigation.

An analysis of the retinal flow at the periphery of our model eye indicates that the characteristics of the longitudinal flow distinguish *hit* from *miss* situations (in the sense described above). The magnitude of the longitudinal flow on the periphery (from a point P in the world) depends on how far the point is from the eye as well as how fast the observer moves. On the other hand, the direction — or sign — of the longitudinal flow *within a quadrant of the retina* only depends on whether the

[9]This holds for an idealized case where the observer is a point. For a practical situation in which the observer has finite size, determining a *hit* situation is more involved.

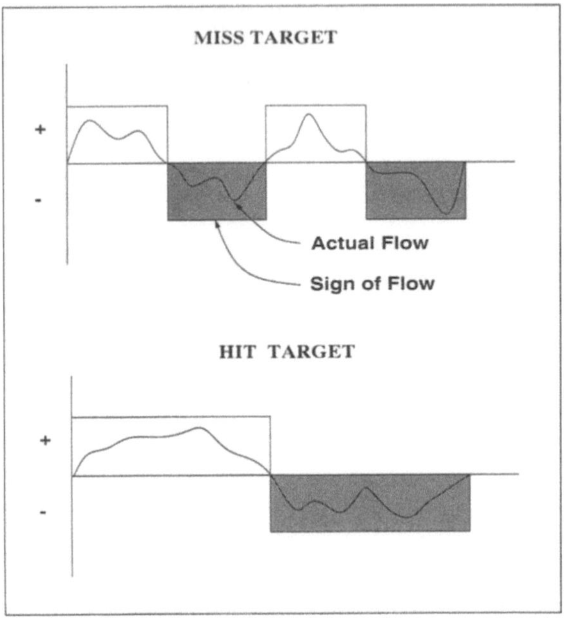

FIGURE 7.7. The sign of the retinal flow at the periphery. (See text for description of "at the periphery".)

observer is heading towards the target. The quadrants are defined by the location of the north and south poles on the periphery.

In the *miss* situation, the sign of the longitudinal flow (sign of $\vec{\omega}_y$ in Eq. 7.17) switches exactly four times as one traces along the periphery (i.e., once for each quadrant). Fig. 7.6(A) indicates the sign of the longitudinal flow along the periphery as well as across the entire retina for the most extreme *miss* situation. Even in less extreme cases the sign changes four times.

Although Eq. 7.17 could be analyzed to derive the number of sign changes, we will turn to Fig. 7.5 for a more intuitive proof. The two planes of zero longitudinal flow divide the entire universe into four quadrants; within each quadrant the longitudinal flow is of the same sign, whereas sign switches at the planes. For example, in the quadrant spanned between the observer's direction of heading and the north pole the longitudinal flow is towards the north pole, whereas in the quadrant on the other side of the critical plane (and on the same side as the observer's direction) the longitudinal flow is towards the south pole. When the observer is not moving directly ahead, these two planes will cut the hemispherical retina into four quadrants, thereby producing the four-way distinction in the sign of the longitudinal flow.

As we continuously move from a *miss* situation towards a *hit* situation, the translation direction gets closer to the direction of gaze. When the translation direction and the direction of gaze coincide, one of the planes with zero longitudinal flow (Fig. 7.5) contains the entire periphery. Thus the longitudinal flow of all the points on the periphery is zero in the *hit* situation. Immediately adjacent to this infinitely thin periphery, a different picture emerges. As one traces the longitudinal flow just inside of the periphery, the sign of the longitudinal flow changes *exactly twice*. Fig. 7.6(B) shows the sign of the longitudinal flow across the entire retina in the *hit* situation.

Thus, in theory, the infinitely thin line of the periphery will contain the four-way change for the *miss* situation and zero flow for the *hit* situation. However, in practice, the observable difference at a periphery with finite thickness involves a four-way change in the *miss* situation and a two-way change in the *hit* situation. This situation is illustrated in Fig. 7.7.

7.6 Experiment I: Simulated Image Sequence

In a simulated experiment we attempted to obtain the sign of the longitudinal flow on the retina as in Fig. 7.6. The number of times the sign changes along the periphery (two vs. four times) allows us to distinguish between the *hit* and the *miss* situations.

In the simulation we created a rigid world consisting of 1024 points. The points were uniformly and randomly distributed in all directions around the initial position of the eye. The points were also randomly (but nonuniformly) distributed in

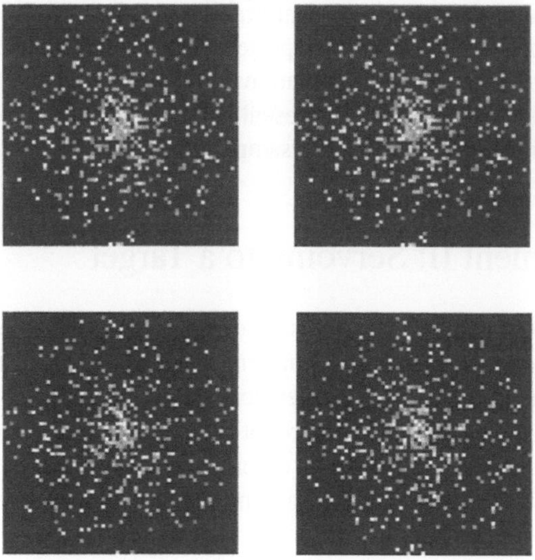

FIGURE 7.8. Image sequences depicting the retina for the *hit* situation (left column) and the *miss* situation (right column). The first image is common to both sequences. The world consists of 3-D points randomly distributed in position (see text for details). In all cases, the camera is fixating the same dot that projects to the center of the image.

distance, with more points closer.[10] Due to the hemispheric nature of the retina, only those points in front of the eye were projected on the retina; the visual field was 180° along any diameter of the retina.

Two pairs of images were generated containing the projections of points in the world onto the retina. The target on which the eye fixated was located at a distance equal to thrice the radius of the eye, directly ahead of the initial position of the eye. The two pairs contained the same first image: a view from an initial eye position. The second image of the first pair was generated to simulate a sideways movement of one-fourth of the radius of the eye (*miss* situation). The second image of the second pair simulated a movement of one-fourth of the radius of the eye towards the target (*hit* situation) from the initial position. The two image sequences are shown in Fig. 7.8.

To test the sign of the longitudinal flow on the retina, the optical flow in both image sequences was estimated, using a multi-scale gradient-based algorithm described by Simoncelli (1993). The longitudinal flow was then extracted from the total optical flow.[11]

Fig. 7.9 shows the sign of the longitudinal flow across the entire retina for the two sequences. The longitudinal flow in one direction is shown in white, and in the opposite direction in black. For the *hit* situation these regions approximately divide the retina into two halves (shown on the left in Fig. 7.9) while in the *miss* situation the retina is divided into quarters, shown on the right. In each row of images, an increasing range of estimated longitudinal flow speeds are treated as zero (shown as grey in each panel).

Although the overall pattern of the sign of the longitudinal flow on the simulated retina is in agreement with the theoretical prediction (Fig. 7.6), there are certain regions on the retina that have an unexpected sign. We attribute this error due to imperfect optical flow; a set of simulations with perfect optical flow obtained perfect patterns (Fig. 7.10). The errors present in the optical flow estimates are due to blank regions of the images and undersampling in time.

7.7 Experiment II: Servoing to a Target

This experiment demonstrates that the sign of noisy retinal flow is sufficient to determine the direction of heading of a moving robot that fixates on a target. The task is to guide the robot toward the target from any starting position and velocity using only visual information. As the robot moves, its direction of heading is continuously computed based on the sign of longitudinal flow. Based on the current direction of heading, the robot's direction of motion is continuously changed until

---

[10] The distance (d) of the points was chosen randomly such that $\frac{1}{d}$ was uniformly distributed in the range $[0, 0.5]$.

[11] The optical flow contained both the latitudinal and longitudinal flow. For the purposes of this demonstration only the longitudinal flow was needed. Simoncelli's optical flow algorithm was easily modified to obtain the longitudinal component of flow.

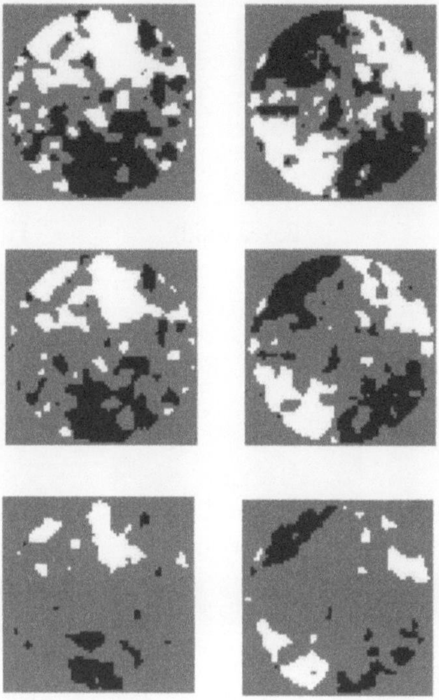

FIGURE 7.9. Simulated experiment: sign of the longitudinal flow for *hit* (left) and *miss* (right) situations. In the first row the points with less that 0.125 pixels of longitudinal flow were ignored. The comparable figure for the second row was 0.25 pixels and for the third row was 0.5 pixels.

FIGURE 7.10. Simulated experiment with perfect optical flow. The sign of the longitudinal flow is depicted for different directions of heading. From left to right, the directions of heading are -90°, -45°, 0°(direct ahead), 45° and 90°.

it heads directly towards the target.[12] It is assumed that there are no obstacles, since obstacle avoidance is beyond the scope of this work.

[12]Note that fixation corresponds merely to having the robot *look* in the direction of the target. We must determine which way the robot is heading relative to the target. Although such information could potentially be obtained through positional encoders (corresponding to vestibular or muscular feedback in humans), the model developed in this paper solves for direction of heading using visual information alone.

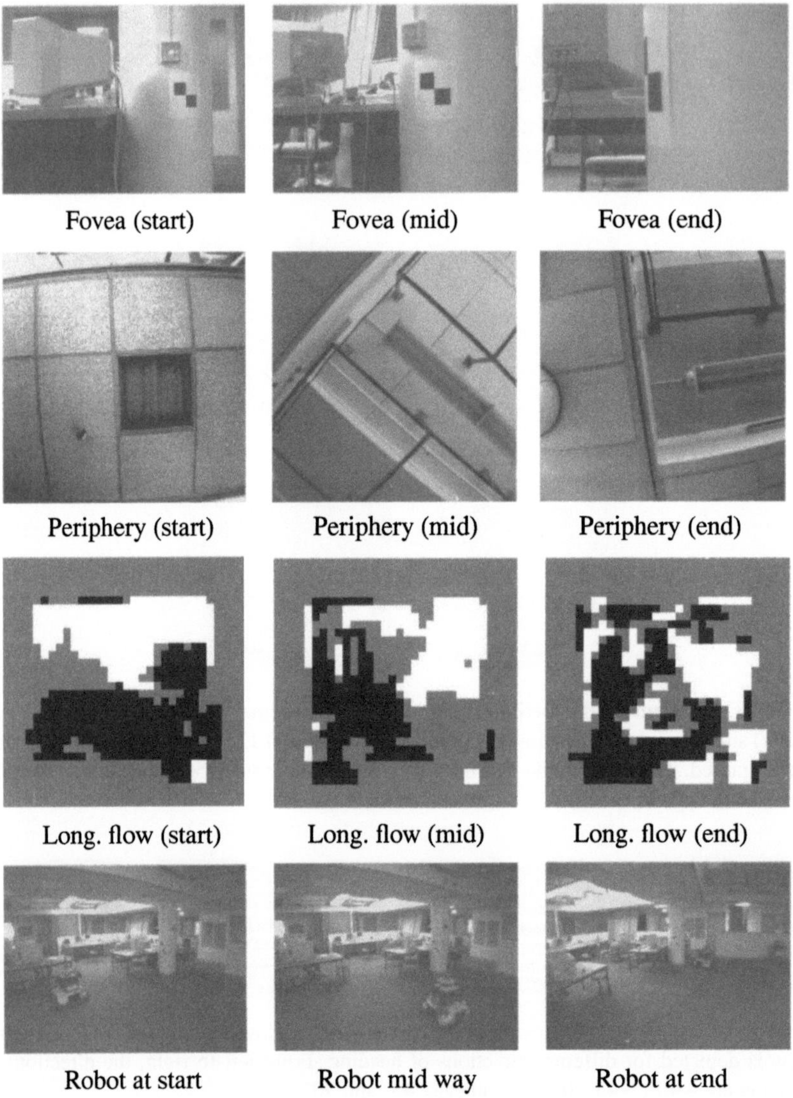

FIGURE 7.11. Each column of images corresponds to one position in the robot's path towards the target (a black-and-white checkerboard pattern hung on a pillar). The first column is at the robot's starting position, the last column is at the robot's finish position and the middle column is about midway between start and finish. At each position, the first row is the foveal image, the second row is the peripheral image. The third row is the sign of the longitudinal flow in the periphery image: these images are used to move the robot towards the target. The fourth row provides an external view of the robot at different positions along its path.

To simulate a wide field of view,[13] we employ two cameras in the present work: (1) a camera pointed at the target, which we denote as the *fovea*, and (2) a large field-of-view camera pointed upwards in a direction orthogonal to the foveal camera, which we denote as the *peripheral* camera. Note that although the second camera obtains a view of just a portion of the periphery, this portion is sufficient to calculate the direction of heading (as indicated by, for example, the top portion of the different situations shown in Fig. 7.10). The two cameras are held rigidly with respect to each other.

The purpose of the foveal camera is to track the target such that the target is always in the center of the image. While the robot is in motion, and the target moves from the center of the image, the fovea is rotated to compensate for the target motion, thereby achieving fixation. The fovea is mounted on a turntable whose stepper motors are constantly servoed using the amount of slip of the target from the image center. The first column in Fig. 7.11 shows the initial position of the robot, and the target being fixated. The target in this case is chosen to be a simple pattern that can be easily tracked in order to maintain fixation. However, since the fixation is performed by a coarse-to-fine correlation scheme, the target could be any arbitrary textured pattern.

The sign of the longitudinal flow provided by the peripheral camera is used to compute the direction of heading. Recall that for this particular camera arrangement, the peripheral camera provides sign of flow only in a circular patch around the north pole. The problem of finding the direction of heading is reduced to the problem of locating the line that divides the positive flow from the negative flow at the north pole; as was seen in Fig. 7.10, this line rotates about the north pole. In practice, the orientation of this line is computed as the phase of the first harmonic of the flow pattern. As is standard, the phase is computed by first projecting the pattern onto two functions, a sine and a cosine, and then calculating the arctangent of the ratio of the two projections.

The third row in Fig. 7.11 shows the sign of the longitudinal flow at three different positions of the robot along its trajectory towards the target (white is one direction and black the opposite direction of sign). The first row of Fig. 7.11 shows the images obtained from the fovea and the second row from the periphery at the three robot locations. The computed value of the direction of heading (with respect to the direction of gaze/target) at any given position of the robot is used to servo the robot to move towards its direction of gaze.

The current implementation of the system is in C running on a Sun Workstation. Each run of the experiment takes under 2 minutes, and the initial distance from the robot to the target is 3.1 meters. The experiment was repeated 10 successive times. In every trial the robot started by heading in a direction perpendicular to the target direction. In two of the ten trials the robot arrived within 40 cm of the pillar containing the target; in the remaining eight trials the base of the robot ended up in contact with the pillar, directly underneath the target. Fig. 7.11 is an example

[13] We are currently testing a recently-acquired 180° field-of-view camera.

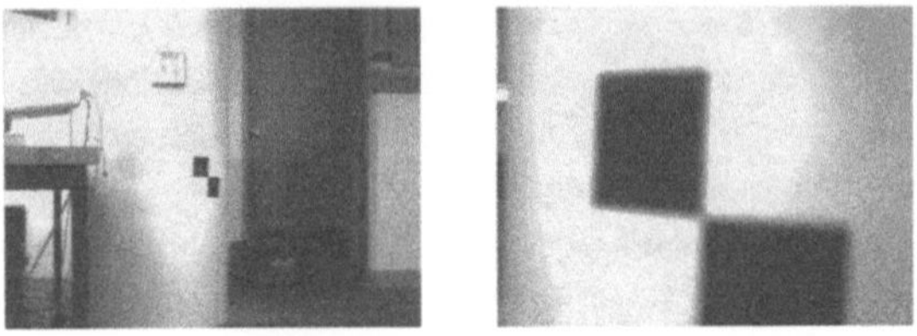

Fovea (start) Fovea (end)

FIGURE 7.12. Foveal images for a typical uninterrupted run of the robot, where the robot reaches the target.

of a case where the robot's base does not touch the pillar. Due to the fact that the robot was stopped in the middle of its run to take the intermediate pictures in Fig. 7.11, the servoing algorithm in effect had to start again, resulting in an overshoot. Pictures from the fovea of a typical uninterrupted sequence, where the robot base touched the pillar, are provided in Fig. 7.12.

7.8 Conclusion

In this paper, we have formulated the precise relationship between retinal flow, the movement of a fixating observer and the geometry of the physical world in a simple and systematic way. The simplicity of the formulation derives from the use of a spherical imaging surface, coupled with the constraints imposed by fixation.

Our theoretical analysis and subsequent experiments reveal that the information along the periphery of the retina appears to be sufficient for determining whether the observer does eventually hit a target if it continues moving in its current direction (assuming the target does not move). This is possible precisely because the observer activelyfixates on the target while moving. We believe this formulation will prove useful in many problems involving optical flow computation.

Acknowledgments: A preliminary version of this work has been published previously (Thomas, Simoncelli & Bajcsy, 1994). The first author is indebted to Prof. V. Aurich, Computer Science, Univ. of Düsseldorf, for triggering this research by bringing to the author's attention interesting patterns on the retinal periphery. Thanks to Dr. A. Vainikka who helped with the style and logic of presentation, to U. Cahn von Seelen for the tracking algorithm, and to J. Kosecka for the robot control program. This work was supported by ARPA Grant N00014-92-J-1647, ARO Grant DAAL03-89-C-0031PRI, NSF Grant CISE/CDA-88-22719 and NSF Grant STC SBR8920230.

7.9 References

Abraham, R. H. & Shaw, C. D. (1993). *Dynamics: The Geometry of Behavior*. Redwood City, CA: Addison-Wesley.

Bajcsy, R. (1985). Active perception vs. passive perception. In *Proceedings of the IEEE Workshop on Computer Vision: Representation and Control* (pp. 55-59). Washington, DC: IEEE Computer Society.

Bajcsy, R. (1988). Active perception. *Proceedings of the IEEE, Special Issue on Computer Vision, 76*, 996-1005.

Fermüller, C. (1993). Navigational preliminaries. In Y. Aloimonos (Ed.), *Active Perception* (pp. 103-150). Hillsdale, NJ: Lawrence Erlbaum.

Hager, G. D. (1988). *Active Reduction of Uncertainty in Multi-Sensor Systems*. PhD thesis, Department of Computer and Information Science, University of Pennsylvania, Philadelphia, PA. Technical Report MS-CIS-88-47, GRASP LAB 147.

Krotkov, E. P. (1989). *Active Computer Vision by Cooperative Focus and Stereo*. New York: Springer Verlag.

Maver, J. & Bajcsy, R. (1993). Occlusions as a guide for planning the next view. *IEEE Transactions on Pattern Analysis and Machine Intelligence, 15*, 417-433.

Pahlavan, K., Uhlin, T. & Eklundh, J. O. (1993). Dynamic fixation. In *Proceedings of the Fourth International Conference on Computer Vision* (pp. 412-419). Berlin: Springer-Verlag.

Raviv, D. & Herman, M. (1990). Towards an understanding of camera fixation. In *Proceedings of the 1990 IEEE International Conference on Robotics and Automation* (pp. 28-33). Los Alamitos, California: IEEE Computer Society.

Simoncelli, E. (1993). *Distributed Representation and Analysis of Visual Motion*. PhD thesis, Department of Electrical Engineering, Massachusetts Institute of Technology, Cambridge, MA. Vision and Modeling Group Technical Report 209, MIT Media Laboratory.

Thomas, I. (1993). *Reducing Noise in 3D Models Recovered from a Sequence of 2D Images*. PhD thesis, Department of Computer Science, University of Massachusetts, Amherst, MA.

Thomas, I., Simoncelli, E. & Bajcsy, R. (1994). Spherical retinal flow for a fixating observer. In *Proceedings of the Workshop on Visual Behaviors* (pp. 37-44). Los Alamitos, California: IEEE Computer Society.

Yarbus, A. L. (1967). *Eye Movements and Vision*. New York: Plenum Press.

8

Local Qualitative Shape from Active Shading

Michael S. Langer[1]
Steven W. Zucker[1]

ABSTRACT We show how to actively compute local qualitative shape properties directly from image intensities, that is, without having to first reconstruct the surface. Our approach differs from classical active shape-from-shading (photometric stereo, Woodham, 1980), in two ways. First, we do not attempt to compute a dense depth map of the surface. Instead, we detect the presence of certain local qualitative geometric features. Second, we use two different *types* of lighting conditions rather than two different instances of a single type. Our two types of lighting conditions are a diffuse source and a point source positioned at the camera. The major result of this paper is that surface concavities such as orientation discontinuities and smooth valleys typically share the same shading signatures. Under diffuse conditions, both produce local minima in the image intensity; under the point source condition, both produce local maxima in the image intensity. These 1-D features may be detected, although not discriminated, by alternating between the two lighting conditions.

8.1 Introduction

Shape-from-shading is the problem of computing geometric properties of a piecewise smooth surface from measurements of light reflected off the surface. Obviously, the distribution of light reflected off a surface depends on how the surface is illuminated. For example, if a surface were illuminated by a diffuse light source such as the sky on a cloudy day, then it would have a different luminance profile than if it were illuminated by a point source. We show in this paper that, by alternating the lighting conditions in the scene from a diffuse source to a point source, certain local qualitative shape properties of the surfaces become salient.

There are three separate components to this paper. First, we compute local qualitative shape properties from the image intensities themselves. We do not attempt to "reconstruct" the surface (Blake & Zisserman, 1987) before computing its local qualitative shape. The second component is an examination of the relationship between local qualitative shape and shading under two different lighting conditions: a diffuse source, and a point source at the camera. We show how local qualitative

---

[1]Center for Intelligent Machines, McGill University

shape is revealed under these two conditions. In particular, we introduce a model of diffuse shading which is quite different from the point source model. The third component is a proposal for actively manipulating the light sources so that features of interest will be salient.

8.2 Local Qualitative Shape

We assume that our scene is composed of a set of piecewise smooth surfaces. In this domain, there exist two commonly held notions of local qualitative shape (see Fig. 8.1). One notion applies to undulating drapery surfaces, such as a blanket or a loose shirt. Here, local qualitative shape refers to the spatial arrangement of surface hills and valleys, and is typically represented with a coarse coding of the surface curvatures in classes such as convex, concave and hyperbolic (Besl & Jain, 1988; Leyton, 1988; Richards, Koenderink & Hoffman, 1987). A second notion applies to surfaces which seem to be best described as a composition of a set of adjacent or interpenetrating convex objects (Hoffman & Richards, 1985). Examples include a pile of apples or the fingers of a hand. Here, local qualitative shape refers to a set of parts and how these parts are connected.

FIGURE 8.1. There are two notions of local qualitative shape. One notion is that undulating surfaces can be described as a set of hills and valleys. A second notion is that surfaces can be described as a set of adjacent or interpenetrating parts. In this scene, the folds of the blanket are best described using the former while the apples are best described using the latter.

The distinction between these two notions is unclear, since blurring a set of parts will produce an undulating surface. For example, how much does a pile of rocks have to erode before it becomes a set of hills and valleys? We take a unified position with respect to the two notions: *hills* and *parts* are surface regions which protrude from the surface, while *valleys* and *part boundaries* are surface regions which are typically hidden by the surface.

Protruding surface regions usually receive direct illumination from the light source, whereas hidden regions usually lie in shadow. Stated another way, protruding surface regions see a larger percentage of the sky than the hidden regions. We introduce a new geometric property of surfaces to capture this last idea. The *aperture* of a surface element is defined as the percentage of the sky which is visible from that element. Thus, protruding surface regions have higher aperture than hidden regions. We will take advantage of a strong relationship between surface aperture and surface luminance under diffuse lighting conditions.

In Sections 8.3 and 8.4, we argue that concavities, or small aperture regions of a surface, typically have a different shading signature under diffuse source conditions than under point source conditions. Under diffuse conditions, surface luminance is typically a local minimum across a concavity, while under point source conditions, surface luminance is typically a local maximum across a concavity. This latter is well-known (Forsyth & Zisserman, 1991; Horn, 1977; Koenderink & van Doorn, 1983). In Section 8.5, we discuss the example of a drapery surface viewed under the two lighting conditions.

8.3 Diffuse Shading

Most shape-from-shading research is based on the assumption that the intensity variations in an image depict unit surface normal variations in a scene (Horn, 1977). This assumption is not always applicable, however, as can be seen in Fig. 8.2. Most people interpret this image as a set of of *flat* surfaces protruding from the ground plane. This percept is in conflict with standard shape-from-shading, which would interpret the surfaces as cylinders.

Standard shape-from-shading methods were designed for scenes in which it makes sense to talk about *the* illuminant direction. However, for a scene illuminated by a diffuse source, the question "Where is *the* illuminant direction?" (Pentland, 1982) is meaningless, since the illumination comes from many directions. We continue by developing a diffuse shading model.

8.3.1 A model of diffuse shading

Let $I(x)$ be the luminance of the surface at x, that is, the luminance scattered from a surface element at x. Let $I_{\mathcal{D}}$ be the *average* luminance from the diffuse light source. (On overcast days, the luminance of the sky is not exactly uniform (Koenderink & Richards, 1992).) Both $I(x)$ and $I_{\mathcal{D}}$ are expressed in lumens per steradian per

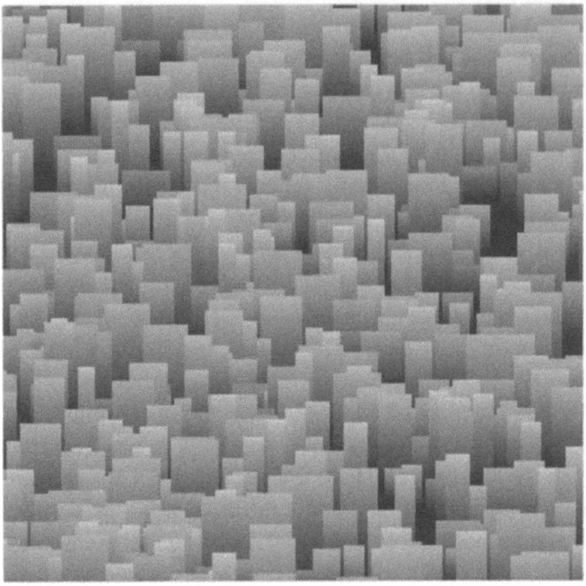

FIGURE 8.2. This image was formed by overlapping a set of rectangles, each having a vertical intensity gradient. Most people perceive a set of flat surfaces protruding from a ground plane. A skyline or a graveyard are two typical interpretations. These percepts cannot be explained by the standard shape-from-shading assumption that smooth variations in image intensity depict smooth variations in surface normal, since the standard assumption would suggest that the surfaces are cylinders.

projected unit surface area. Let $\rho(x)$ be the surface albedo, that is, the fraction of incident light energy which is reflected by the surface, so that $0 < \rho(x) < 1$.

Let S^2 denote a unit sphere and let $\mathcal{H}(x) \subset S^2$ be the hemisphere of directions from which the surface element at x can receive incident illumination (see Fig. 8.3). For any $\mathbf{L} \in \mathcal{H}(x)$, $\mathbf{L}$ points either to the diffuse source (a region of the sky, in the example of the cloudy day) or to another surface element in the scene. Let $\mathcal{V}(x) \subset \mathcal{H}(x)$ denote the set of directions at x in which the diffuse source is visible, so that $\mathcal{H}(x)\backslash\mathcal{V}(x)$ is the set of directions at x in which other surface elements in the scene, for example the ground plane, are visible.

Gibson (1966) observed that the ambient light from the sky has greater luminance than the ambient light from surfaces in the scene. In terms of the above definitions, the luminance from the directions $\mathcal{V}(x)$ is greater than that coming from the directions $\mathcal{H}(x)\backslash\mathcal{V}(x)$. Langer and Zucker (1994) derived bounds on the relation between the surface luminance, the luminance of a sky, and the surface albedo. In particular, if the sky has uniform luminance and if the surface is matte (i.e., Lambertian), then the surface luminance is bounded above as follows,

$$I(x) \leq \rho(x)\, I_{\mathcal{D}}\,,$$

which we call the *dominating sky principle.*

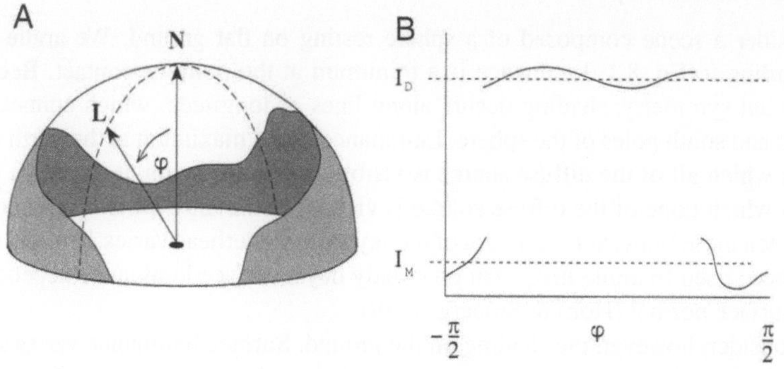

FIGURE 8.3. (A) The hemisphere of incident directions above a point on a surface can be partitioned into two sets. $\mathcal{V}(x)$ denotes the set of directions pointing to the diffuse source. $\mathcal{H}(x)\backslash\mathcal{V}(x)$, shaded in the figure, denotes the set of directions pointing to other surface elements in the scene. (B) Incident luminance is defined on $\mathcal{H}(x)$. Directions on an arbitrary great circle, such as the dashed curve, can be indexed by the angle φ. Notice that the incident luminance is much higher on $\mathcal{V}(x)$ than on $\mathcal{H}(x)\backslash\mathcal{V}(x)$.

The above definitions suggest the following model of surface luminance. This model is similar in form to classical models which use only a single point source. The difference is that we have to integrate over those regions of the sky which are directly illuminating the surface element (Nishita & Nakamae, 1986). We use the symbol $\mathcal{M}$ to denote "mutual illumination". Let $d\mathbf{L}$ be an infinitesimal region of $\mathcal{H}(x)$ which is centered at direction $\mathbf{L}$. Then surface luminance may be approximated by

$$I(x) \;=\; \frac{\rho(x)}{\pi}\int_{\mathcal{V}(x)} (I_{\mathcal{D}} - I_{\mathcal{M}})\, \mathbf{N}(x)\cdot\mathbf{L}\, d\mathbf{L} \;+\; \rho(x)\, I_{\mathcal{M}} \tag{8.1}$$

The first term in this model is due to the average luminance, $I_{\mathcal{D}} - I_{\mathcal{M}}$, of the diffuse source that is *in excess of* the average luminance $I_{\mathcal{M}}$ of the surfaces in the scene. The second term is due to the average luminance from other surface elements in the scene. Although this model is not exact, it does illustrate how the surface luminance depends on the amount of the diffuse source that is visible.

This model motivates the following definition. The *aperture* $A(x)$ of a surface point x is the ratio of the solid angle of $\mathcal{V}(x)$ to the solid angle of $\mathcal{H}(x)$. That is,

$$A(x) = \frac{1}{2\pi}\int_{\mathcal{V}(x)} d\mathbf{L}\,. \tag{8.2}$$

Intuitively, $A(x)$ is the fraction of $\mathcal{H}(x)$ for which the diffuse source (e.g., the sky on a cloudy day) is visible from x. Notice that for all x, $0 \leq A(x) \leq 1$. It is important to observe that aperture is a geometric property which cannot be defined locally, since the amount of source that is visible from a point depends on objects which may be distant from that point.

8.3.2 An example

Consider a scene composed of a sphere resting on flat ground. We argue that, according to Eq. 8.1, luminance is a minimum at the point of contact. Because of radial symmetry, shading occurs along lines of longitude, which connect the north and south poles of the sphere. Luminance is at a maximum at the north pole, from which all of the diffuse source is visible, and a minimum at the south pole, from which none of the diffuse source is visible. Luminance varies continuously between these limits as the amount of the sky visible overhead varies. This example has been used to argue that, even on cloudy days, surface luminance depends on the surface normal (Horn & Sjoberg, 1979).

Consider, however, the shading on the ground. Surface luminance varies along straight lines through the point of contact with the sphere. In particular, luminance decreases continuously toward the point of contact, since the sphere occludes a greater region of the diffuse source for points nearby. As in Fig. 8.2, luminance variations occur in the absence of surface curvature, and the classical shading model fails to explain this shading effect.

8.3.3 Diffuse shading in concavities

We claim that under diffuse lighting conditions, surface luminance typically attains a 1-D minimum across concavities. To see why, consider first the case of two interpenetrating convex surface regions (Fig. 8.4). We claim that, according to Eq. 8.1, surface luminance and aperture are at a local minimum across the locus of points where the two surfaces meet. Formally, let $\vec{\alpha}(t)$ be a unit speed curve passing perpendicularly through the meeting point ($t = 0$), at which there is a concave orientation discontinuity, so that the tangent vector of $\vec{\alpha}(t)$ is discontinuous at $t = 0$. We argue that, according to Eq. 8.1, $I(\vec{\alpha}(t))$ is a minimum at $t = 0$.

The proof of this claim goes as follows. Imagine walking along the curve $\vec{\alpha}(t)$, and consider what you see in the hemisphere of directions overhead. To be precise about the term "overhead", define a moving coordinate frame at $\vec{\alpha}(t)$ by $\{\vec{\alpha}'(t),\ \mathbf{N}(t),\ \mathbf{N} \times \vec{\alpha}'(t)\}$. Let $\mathcal{H}^*$ denote the hemisphere of directions as expressed in this moving frame. $\mathcal{H}^*$ does not depend on t since the coordinate axes are intrinsic. Let $\partial\mathcal{H}^*$ be the boundary of $\mathcal{H}^*$, that is, the "horizon". Let $\mathbf{N}^* \in \mathcal{H}^*$ be the unit surface normal, and let $\mathcal{V}^*(t) \subset \mathcal{H}^*$ be the directions overhead in which the diffuse source is visible from $\vec{\alpha}(t)$. The variations in surface aperture and luminance along $\vec{\alpha}(t)$ are entirely a result of the deformation in $\mathcal{V}^*(t)$. To see this, rewrite Eqs. 8.1 and 8.2 as

$$I(\vec{\alpha}(t)) = \frac{1}{\pi}\rho(\vec{\alpha}(t))(I_{\mathcal{D}} - I_{\mathcal{M}}) \int_{\mathcal{V}^*(t)} \mathbf{N}^* \cdot \mathbf{L}^* d\mathbf{L}^* + \rho(\vec{\alpha}(t)) I_{\mathcal{M}} \tag{8.3}$$

and

$$A(\vec{\alpha}(t)) = \frac{1}{2\pi} \int_{\mathcal{V}^*(t)} d\mathbf{L}^*. \tag{8.4}$$

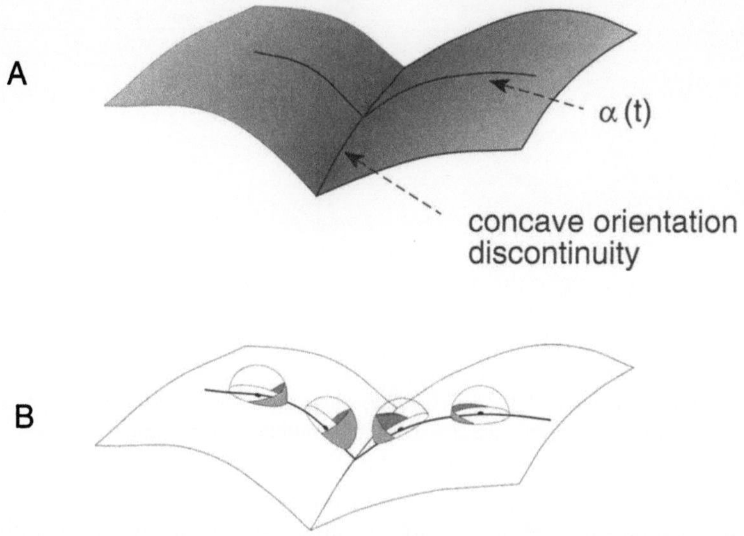

FIGURE 8.4. (A) $\vec{\alpha}(t)$ is a curve which is perpendicular to the concave orientation discontinuity at $t = 0$. We wish to show that $I(\vec{\alpha}(t))$ is at a local minimum at $t = 0$. (B) The idea behind the proof is that the amount of the diffuse source visible from $\vec{\alpha}(t)$ is at a local minimum at $t = 0$. The dominating sky principle guarantees that if a region of the diffuse source is occluded by a surface in the scene, then the incident illumination from that direction is decreased.

These equations are the integral of fixed scalar functions over a domain which varies with t, so luminance and aperture variations are determined entirely by variations in the domain $\mathcal{V}^*(t)$.

We can now complete the argument that luminance is a minimum at the concave orientation discontinuity. As we move along $\vec{\alpha}(t)$ toward $t = 0$, we see nothing but clear sky to the rear, while ahead we see a surface rising up and blocking regions of the diffuse light source. It follows immediately from Eq. 8.3 that $L(\vec{\alpha}(t))$ decreases as $t \rightarrow 0$. Exactly the same argument shows that surface aperture reaches a minimum across the concave orientation discontinuity.

An example of this shading effect is shown in Fig. 8.5. The surface is planar with a wedge cut from it. The luminance profile, computed exactly using a 1-D version of the radiosity equation (Forsyth & Zisserman, 1991), is shown for a set of five different albedos and under a diffuse light source of unit luminance. Observe that luminance within the concavity is a minimum at the orientation discontinuity.

To see a real example of how surface luminance reaches a minimum at concave orientation discontinuities, perform the following experiment. Curl the palm of your hand slightly to make a cup. The skin of your hand will fold along wrinkles, and dark lines will emerge along these wrinkles. Also observe that dark lines emerge where the fingers are pressed together. Dark lines are the luminance minima which separate the skin of the cupped hand into its convex regions.

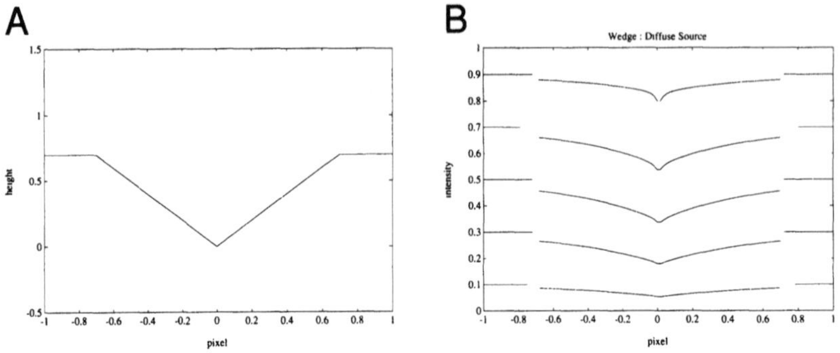

FIGURE 8.5. (A) Height profile of a wedge surface in image coordinates. (B) Image intensity for five different albedos (0.1, 0.3, 0.5, 0.7, 0.9) where the diffuse source has unit luminance. Observe that luminance within the concavity is a minimum at the orientation discontinuity.

These observations are related to ideas mentioned by Koenderink and van Doorn (1980) concerning the stability of surface luminance structure over many different lighting conditions. In particular, surface luminance is stable for surface elements with small aperture in that when a scene is illuminated by a point source, surface elements with a small aperture tend to lie in shadows. Since a diffuse light source can be considered as the average of many different point sources, surface elements with small aperture will have small luminance under diffuse source conditions.

Before moving on to the point source scenario, we should mention that if a concavity is smooth, it may not have a local 1-D luminance minimum across its bottom. To understand why, consider the special case of a hemi-cylindrical groove cut out of a ground plane. The key property of such a groove is that aperture is constant within it. Yet, according to Eq. 8.1, the groove would not have constant luminance: points at the bottom of the groove would have a surface normal centered on the visible region of the sky, whereas points on the side of the groove would have a normal slightly off the center of the visible region of the sky (resulting in a smaller value of $\mathbf{N} \cdot \mathbf{L}$). As a result, the bottom of the groove would have a slightly greater luminance than the side. This surface normal effect is small relative to the aperture effect (Langer & Zucker, 1994).

To summarize, when a concavity is illuminated by a diffuse source overhead, mutual shadowing by the sides of the concavity tends to produce local luminance minima at the bottom of the concavity, although in special cases, surface normal effects may conspire to neutralize these minima.

8.4 Point Source Shading

Quite a different shading effect occurs under point source illumination. For example, when a concave wedge is illuminated by a point source, surface luminance

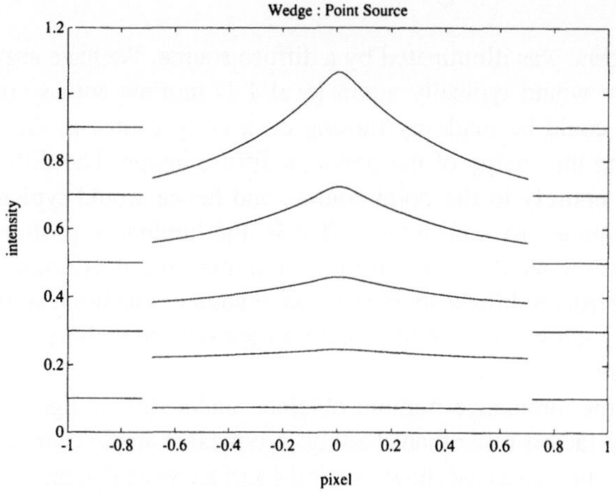

FIGURE 8.6. Image intensity for a point light source overhead, and for five different albedos. Observe the local luminance maximum at the concave orientation discontinuity.

attains a local maximum at the orientation discontinuity (Forsyth & Zisserman, 1991; Horn, 1977; Nayar, Ikeuchi & Kanade, 1991). This maximum is due to the increasing solid angle subtended by the opposite side of the wedge (and hence increase in mutual illumination) as one moves toward the orientation discontinuity. An example is shown in Fig. 8.6.

Such a 1-D local luminance maximum would be produced across the wedge even if the light source were not directly overhead. Because the sides of the wedge are planar, the surface luminance that is due to direct illumination from the point source would be constant along each side of the wedge, while the luminance that is due to mutual illumination would increase toward the orientation discontinuity.

Next, consider the case of a smooth concavity illuminated from above. At the bottom of the concavity, the surface normal would point toward the point source, thus maximizing the luminance from direct illumination. A large mutual illumination component would also be present, as in Fig. 8.6. These effects would give rise to a large luminance maximum in the bottom of the concavity.

There are special cases in which luminance is not a maximum within a concavity. Consider two parallel cylinders, for example extended fingers, in contact and illuminated from above. Along the locus of contact, the surface normals would be perpendicular to the light source direction, giving rise to a local luminance minimum.

To summarize, when a concavity is illuminated by a point source overhead, the walls of the concavity illuminate one another, and this tends to produce 1-D local luminance maxima at the bottom of the concavity. In special cases, however, surface normal effects may neutralize these maxima.

8.5 Active Shading

Suppose the scene was illuminated by a diffuse source. We have argued that surface luminance would typically attain local 1-D minima across concavities. A second image could be made by turning on a point source at the camera, and then subtracting the energy of the previous diffuse image. The difference image would be due entirely to the point source, and hence would typically have luminance maxima across concavities. That is, the luminance profile across each concavity typically would switch from a minimum to a maximum if the source were switched from a diffuse source to a point source. The point source thus provides further shading evidence to support a concavity hypothesis obtained using the diffuse source.

Of course, the luminance minima obtained under diffuse lighting conditions could be due to factors other than shading. A typical example is an albedo change along a surface. In such a case, however, if the surface were flat, then the luminance minimum would be unaffected by the change in light source. The situation becomes more complicated if both albedo variations and shading variations due to shape are present. In the following experiment, we consider the case in which the surface has constant albedo.

Fig. 8.7 shows a drapery surface which was illuminated by a point light source at the camera, and by a diffuse source. Notice how different the images are. In particular, the spatial variation of the shading is much more rapid in the point source image, in which both hills and valleys contain large local luminance maxima, than in the diffuse source image, in which only the hills are luminance maxima.

For the point source image (upper) the locations of luminance maxima were detected with a positive contrast line operator (Iverson & Zucker, 1990). For the diffuse source image (lower), the locations of luminance minima were detected using a negative contrast line operator. These initial estimates should be refined with further processing, as in Parent and Zucker (1989). There are a number of reasons that such processing is needed. First, the valleys are so closely packed that unless the scale of the operators happens to be that of the valley width, the operator may cover more than one valley. Second, in our implementation, the point source was *near* rather than *at* the camera, and this produced unwanted cast shadows.

8.6 Conclusion

In this paper we have discussed a new idea for computing local qualitative shape information directly from the image intensities, that is, without first reconstructing the surface. We have argued that two popular local qualitative shape features — orientation discontinuities and smooth valleys — typically have the same local shading signatures. Under diffuse lighting conditions, local luminance minima are typically produced. Under point source overhead conditions, local luminance maxima are produced. By designing an environment such that images can be

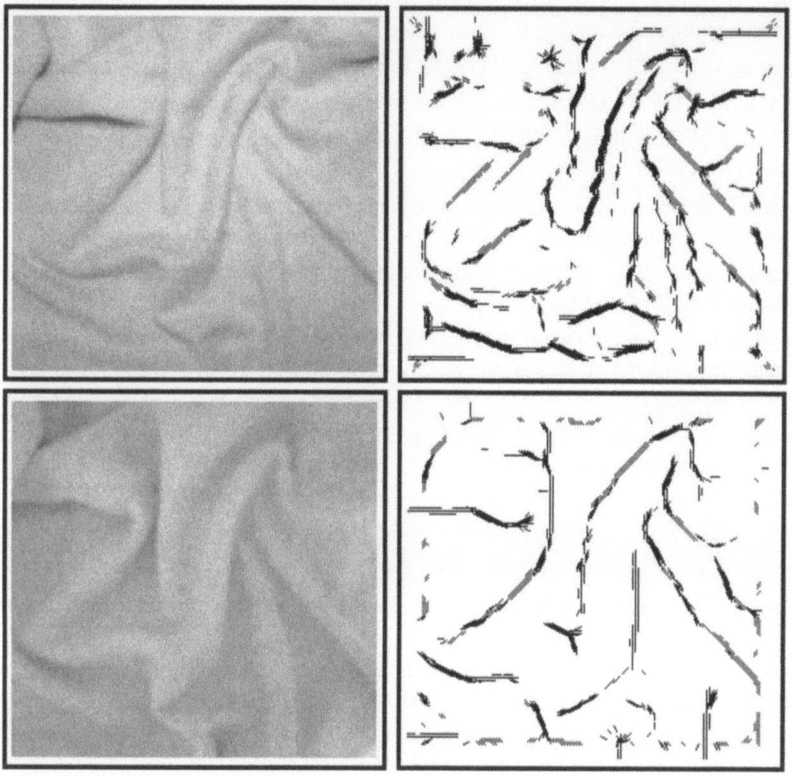

FIGURE 8.7. The top image is a drapery surface illuminated by a point source at the camera. The bottom image is the same surface illuminated by a diffuse source. Initial estimates of the location of the local luminance maxima (in the top image) and minima (in the bottom image) are shown.

obtained under both these lighting conditions, these local qualitative shape features can be more easily detected.

Acknowledgments: The research discussed in this chapter was supported by grants from NSERC and AFOSR. The authors would like to thank Michael Landy for his careful review and comments.

8.7 References

Besl, P. J. & Jain, R. C. (1988). Segmentation through variable-order surface fitting. *IEEE Transactions on Pattern Analysis and Machine Intelligence, 10*, 167-192.

Blake, A. & Zisserman, A. (1987). *Visual Reconstruction.* Cambridge, Mass.: MIT Press.

Forsyth, D. & Zisserman, A. (1991). Reflections on shading. *IEEE Transactions on Pattern Analysis and Machine Intelligence, 13*, 671-679.

Gibson, J. J. (1966). *The Senses Considered as Perceptual Systems*. Boston: Houghton Mifflin Company.

Hoffman, D. D. & Richards, W. A. (1985). Parts of recognition. *Cognition, 18*, 65–96.

Horn, B. K. P. (1977). Understanding image intensities. *Artificial Intelligence, 8*, 201–231.

Horn, B. K. P. & Sjoberg, R. W. (1979). Calculating the reflectance map. *Applied Optics, 18*, 1770–1779.

Iverson, L. & Zucker, S. W. (1990). Logical/linear operators for measuring orientation and curvature. Technical Report TR–CIM–90–06, McGill University, Montréal.

Koenderink, J. J. & van Doorn, A. J. (1980). Photometric invariants related to solid shape. *Optica Acta, 27*, 981–996.

Koenderink, J. J. & van Doorn, A. J. (1983). Geometrical modes as a general method to treat diffuse interreflections in radiometry. *Journal of the Optical Society of America, 73*, 843–850.

Koenderink, J. J. & Richards, W. A. (1992). Why is snow so bright? *Journal of the Optical Society of America A, 9*, 643–648.

Langer, M. S. & Zucker, S. W. (1994). Shape-from-shading on a cloudy day. *Journal of the Optical Society of America A, 11*, 467–478.

Leyton, M. (1988). A process grammar for shape. *Artificial Intelligence, 34*, 213–247.

Nayar, S. K., Ikeuchi, K. & Kanade, T. (1991). Shape from interreflections. *International Journal of Computer Vision, 6*, 173–195.

Nishita, T. & Nakamae, E. (1986). Continuous tone representation of three-dimensional objects illuminated by sky light. In *SIGGRAPH 86* (pp. 125–132).

Parent, P. & Zucker, S. W. (1989). Trace inference, curvature consistency, and curve detection. *IEEE Transactions on Pattern Analysis and Machine Intelligence, 11*, 823–839.

Pentland, A. P. (1982). Finding the illuminant direction. *Journal of the Optical Society of America, 72*, 448–455.

Richards, W. A., Koenderink, J. J. & Hoffman, D. D. (1987). Inferring 3-d shapes from 2-d silhouettes. *Journal of the Optical Society of America, 4*, 1168–1175.

Woodham, R. J. (1980). Photometric method for determining surface orientation from multiple images. *Optical Engineering, 19*, 139–144.

Part III

Robots that Explore

The chapters in this section are concerned with practical and theoretical problems surrounding the design of robots that see as they move.

9

The Synthesis of Vision and Action

Cornelia Fermüller[1]
Yiannis Aloimonos[1]

ABSTRACT
Our efforts to reconstruct the world using visual information have led to the insight that the study of vision should not be separated from the study of a system's actions and purposes. In computational terms this relates to approaching the analysis of perceptual information processing systems through the modeling of the observer and world in a synergistic manner, not through the isolated modeling of observer and world as closed systems. The question still remains: how should such a synergistic model be realized? This chapter addresses the question by providing a methodology for synthesizing vision systems and integrating perception and action. In particular, we outline an architecture for purposive vision systems and present a hierarchy of navigational competences based on computational models of increasing complexity, employing representations of motion, shape, form and space. Pure computational considerations will not tell us what visual competences and representations are important to vision systems performing a set of tasks. Interaction, however, with empirical sciences such as neurobiology, physiology, psychology, ethology, etc., can give us inspiration about the visual categories relevant to systems existing in real world environments. Throughout the chapter, we describe biological findings and how they affect the choice of computational models and representations needed for the synthesis of a hierarchy of navigational competences in a working system.

9.1 Prolegomena

A complete theory of perception must examine a broad set of topics and their interaction, ranging from the environment and the stimuli to sensory organs, the brain and the tasks supported by perception. Many theories for explaining visual perception have been proposed over the centuries. For tractability reasons, these theories concentrated on only one or a few of the topics mentioned above. They range from the theory of Empedocles (440 B.C.) to the computational theory of David Marr (1982); some of them are well known to students of perception (Gestaltist theories (Köhler, 1947), empiricism (von Helmholtz, 1866), Gibson's theory of direct perception (1979)) while others are almost forgotten (e.g., Brunswick's theory of probabilistic functionalism (1956)).

[1]Center for Automation Research, University of Maryland

The development of each theory, as is the case in all disciplines, was influenced by previous ones and other philosophical or scientific ideas prominent at the time. For example, the Gestalt theories were influenced by Kant's (1786/1973) Critique of Pure Reason (although 150 years after its publication) and Marr's computational theory by neuroanatomical developments of his time (Hubel & Wiesel, 1968) as well as by initial results on visual agnosia (Warrington & Shallice, 1984). In our days, more than ever before, our views on the architecture and the structure of the computational mechanisms underlying the behavior of intelligent organisms or robots possessing perception are influenced by advances in various disciplines that study the brain.

During the 1960's and 70's it was commonly supposed that each visual area of the cerebral cortex analyzes all the information in the field of view, but at a more complex level than the antecedent area, the areas forming a sort of hierarchical chain. The computational theories that appeared at these times reflected this doctrine and efforts were made for discovering mechanisms to reconstruct general descriptions of the visible scene (Aloimonos & Shulman, 1989; Horn, 1986; Marr, 1982). Vision was studied in a disembodied manner by concentrating mostly on *stimuli*, *sensory organs* and the *brain*. In some sense, this work argued that "seeing" and "thinking" are distinguishable activities, with "seeing" being a mechanical act that does not originate anything (Kanizsa, 1979; Nalwa, 1993). This in turn contributed to the separation of main stream artificial intelligence (that was about "thinking") and computer vision (that was about "seeing").

During the 1980's with the emergence of "active vision" (Aloimonos, 1993; Aloimonos, Weiss & Bandopadhay, 1988; Bajcsy, 1988; Ballard, 1991), researchers considered, in addition to the topics of stimuli, sensors and brain, the topics of reflexes and motor responses and in particular the motor responses whose goal is to control the image acquisition process. It was noted that active vision systems could, with selective perception, perform a number of interesting tasks. This contributed to a novel view of the global structure of a "seeing" system. Several researchers today view a vision system as consisting of a set of behaviors, programs capable of supporting a set of actions. It has been realized that perception should not be studied in isolation, but in conjunction with the physiology of systems and with the tasks that systems perform. In the discipline of computer vision such ideas caused researchers to extend the scope of their field. Computer vision was originally limited to the study of mappings of a given set of visual data into representations on a more abstract level. Now, it has become clear that image understanding should also include the process of selective acquisition of data in space and time. This has led to a series of studies published under the headings of *Active*, *Animate*, *Purposive* or *Behavioral Vision*. A good theory of vision would be one that can create an interface between perception and other cognitive abilities. However, with a formal theory integrating perception and action still lacking, most studies treated active vision (Aloimonos et al., 1988; Bajcsy, 1988) as an extension of the classical reconstruction theory, employing activities only as a means to regularize the ill-posed classical inverse problems.

9.2 Marr's Theory and Its Drawbacks

Let us summarize the key features of the classical theory of vision to point out its drawbacks as an overall framework for studying and building perceptual systems. In the theory of Marr (1982), the most influential in recent times, vision is described as a reconstruction process, that is, a problem of creating representations of increasingly high levels of abstraction, leading from 2-D images over the primal sketch through the $2\frac{1}{2}$-D sketch to object-centered descriptions ("from pixels to predicates", Pentland, 1986). Marr suggested that visual processes—or any perceptual/cognitive processes—are information processing tasks and thus should be analyzed at three levels: (a) at the computational theoretic level (definition of the problem and its boundary conditions; formulation of theoretical access to the problem), (b) at the level of selection of algorithms and representations (specification of formal procedures for obtaining the solution), and (c) at the implementational level (depending on the available hardware).

In the definition of cognitive processing in the classical theory, vision is formalized as a pure information processing task. Such a formalization requires a well-defined closed system. Since part of this system is the environment, the system would be closed only if it were possible to model all aspects of objective reality. The consequences are well-known: Only toy problems (blocks worlds, Lambertian surfaces, smooth contours, controlled illumination, and the like) were solved successfully.

The strict formalization of representations at different levels of abstraction gave rise to breaking the problems into autonomous subproblems and solving them independently. The conversion of external data (sensor data, actuator commands, decision making, etc.) into an internal representation was separated from the phase of algorithms to perform computations on internal data; signal processing was separated from symbolic processing and action. Processing of visual data was treated, for the most part, in a syntactic manner and semantics was treated in a purely symbolic way using the results of the syntax analysis. This is not surprising, since computer vision was considered as a subfield of artificial intelligence and thus studied using the same methodology, influenced by the ideas about computational theories of the last decades (Ernst & Newell, 1969; Gelernter, 1959; Nilsson, 1980).

The strict hierarchical organization of representational steps in the Marr paradigm makes the development of learning, adaptation and generalization processes practically impossible (no doubt there hasn't been much work on computational "vision and learning"). Furthermore, the conceptualization of a vision system as consisting of a set of modules recovering general scene descriptions in a hierarchical manner introduces computational difficulties with regard to issues of robustness, stability, and efficiency. These problems lead us to believe that general vision does not seem to be feasible. Any system has a specific relationship with the world in which it lives, and the system itself is nothing but an embodiment of this relationship. In the Marr approach the algorithmic level has been separated from the physiology of the system (the hardware) and thus vision was studied in a disembodied transcendental manner.

Of course, many of the solutions developed for disembodied systems may also be of use for embodied ones. In general, however, this does not hold. Having available an infinite amount of resources, every (decidable) problem can be solved in principle. Assuming that we live in a finite world and that we have a finite number of possibilities for performing computations, any vision problem might be formulated as a simple search problem in a very high dimensional space. From this point of view, the study of embodied systems is concerned with the study of techniques to make seemingly intractable problems tractable.

To contribute to the understanding of perceptual systems, one should model the observer and world in a synergistic manner rather than as isolated, closed systems (Sommer, 1994). The question, of course, still remains how such a synergistic modeling should be realized. How can we relate perception and action? What are the building blocks of an intelligent perceptual system? What are the categories into which the system divides its perceptual world? What are the representations it employs? How is it possible to implement such systems in a flexible manner to allow them to learn from experience and improve performance? In this paper we present a formal framework for addressing these questions. Our exposition describes both recent technical results and some of our future research agenda.

9.3 The Architecture

9.3.1 The modules of the system

At the highest level of abstraction, a vision system consists of a set of maps (Fig. 9.1) that relate the observer's space-time representations. For the purpose of a more systematic study, we classify the maps into three categories: the visual competences, the action routines, and the learning procedures. We distinguish two kinds of representations according to the amount of processing performed on them: representations of the perceptual information computed from visual input, and representations of any kind of perceptual information acquired over time and shared and organized in memory. Fig. 9.2 gives a more detailed description of a purposive vision system.

At a different level of abstraction that modularizes the system, Fig. 9.3 describes the basic components of a purposive vision system. The abstract procedures and representations of a vision system are: the procedures for performing visual perceptions, physical actions, learning, and information retrieval, and purposive representations of the perceptual information along with representations of information acquired over time and stored in memory.

At any time a purposive vision system has a goal or a set of goals it approaches as best as it can by means of its available resources. Thus, at any time the system is engaged in executing a task. The visual system possesses a set of visual competences with which it processes the visual information. The competences compute purposive representations. Each of these representations captures some aspect of the total visual information. Thus, compared with the representations of the old paradigm, they are partial. The representations vary in complexity with regard to

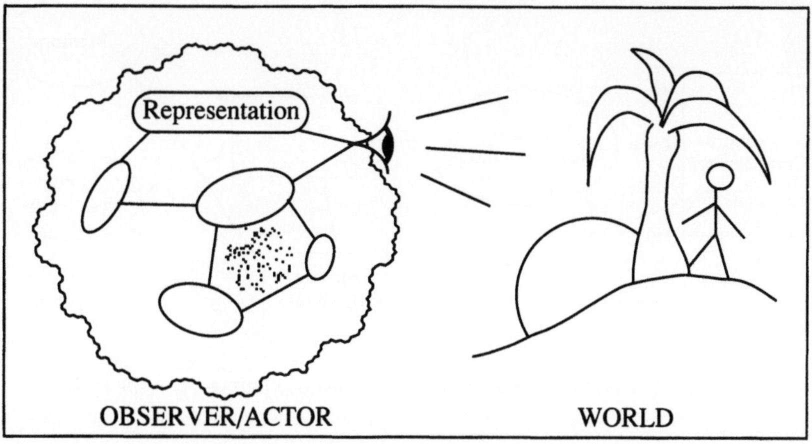

FIGURE 9.1. An intelligent system with vision creates various representations of space-time that it uses to perform various actions.

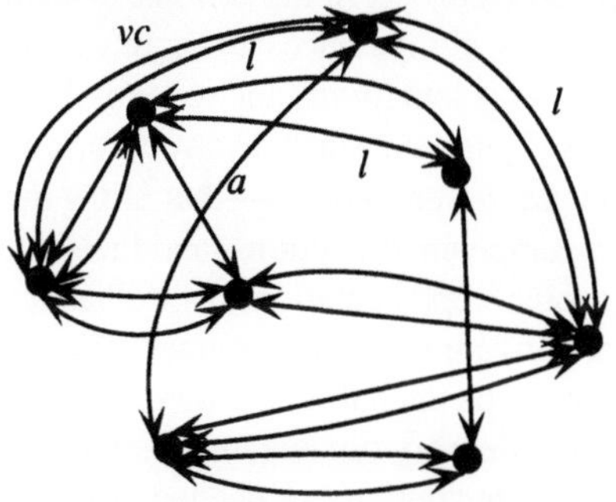

FIGURE 9.2. An intelligent system with vision consists of a map of maps. These maps map different representations of space-time (•) into each other. Space-time includes, of course, the system itself. To study these maps more systematically, we divide them into three categories: the visual competences (*vc*), the action routines (*a*), and the learning procedures (*l*).

the space they describe. The purposive representations themselves are purposive descriptions of the visual information organized in certain data structures. The purposive representations access programs which we call *action routines*. This collective name refers to two kinds of routines; the first kind are the programs that schedule the physical actions to be performed, that is, they initialize motor commands and thus provide the interface to the body, and the second kind schedule the selection of information to be retrieved from the purposive representations and stored in long-term memory. An important aspect of the architecture is that the access of the visual processes to the actions is on the basis of the contents of the purposive representations, that is, the contents of the purposive representations serve as addresses to the actions. Another class of programs is responsible for

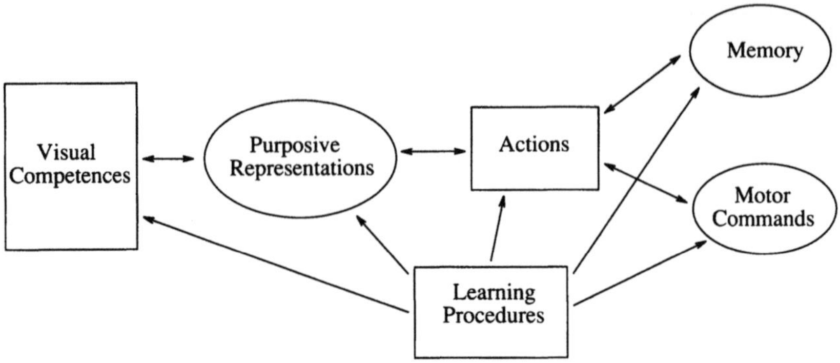

FIGURE 9.3. Working model: basic components of a purposive vision system.

learning by providing the actions, the competences, and the representations with the means to change and adjust parameters.

As can be seen from Fig. 9.3, learning takes place at various levels of, as well as in between, the modules of the system. For a flexible vision system, it should be possible to learn the parameters describing actions, to acquire new actions, to learn parameters describing visual competences, to acquire new visual competences that compute new purposive representations, and to learn the sequences of actions and perceptual competences to perform a task. In any case, learning is accomplished by means of programs—learning procedures—that allow the change and adaptation of parameters to learn competences, actions, and their interrelationships.

The purposive perceptual representations, as well as representations containing other kinds of information, are stored in memory. The storage must happen in an efficient way according to the available memory space. Different representations share common elements. Memory organization techniques that allow you to store information according to its content are needed. Also, designing a memory for representations includes designing the procedures necessary for fast and reliable access. In this chapter we focus our discussion on the visual competences.

Let us summarize in which way the above model captures the study of perception and action in a synergistic way, and address some of the questions posed in Section 9.2. In this model the intelligence of a purposive system is embodied in its visual competences and its actions. Thus, competences and actions are considered to be the building blocks of an intelligent system. To meet a purpose (a task which is stated in the form of events that can be perceived by means of the perceptual processes), a system executes behaviors. Thus, behaviors, which are an emergent attribute of the system, couple perception and action. They constitute some form of structural adaptation which might either be visible externally or take place only internally in the form of parameter adaptation.

9.3.2 *Outline of the approach*

If we aim to understand perception, we have to come up with some methodology to study it. The ideal would be to design a clearly defined model for the architecture of vision systems and start working on its components. However, we have few answers available when it comes to actually talking about the visual categories that are relevant for visual systems. The kinds of representations needed to perform a task depend on the embodiment of the system and the environment in which it lives. Answers to these questions cannot come only as insight gained from the study of mathematical models. There must be empirical studies investigating systems (biological and artificial ones) that will tell us how to couple functionality, visual categories and visual processes. If we haven't understood how we actually could develop visual competences for systems that work in environments as complex as our own, we won't be able to obtain a global view of the overall architecture and functionality of vision systems. At this time it also wouldn't contribute much to the development of our understanding to just develop particular systems that perform particular tasks, for example, a system that recognizes tables. Even if we were able to create such a system having a success rate of 99%, it would have the capacity of recognizing many things that are unknown to us, and not just tables. Thus, by aiming to build systems that recognize certain categories that seem relevant to our symbolic language repertoire, we wouldn't gain much insight into perception.

It seems somehow natural that the only way out of this problem of where to start is to approach the study of vision systems in an "evolutionary" way. We call such an approach the synthetic (evolutionary) approach. We give here a short outline of the ideas behind this approach, which we discuss in detail in the remainder of the paper. It means we should start by developing primitive visual operations and provide the system in this way with visual capabilities (or competences). As we go on, the competences will become more and more complex. At the same time, as soon as we have developed a small number of competences, we should work on their integration. Such an endeavor throws us immediately into the study of two other major components of the system. How is visual information related to action and how is the information represented? How is it organized and how it is coordinated with the object recognition space? We are confronted on the one hand with the study of activities and the integration of vision and action, and on the other hand with the study of the memory space with all its associated problems of memory organization, visual data representation, and indexing—the problem of associating data stored in the memory with new visual information. Furthermore, we also have to consider the problem of learning from the very beginning.

9.4 The Competences

9.4.1 Computational principles

9.4.1.1 Model complexity

Our goal is to analyze, in order to design, a system from a computational point of view. We argued earlier that the study of visual systems should be performed in a hierarchical manner according to the complexity of the visual processes. As a basis for its computations a system has to utilize mathematical models, which serve as abstractions of the representations employed. Thus, when referring to the complexity of visual processes, we mean the complexity of the mathematical models involved.

The synthetic approach calls first for studying capabilities whose development relies on only simple models and then going on to study capabilities requiring more complex models. Simple models do not refer to environment- or situation-specific models which are of use in only a limited number of situations. Each of the capabilities requiring a specified set of models should be used for solving a well-defined class of tasks in every environment and situation to which the system is exposed. If our goal is to pursue the study of perception in a scientific way, as opposed to industrial development, we have to accept this requirement as one of the postulates, although it is hard to achieve. Whenever we perform computations, we design models on the basis of assumptions, which in the case of visual processing are constraints on the space-time in which the system is acting, on the system itself, and on their relationship. An assumption, however, can be general with regard to the environment and situation, or very specific.

For example, the assumption about piecewise planarity of the world is general with regard to the environment (every continuous differentiable function can be approximated in a sufficiently small area by a linear function). However, to use this assumption for visual recovery, additional assumptions regarding the number of planar patches have to be made; these are environment-specific assumptions. Similarly, we may assume that the world is smooth between discontinuities; this is general with regard to the environment. Again, for this assumption to be utilized we must make some assumptions specifying the discontinuities, and then we become specific. We may assume that an observer only translates. If indeed the physiology of the observer allows only translation, then we have made a general assumption with regard to the system. If we assume that the motion of an observer in a long sequence of frames is the same between any two consecutive frames, we have made a specific assumption with regard to the system. If we assume that the noise in our system is Gaussian or uniform, again we have made a system-specific assumption.

Our approach requires that the assumptions used be general with regard to the environment and the system. Scaled up to more complicated systems existing in various environments, this requirement translates to the system's capability to decide whether a model is appropriate for the environment in which the system is acting. A system might possess a set of processes that together supply the system with one competence. Some of the processes are limited to certain environmental specifications. Thus, the system must be given the capability to acquire knowledge about what processes to apply in a specific situation.

The motivation for studying competences in a hierarchical way is to gain increasing insight into the process of vision, which is extremely complex. Therefore, the capabilities which require more complex models should be based on "simpler," already developed capabilities. The complexity of a capability is given by the complexity of the assumptions employed; what has been considered a "simple" capability might require complex models, and vice versa.

9.4.1.2 Qualitative models

The basic principle concerning the implementation of processes subserving the capabilities, which is motivated by the need for robustness, is the quest for algorithms which are qualitative in nature. We argue that visual competences should not be formulated as processes that reconstruct the world but as recognition procedures. Visual competences are procedures that recognize aspects of objective reality which are necessary to perform a set of tasks. The function of every module in the system should constitute an act of recognizing specific situations by means of primitives which are applicable in general environments. Each such entity recognized constitutes a category relevant to the system. We next discuss some examples from navigation.

The problem of independent motion detection by a moving observer usually has been addressed with techniques for segmenting optical flow fields. But it also may be tackled through the recognition of nonrigid flow fields for a moving observer partially knowing its motion (Aloimonos, 1990; Nelson, 1991; Thompson & Pong, 1990). Pursuing a target amounts to recognizing the target's location on the image plane along with a set of labels representing aspects of its relative motion sufficient for the observer to plan its actions. Motion measurements of this kind could be relative changes in the motion such as a turn to the left, right, above, or down; or focusing further away, or closer. In the same way, the problem of hand/eye coordination can be dealt with using stereo and other techniques to compute the depth map and then solve the inverse kinematics problem in order to move the arm. While the arm is moving the system is blind (Brady, Hollerbach, Johnson, Lozano-Perez & Mason, 1983); however, the same problem can be solved by creating a mapping (the perceptual kinematic map) from image features to the robot's joints. The positioning of the arm is achieved by recognizing the image features (Hervé, 1993).

Instead of reconstructing the world, the problems described above are solved through the recognition of entities that are directly relevant to the task at hand. These entities are represented by only those parameters sufficient to solve the specific task. In many cases, there exists an appropriate representation of the space-time information that allows us to derive directly the necessary parameters by recognizing a set of locations on this representation along with a a set of attributes. Since recognition amounts to comparing the information under consideration with prestored representations, the described approaches to solving these problems amount to matching patterns.

In addition, image information should be utilized globally whenever possible. Since the developed competences are meant to operate in real environments, the computations have to be insensitive to errors in the input measurements. This postulates a requirement for redundancy in the input used. In most cases, the partial information about the scene, which we want to recognize, will be globally encoded in the image information. The computational models we are using should be such that they map global image information into partial scene information. Later in this section we will demonstrate our point by means of the rigid motion model.

If we wish to speak of an algorithm as qualitative, the primitives to be computed should not rely on explicit, unstable, quantitative models. Qualitativeness can be achieved in a number of ways. The primitives might be expressed in qualitative terms. Their computation might be derived from inexact measurements and pattern recognition techniques. Or, the computational model itself might be proved stable and robust in all possible cases.

At the philosophical level, the synthetic approach is similar to the proposal of Brooks (1986) for understanding intelligent behavior through the construction of working mechanisms. In proposing the subsumption architecture, Brooks suggested a hierarchy of competences such as avoiding contact with objects, exploring the world by seeing places, reasoning about the world in terms of identifiable objects, etc. This proposal, however, did not provide a systematic way of creating a hierarchy of competences by taking into account the system's purpose and physiology. This is the relevant question.

9.4.2 Biological hierarchy

It remains to be discussed what these simple capabilities actually are on which we should concentrate our first efforts. Other scientific disciplines give us some answers. Much simpler than the human visual system are the perceptual systems of lower animals like medusae, worms, crustaceans, insects, spiders and molluscs. Researchers in neuroethology study such systems and by now have gained some understanding of them. Horridge (1987, 1991), working on insect vision, studied the evolution of visual mechanisms and proposed hierarchical classifications of visual capabilities. He argued that the most basic capabilities found in animals are based on motion. Animals up to the complexity of insects perceive objects entirely by relative motion. His view concerning the evolution of vision is that objects are first separated by their motions and, with the evolution of a memory for shapes, form vision progressively evolves. The importance of these studies on lower animals becomes very clear when we take into account the view commonly held by leaders in this field that the principles governing visual motor control are basically the same in lower animals and humans—whereas, of course, we humans and other primates can see without relative motion between ourselves and our surroundings.

9.4.2.1 Parallel visual pathways

In the last decades the part of the brain in primates responsible for visual processing—the visual cortex—has been studied from an anatomical, physiological and also behavioral viewpoint. Different regions of visual cortex have been identified and most of their connections established. Most scientists subscribe to the theory that the different parts perform functionally specialized operations. It remains to clarify the precise function of these regions. In particular, opinions diverge about the specialization and the interconnections involved in later stages of processing of visual data. Much more is known about the earlier processes. The vi-

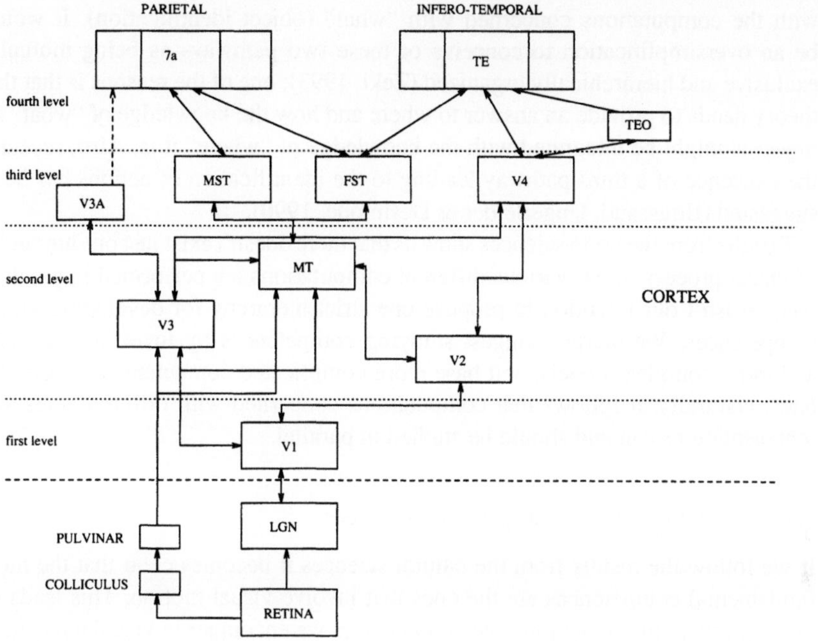

FIGURE 9.4. Diagram of the primate visual system indicating the subcortical structure as well as the four tentative levels of cortical visual processing (from Orban, 1992).

sual signal reaches the cortex at the primary visual cortex (also called V1 or striate cortex) via the retina and the lateral geniculate body. From the primary visual cortex the visual signals are sent to about 30 extrastriate or higher-order visual cortical areas, among which about 300 connections have been reported. Fig. 9.4, taken from Orban (1992) shows the major areas involved in visual processing. According to Orban, the modules in the primate visual cortex can be divided into four hierarchical levels of processing. It seems to be pretty well accepted that there exist lower areas that are specialized for the processing of either static or dynamic imagery. MT (also called V5), MST, and FST seem to be involved in motion processing, and V4 in color processing. Form vision seems to be accomplished by different lower modules, which use both static and dynamic information. Zeki (1993), for example, suggests that V3 is responsible for the understanding of form from motion information, and V4 derives form and color information. At later stages the modules process both kinds of information in a combined way.

On the basis of anatomical evidence and behavioral studies (studies on patients with lesions of specific cortical areas), the hypothesis has been brought forward that there exist two visual pathways originating from V1; a dorsal one leading to the parietal cortex and a ventral one leading to the infero-temporal cortex (Ungerleider & Mishkin, 1982). The dorsal pathway is concerned with either the computations concerned with "where" (object localization) or "how" (the visual guidance of movements, Goodale, Milner, Jacobson & Carey, 1991), and the ventral pathway with the computations concerned with "what" (object identification). It would be an oversimplification to conceive of these two pathways as being mutually exclusive and hierarchically organized (Zeki, 1993); one of the reasons is that this theory needs to provide an answer to where and how the knowledge of "what" an object is might be integrated with the knowledge of "where" it is. Also, recently the existence of a third pathway leading to the identification of actions has been suggested (Boussaud, Ungerleider & Desimone, 1990).

Results from the brain sciences show us that there doesn't exist just one hierarchy of visual processes, but various different computations are performed in parallel. Also, it isn't our intention to propose one strict hierarchy for developing visual competences. We merely suggest studying competences by investigating more and more complex models, and base more complicated competences on simpler ones. Naturally, it follows that computations concerned with different cues and representations can and should be studied in parallel.

9.4.2.2 Motion, shape and space competences

If we follow the results from the natural sciences it becomes clear that the most fundamental competences are the ones that involve visual motion. This leads us to the problem of navigation. The competences we encounter in visual navigation encompass representations of different forms. To elucidate the synthetic approach, in the next section we will discuss a series of competences of increasing complexity employing representations of motion, shape, and space. In the following section we will then outline our realizations of the most basic competences in visual navigation, which only require motion information.

Next in the hierarchy follow capabilities related to the understanding of form and shape and the learning of space. Concerning form and shape, our view is that we should not try to adopt the classical idea of computing representations that capture the 3-D world metrically. Psychological studies on the role of the eye movements suggest that fixations play an important role in our understanding of space. It seems that the level on which information from successive fixations is integrated is relatively abstract and that the representations from which organisms operate on the world are 3-D only locally. Therefore, it will be necessary to study new forms of shape representation. In nature too, there doesn't exist just one method of shape representation. As results from neurobiology show, form perception in the human visual system takes place in more than just one part of the cortex and is realized with different kinds of hardware.

Space is also understood from the processing of various cues in a variety of ways. Furthermore, different tasks will require representations of space with regard to different reference systems—not just one, as often has been debated in the past. Representations might be object-centered, ego-centered, or action-driven.

Actions can be very typical for objects. Early perceptual studies have shown that humans are able to interpret moving scenes correctly, even when the static view does not contain any information about the structure. In the experiments of Johansson (1973) subjects were able to recognize animals, as well as specific human beings, given only the motions of light bulbs mounted on the object's joints. Since our viewpoint is that we should formulate competences as recognition procedures, the study of navigation also leads us to the study of action-driven visual processing.

9.4.3 A hierarchy of models for navigational competences

Navigation, in general, refers to the performance of sensory mediated movement, and visual navigation is defined as the process of motion control based on an analysis of images. A system with navigational capabilities interacts adaptively with its environment. The movement of the system is governed by sensory feedback which allows it to adapt to variations in the environment. By this definition visual navigation comprises the problem of navigation where a system controls its various components relative to the environment and relative to each other.

Visual navigation encompasses a wide range of perceptual competences, including tasks that every biological species possesses such as motion segmentation and kinetic stabilization (the ability of a single compact sensor to understand and control its own motion), as well as advanced specific hand-eye coordination and servoing tasks.

To explain the principles of the synthetic approach, we describe six such competences, all of which are concerned only with the movement of a single compact sensor. These are: egomotion estimation, partial object-motion estimation, independent motion detection, obstacle avoidance, target pursuit, and homing. These particular competences allow us to demonstrate a hierarchy of models concerned with the representation of motion, form and shape. Table 9.1 describes these competences and formulates them as recognition procedures that rely on increasingly more complex models.

In the past, navigational tasks, since they inherently involve metric relationships between the observer and the environment, have been considered as subproblems of the general "structure-from-motion" problem (Ullman, 1979). The idea was to recover the relative 3-D motion and the structure of the scene in view from a given sequence of images taken by an observer in motion relative to its environment. Indeed, if structure and motion can be computed, then various subsets of the computed parameters provide sufficient information to solve many practical navigational tasks. However, although a great deal of effort has been spent on the subject, the problem of structure from motion still remains unsolved for all practical purposes. The main reason for this is that the problem is ill-posed, in the sense that its solution does not continuously depend on the input.

The most simple navigational competence, according to our definition, is the estimation of egomotion. The observer's sensory apparatus (eye or camera), independent of the observer's body motion, is compact and rigid and thus moves rigidly with respect to a static environment. As we will demonstrate, the estimation of an observer's motion can indeed be based on only the rigid motion model. A geometric analysis of motion fields reveals that the rigid motion parameters manifest themselves in the form of patterns defined on partial components of the motion

TABLE 9.1.

Egomotion estimation	Recognizing locations of intersection of axis of rotation and axis of translation with the image plane by locating patterns on the flow field. Rigid motion model applied globally.
Object-motion estimation	Recognition of tracking acceleration. Rigid motion model applied locally.
Independent motion detection	Recognition of locations whose flow vectors do not originate from rigid motion. Various motion models responding to nonrigidity.
Obstacle avoidance	Recognition of locations that represent parts of the 3-D world on a collision course with observer. Models for time-to-contact.
Target pursuit	Recognizing target's location along with label sufficient to plan pursuit. Models of operational space and the motion of the target.
Homing	Recognition of routes connecting different locations. Models of shape, form and space.

fields (Fermüller, 1993). Algorithmically speaking, the estimation of motion thus can be performed through pattern recognition techniques.

Another competence, the estimation of partial information about an object's motion (its direction of translation), can be based on the same model; but, whereas for the estimation of egomotion the rigid motion model could be employed globally, for this competence only local measurements can legitimately be employed. Following our philosophy about the study of perception, it makes perfect sense to define such a competence which appears very restricted. Since our goal is to study visual problems in the form of modules directly related to the visual task in which the observer is engaged, we argue that in many cases when an object is moving in an unrestricted manner (translation and rotation) in the 3-D world we are only interested in the object's translational component, which can be extracted using dynamic fixation (Fermüller & Aloimonos, 1992).

Next in the hierarchy follow the capabilities of independent motion detection and obstacle avoidance. Although the detection of independent motion seems to be a very primitive task, it can easily be shown by a counterexample that in the general case it cannot be solved without any knowledge of the system's own motion. Imagine a moving system that takes an image showing two areas of different rigid motion. From this image alone, it is not decidable which area corresponds to the static environment and which to an independently moving object.

However, such an example shouldn't discourage us and drive us to the conclusion that egomotion estimation and independent-motion detection are "chicken and egg" problems, and that unless one of them has been solved, the other can't be addressed either. Have you ever experienced the illusion that you are sitting in front of a wall which covers most of your visual field, and suddenly this wall (which actually isn't one) starts to move? You seem to experience yourself moving. It seems that vision alone does not provide us (humans) with an infallible capability of estimating motion. In nature the capability of independent motion detection appears at various levels of complexity. We argue that to achieve a very sophisticated mechanism for independent motion detection, various different processes have to be employed. Another glimpse at nature should give us some inspiration. We humans also do not perceive everything moving independently in our visual field. We usually concentrate our attention on the moving objects in the center of the visual field (where the image is sensed with high resolution) and pay attention only if something is moving fast in the periphery. It thus seems to make sense to develop processes that detect anything moving very fast (Nelson, 1991). If some upper bound on the observer's motion is known (maximal speed), it is possible to detect even for small areas where motions above the speed threshold appear. Similarly, for specific systems, processes that recognize specific types of motion may be devised by employing filters that respond to these motions (of use, for example, when an enemy moves in a particular way). To cope with the "chicken and egg" problem in the detection of larger independently moving objects, we develop a process based on the same principle as the estimation of egomotion, which for an image patch recognizes whether the motion field within the patch originates from only rigid motion, or whether the constraint of rigidity does not hold. Having some idea about the egomotion or the scene (for example, in the form of bounds on the motion, or knowing that the larger part of the scene is static) we can also decide where the independently moving objects are.

To perform obstacle avoidance it is necessary to have some representation of space. This representation must capture in some form the change of distance between the observer and the scene points which have the potential of lying in the observer's path. An observer that wants to avoid obstacles must be able to change its motion in a controlled way and must therefore be able to determine its own motion and set it to known values. As can be seen, the capability of egomotion estimation is a prerequisite for obstacle avoidance mechanisms, and general independent motion detection will require a model which is as complex as that used in egomotion estimation in addition to other simple motion models.

Even higher in the hierarchy are the capabilities of target pursuit and homing (the ability of a system to find a particular location in its environment). Obviously, a system that possesses these capabilities must be able to compute its egomotion, and avoid obstacles while detecting independent motion. Furthermore, homing requires knowledge of the space and models of the environment (for example, shape models), whereas target pursuit relies on models for representing the operational space and the motion of the target. These examples should demonstrate the principles of the synthetic approach, which argues for studying increasingly complex visual capabilities and developing robust (qualitative) modules in such a way that more complex capabilities require the existence of simpler ones.

9.4.4 Motion-based competences

In this section we describe the ideas behind some of the modules we have developed to realize the most basic competences for visual navigation: the competence of egomotion estimation, a process for partial object motion estimation and a process for independent motion detection. This description should merely serve to demonstrate our viewpoint concerning the implementation of qualitativealgorithms; more detailed outlines and analyses are found elsewhere.

First, let us state some of the features that characterize our approach to solving the above mentioned competences, and distinguish it from most existing work.

In the past, the problems of egomotion recovery for an observer moving in a static scene and the recovery of an object's 3-D motion relative to the observer, since they both were considered as reconstruction problems, have been treated in the same way. The rigid motion model is appropriate if only the observer is moving, but it holds only for a restricted subset of moving objects—mainly man-made ones. Indeed, all objects in the natural world move nonrigidly. However, considering only a small patch in the image of a moving object, a rigid motion approximation is legitimate. For the case of egomotion, data from all parts of the image plane can be used, whereas for object motion only local information can be employed.

Most current motion understanding techniques require the computation of exact image motion (optical flow in the differential case or correspondence of features in the discrete case); however, this amounts to an ill-posed problem and additional assumptions about the scene have to be employed. As a result, in the general case, the computed image displacements are imperfect. In turn, the recovery of 3-D motion from noisy flow fields has turned out to be a problem of extreme sensitivity with small perturbations in the input causing large amounts of error in the motion parameter estimation. To overcome this problem, in our approach to the development of motion related competences, we skip the first computational step. All the techniques developed are based on the use of only the sign of the projection of the motion vector along some directions. That is, we assume that the system has the capability to estimate the direction (positive or negative) of the projection of the motion vector along a set of directions. The minimum a system can accomplish is to estimate the direction of retinal motion in at least one direction, namely the one perpendicular to the local edge, known as the direction of normal flow. It should be mentioned that a few techniques using normal flow have appeared in the literature; however, they deal with restricted cases (only translation or only rotation, Aloimonos & Brown, 1984; Horn & Weldon, 1987).

Another characteristic is that the constraints developed for the motion modules, for which the rigid motion module is the correct one globally, are such that the input also is utilized globally. The basis of these computations forms global constraints which relate the spatiotemporal derivatives of the image intensity function to the 3-D motion parameters.

If we consider a spherical retina translating with velocity $\vec{t}$, then the motion field is along the great circles connecting the two antidiametric points, the focus of

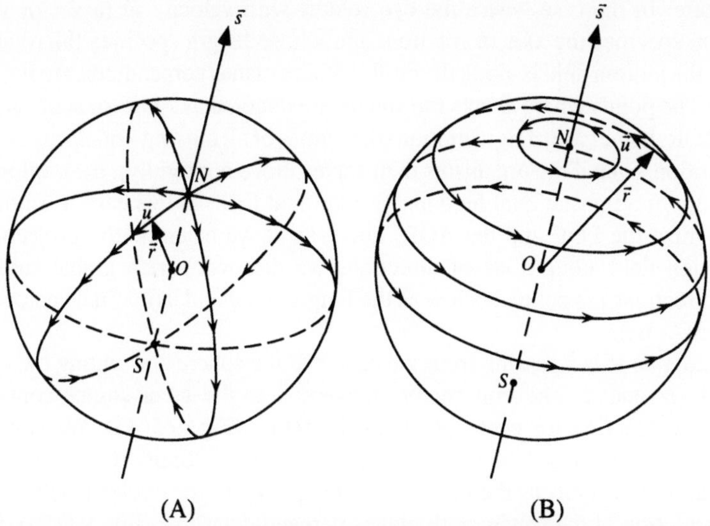

FIGURE 9.5. (A) An axis $\vec{s}$ passing from the center of the sphere and cutting the sphere at points S and N defines a longitudinal vector field. At each point we consider the unit vector tangent to the geodesic connecting S and N. The value of the vector $\vec{u}$ at point $\vec{r}$ is: $u = (\vec{s} - (\vec{s} \cdot \vec{r})\vec{r})/\|\vec{s} - (\vec{s} \cdot \vec{r})r\|$. (B) An axis $\vec{s}$ passing from the center of the sphere and cutting the sphere at points S and N defines a latitudinal vector field. At each point we consider the unit vector tangent to the circle which is the intersection of the sphere with the plane perpendicular to $\vec{s}$. The value of vector $\vec{u}$ at point $\vec{r}$ is $\vec{u} = (\vec{s} \times \vec{r})/\|\vec{s} \times \vec{r}\|$.

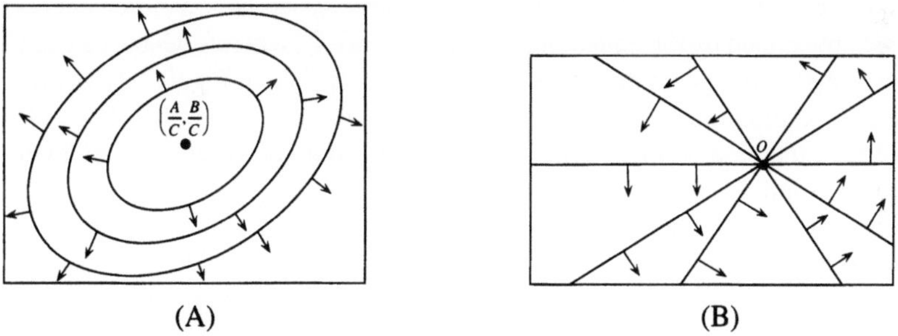

FIGURE 9.6. (A) A longitudinal flow field projected onto a planar retina, with the vectors perpendicular to a set of conic sections, as defined below. An axis $\vec{s} = (A, B, C)$ passing from the nodal point of the eye cuts the image plane at the point $(\frac{A}{C}, \frac{B}{C})$. The family of cones with $\vec{s}$ as their axis intersects the image plane at the set of conics. We call this a *co-axis* field (as it is defined by an axis). (B) A latitudinal field projected onto a planar retina, with the vectors perpendicular to lines passing from a single point O, which defines the field. We call this a *co-point* field.

expansion (FOE) and the focus of contraction (FOC), where the vector $\vec{t}$ intersects the sphere. In the case where the eye rotates with velocity $\vec{w}$ (a vector whose direction specifies the axis of rotation, and whose length specifies the rotational speed), the motion field is along the circles where planes perpendicular to $\vec{w}$ cut the sphere. The points where $\vec{\omega}$ cuts the sphere are denoted as AOR (axis of rotation) and $-$AOR. In the case of rigid motion (the retinas of all moving organisms undergo rigid motion, even if the organisms themselves move nonrigidly), the motion field is the sum of a translational field and a rotational field. In this case it is not easy to recognize the FOE and the AOR; however, if we examine the projection of the motion field along a set of directions we discover a rich global structure. These directions are defined below as the longitudinal and latitudinal vector fields (Fig. 9.5(A–B)).

Consider an axis $\vec{s}$ passing from the center of the sphere and cutting the sphere at points N and S. The unit vectors tangential to the great circles containing $\vec{s}$ define a direction for every point on the retina (Fig. 9.5(A)). We call these directions $\vec{s}$-longitudinal, as they depend on the axis $\vec{s}$. Similarly, we define the $\vec{s}$-latitudinal directions as the unit vectors tangential to the circles resulting from the intersection of the sphere with planes perpendicular to $\vec{s}$ (Fig. 9.5(B)). In the case of a planar retina the longitudinal and latitudinal vector fields become those shown in Fig. 9.6(A–B).

We introduce here a property of these directions that will be of use later. Consider two axes $\vec{s}_1$ and $\vec{s}_2$ cutting the sphere at N_1, S_1 and N_2, S_2 respectively. For every point, each axis defines a longitudinal and a latitudinal direction. We ask the question: Where on the sphere are the $\vec{s}_1$ longitudinal (or latitudinal) directions perpendicular to the $\vec{s}_2$ longitudinal (or latitudinal) directions? Considering the $\vec{s}_1$ and $\vec{s}_2$ longitudinal (or latitudinal) directions, this question translates to: Where on the sphere will a great circle containing $\vec{s}_1$ be perpendicular to a great circle containing $\vec{s}_2$? In general, the locus of such points consists of two closed curves on the sphere defined by the equation $(\vec{r}.\vec{s}_1)(\vec{r}.\vec{s}_2) = \vec{s}_1.\vec{s}_2$, where $\vec{r}$ denotes a position on the sphere. The geometry of these curves is described in Fig. 9.7. Considering now the longitudinal directions of one axis and the latitudinal directions of the other axis, they are perpendicular to each other along the great circle defined by the axes $\vec{s}_1$ and $\vec{s}_2$. (Fig. 9.8).

We now examine the structure of the projection of a rigid motion field on an $\vec{s}$ (NS) longitudinal set of directions. Since a rigid motion field is the sum of a translational and a rotational field, we first study the cases of pure translation and pure rotation.

If we project a translational motion field on the $\vec{s}$ longitudinal vectors, the resulting vectors will either be zero, positive (pointing towards S) or negative (pointing towards N). The vectors will be zero on two curves (symmetric around the center of the sphere) whose shape depends on the angle between the vectors $\vec{t}$ and $\vec{s}$ as in Fig. 9.7. Inside the curves the vectors will be negative and outside the curves positive (Fig. 9.9).

If we project a rotational motion field on the $\vec{s}$ (NS) longitudinal vectors, the projections will be either zero (on the great circle defined by $\vec{\omega}$ and $\vec{s}$), positive (in the one hemisphere) or negative (in the other hemisphere) (Fig. 9.10).

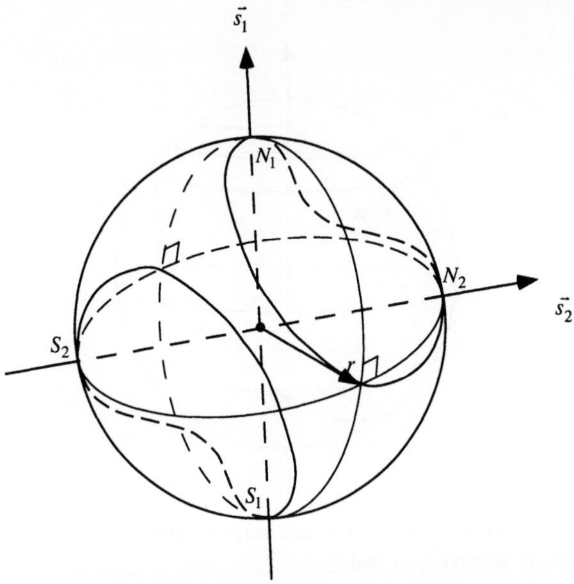

FIGURE 9.7. The great circles containing $\vec{s}_1$ and $\vec{s}_2$ are perpendicular at points of the sphere lying on two closed curves. If $\vec{r}$ denotes a point on the curves, then $(\vec{s}_1 \cdot \vec{r})(\vec{s}_2 \cdot \vec{r}) = \vec{s}_1 \cdot \vec{s}_2$. The shape of the curves depends on the angle between $\vec{s}_1$ and $\vec{s}_2$.

If the observer now translates and rotates with velocities $\vec{t}$ and $\vec{\omega}$, it is possible to classify some parts of the projection of the general motion field on any set of $\vec{s}$ longitudinal vectors by intersecting the patterns of Figs. 9.9 and 9.10. If at a longitudinal vector the projection of both the translational and rotational vectors is positive, then the projection of the image motion vector (the sum of the translational and rotational vectors) will also be positive. Similarly, if the projections of both the translational and rotational vectors on a longitudinal vector at a point are negative, so also will the projection of the motion vector at this point. In other words, if we intersect the patterns of Figs. 9.9 and 9.10, whenever positive and positive come together the result will be positive and whenever negative and negative come together the result will be negative. However, whenever positive and negative come together, the result cannot be determined without knowledge of the environment.

Thus, if we project a rigid motion field on an $\vec{s}$ longitudinal vector field, then the projections will be strictly negative or strictly positive in the areas identified in Fig. 9.11. In the rest of the sphere the projections can be negative, positive or zero. The pattern of Fig. 9.11 is defined by one great circle containing $\vec{\omega}$ and $\vec{s}$ and by two curves containing the points FOE, FOC, N and S.

It is worth pointing out that the pattern of Fig. 9.11 depends only on the directions of vectors $\vec{s}$ (that defines the longitudinal vectors), $\vec{t}$ and $\vec{\omega}$; and is independent of the scene in view. Also, the pattern is different for a different choice of the vector $\vec{s}$.

If we consider the projection of a rigid motion field on the $\vec{s}$ latitudinal directions (defined by the vector $\vec{s}$ (NS)), we obtain a pattern which is dual to the one of

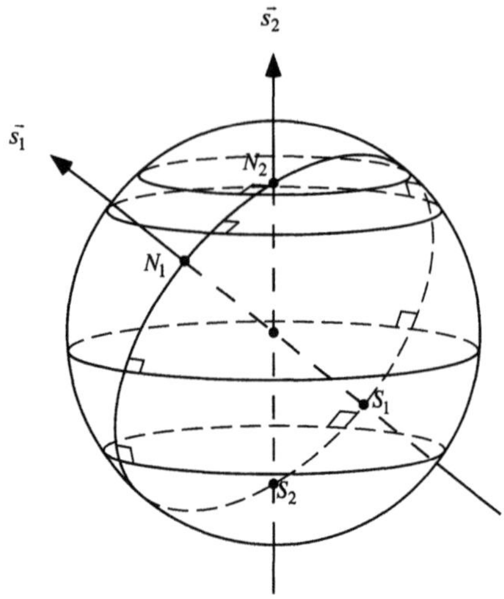

FIGURE 9.8. The $\vec{s}_1$-longitudinal vectors are perpendicular to the $\vec{s}_2$-latitudinal vectors along the great circle defined by $\vec{s}_1$ and $\vec{s}_2$.

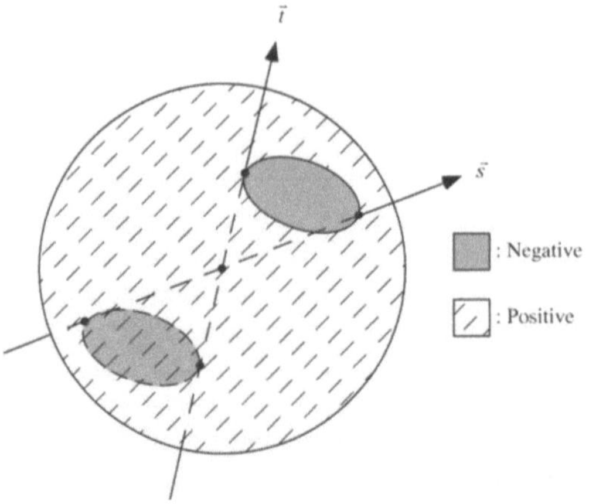

FIGURE 9.9. Projection of a translational motion field on an $\vec{s}$ longitudinal pattern. It is zero on two curves on the sphere (symmetric with regard to the center of the sphere). The points where $\vec{t}$, $\vec{s}$, $-\vec{t}$ and $-\vec{s}$ intersect the sphere lie on the curves. The values are negative inside the curves and positive outside them.

Fig. 9.11. This time, the translational flow is separated into positive and negative by a great circle and the rotational flow by two closed curves passing from the points AOR, −AOR, N and S, as in Fig. 9.7.

The geometric analysis described above allows us to formulate the problem of egomotion estimation as a pattern recognition problem. If the system has the capability of estimating the sign of the retinal motion along a set of directions at each point, then this means that the system can find the sign of the longitudinal and

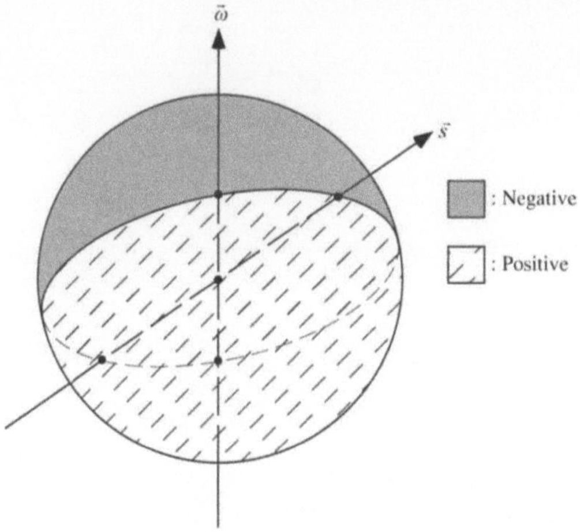

FIGURE 9.10. Projection of a rotational motion field on an $\vec{s}$ longitudinal pattern. The values are zero on the great circle defined by the plane of $\vec{\omega}$ and $\vec{s}$. In the one hemisphere the values are positive and in the other they are negative.

latitudinal vectors for a set of axes $\vec{s}_i$, $i = 1, \ldots, n$. If the system can now locate the patterns in each longitudinal and latitudinal vector field, then it has effectively recognized the directions $\vec{t}$ and $\vec{\omega}$. If, however, the system has less power and can only compute the motion in at most one direction at every point, namely the one perpendicular to the local edge, then the solution proceeds exactly as before. The difference is that for each longitudinal or latitudinal set of directions we do not have information (positive, negative or zero) at every point of the sphere.

Considering a planar retina instead of a spherical retina, we have co-point vectors instead of latitudinal vectors and co-axis vectors instead of longitudinal vectors (Fermüller, 1993). The curves separating positive from negative values become a second order curve and a line in the plane. Fig. 9.12 shows the pattern of co-point (latitudinal) vectors for a planar retina.

Thus, we see that utilizing the geometry of the motion field globally, we can get a lot of information from only a part of the image: the part where we know that the vectors are only negative or only positive. Recall that to find the pattern of Fig. 9.11, we had to intersect the patterns of Figs. 9.9 and 9.10. At the intersection of positive and negative parts, the sign depends on the depth. It is only in these areas that the value along the longitudinal or latitudinal vectors can become zero. The distribution of the image points where the normal flow in some direction becomes zero has again a rich geometric structure containing egomotion information. The interested reader is referred to Fermüller and Aloimonos (1994).

Finally, based on the same basic constraints, a process for the detection of independent motion has been designed. Since the observer is moving rigidly, an area with a motion field not due only to one rigid motion must contain an independently moving object. The constraints are defined for the whole visual field, but the motion vectors in every part of the image plane must obey a certain structure.

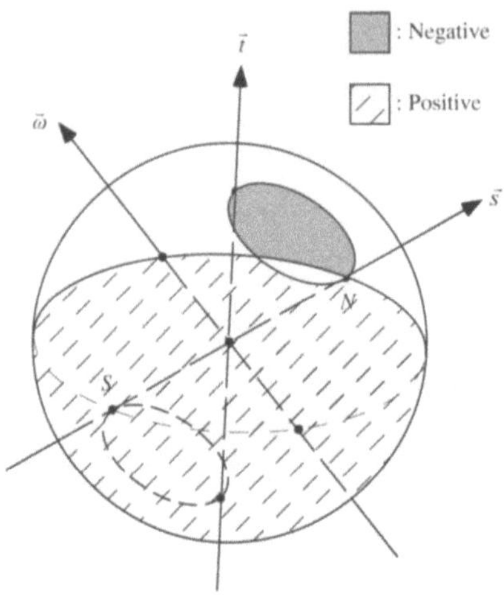

FIGURE 9.11. The projection of a rigid motion field on an $\vec{s}$ longitudinal pattern. The sphere is divided in two halves with the great circle of the plane defined by $\vec{\omega}$ and $\vec{s}$. There are also two curves (the ones of Fig. 9.9) passing from the points where $\vec{t}$, $\vec{s}$, $-\vec{t}$ and $-\vec{s}$ intersect the sphere. Whatever the motion $\vec{t}$ and $\vec{\omega}$ is, there exists a pattern of positive and negative longitudinal vectors in a part of the sphere. (The intersection of the negative parts of Figs. 9.9 and 9.10 provides the negative part and the intersection of the positive parts provides the positive.)

Our approach consists of comparing the motion field within image patches with prestored patterns (which represent all possible rigid motions).

By considering patches of different sizes and using various resolutions, the patterns may also be of use in estimating the motion of objects. Differently sized filters can first be employed to localize the object and then an appropriately sized filter can be used to estimate the motion. However, objects do not always move rigidly. Furthermore, in many cases the area covered by the object will not be large enough to provide sufficiently accurate information. In the general case, when estimating an object's motion, only local information can be employed. In such a case, we utilize the observer's capability to move in a controlled way. We describe the object's motion with regard to an object centered coordinate system. From fixation on a small area on the object the observer can derive information about the direction of the object's translation parallel to its image plane. By tracking the object over a small amount of time, the observer derives additional information about the translation perpendicular to the image plane. Combining the computed values allows the observer to derive the direction of an object's translation (Fermüller & Aloimonos, 1993).

9.4.5 *A look at the motion pathway*

There is a very large literature on the properties of neurons involved in motion analysis (e.g. Duffy & Wurtz, 1991; Maunsell & van Essen, 1983; Tanaka & Saito,

1989; Ungerleider & Desimone, 1986). The modules which have been found to be involved in the early stages of motion analysis are the retinal M-cells, the magnocellular neurons in the LGN, layers 4Cα, 4B, and 6 of V1, the thick bands of V2 and MT. These elements are referred to as the early motion pathway. Among others, they feed further motion processing modules (MST and FST) which in turn have connections to the parietal lobe. Here, we present a hypothesis, based on the computational model described earlier, about how motion is handled in the cortex.

Fig. 9.13 (from Movshon, 1990) shows an outline of the process to be explained which involves four kinds of cells with different properties. In the early stages, from the retinal M-cells through the magnocellular LGN cells, the cells appear functionally homogeneous and respond almost equally well to the movement of a bar (moving perpendicularly to its direction) in any direction (Fig. 9.13(A)). The receptive fields of some neurons in layer 4Cα of V1 are divided into separate excitatory and inhibitory regions (these are called "simple cells"). The regions are arranged in parallel stripes and this arrangement provides the neurons with a preference for a particular orientation of a bar target. In layer 4Cα, a subset of the cells also display directional selectivity to moving bars (which is displayed in the polar diagram, Fig. 9.13(B)). In layer 4B of V1 another major transformation takes place as we begin to see "complex cells", which respond to oriented bars at any location within the receptive field. In addition, these neurons respond better or solely to one direction of motion of an optimally oriented bar target, and less or not at all to the other (Fig. 9.13(C)). Finally, in MT neurons have large receptive fields and in general the precision of the selectivity for direction of motion that the neurons exhibit is typically less than in V1 (Fig. 9.13(D)).

One can easily envision an architecture that, using neurons with the properties listed above, implements a global decomposition of the normal motion field. Neurons of the first kind could be involved in the estimation of the local retinal motion perpendicular to the local edge (normal flow). Neurons at this stage could be thought of as computing whether the projection of retinal motion along some direction is positive or negative. Neurons of the second kind could be involved in the selection of local vectors in particular directions as parts of the various different patterns discussed in the previous section, while neurons of the third kind could be involved in computing the sign (positive or negative) of pattern vectors for areas in the image; that is, they might compute, for patches of different sizes, whether the normal flow in certain directions is positive or negative. Finally, neurons of the last kind could be the ones that piece together the parts of the patterns developed already into global patterns that are matched with prestored global patterns. Matches provide information about egomotion and mismatches provide information about independent motion.

In this architecture we are not concerned with neurons that possibly estimate the motion field (optic flow). This is not to say that optic flow is not estimated in the cortex; several neurons could be involved in approximating the motion field. However, if the cortex is capable of solving some motion problems without the use of optic flow, whose estimation amounts to the solution of an optimization problem, it is quite plausible to expect that it would prefer such a solution. After all, it is

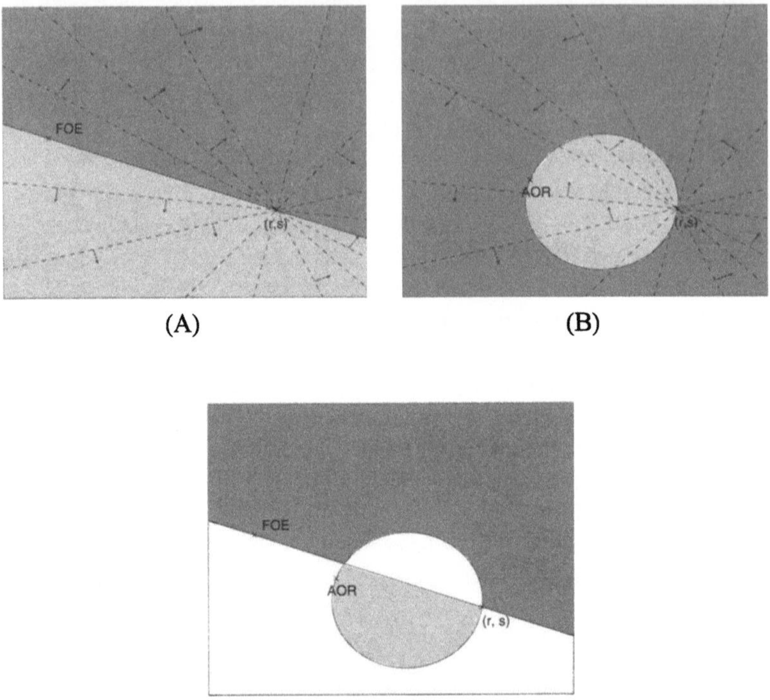

(A) (B)

(C)

FIGURE 9.12. (A) The translational (r, s) co-point vectors are separated by a line that passes through the FOE (the point which denotes the direction of translation); in one half-plane all vectors have positive values (light grey), in the other half-plane negative values (dark grey). (B) The rotational (r, s) co-point vectors are separated by a second order curve that passes through the AOR (the point where the rotation axis pierces the image plane). (C) A general rigid motion separates the (r, s) co-point vectors into an area of negative vectors, an area of positive vectors, and an area that may contain vectors of any value (white).

important to realize that at the low levels of processing the system must utilize very reliable data, such as the sign of the motion field along some direction. It is worth noting that after deriving egomotion from normal flow, information about 3-D motion is available, and the cortex could involve itself with approximating optic flow, because in this way the problem is not ill-posed any longer (at least for background scene points).

9.4.6 Form-based competences

Since computer vision was considered to be best approached through the construction of 3-D descriptions of the world, a lot of effort was spent on developing techniques for computing metric shape and depth descriptions from 2-D imagery. Studies concerned with this kind of work are collectively referred to as "shape

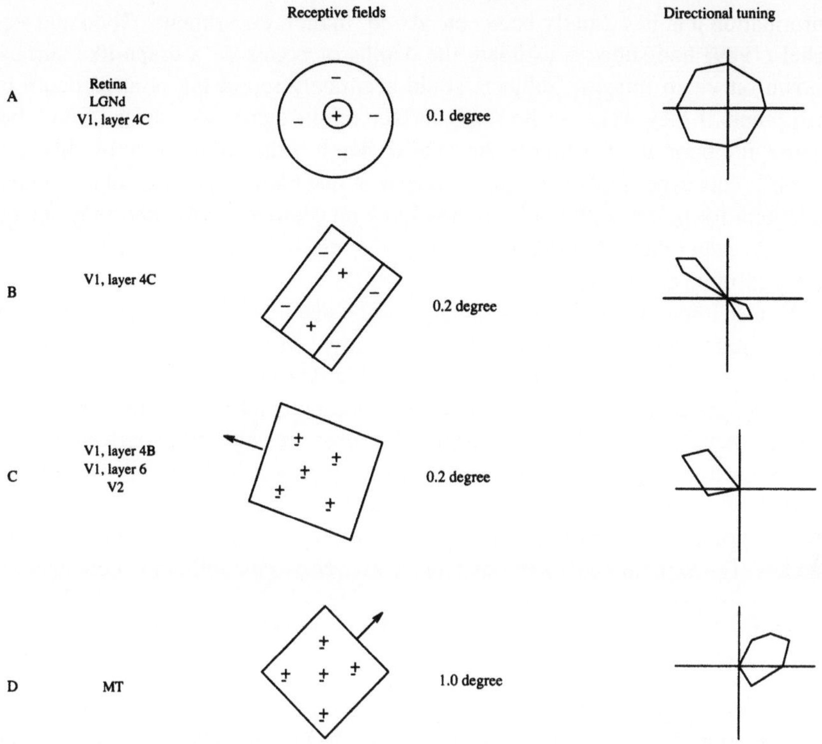

FIGURE 9.13. The spatial structure of visual receptive fields and their directional selectivity at different levels of the motion pathway (from Movshon, 1990). The spatial scales of the receptive fields (0.1 degree, etc.) listed here are for neurons at the center of gaze; in the periphery these dimensions would be larger. The polar diagrams illustrate responses to variation in the direction of a bar target oriented at right angles to its direction of motion. The angular coordinate in the polar diagram indicates the direction of motion and the radial coordinate the magnitude of the response.

from X" computations, where X refers to cues such as shading, texture, pattern, motion, or stereo. Exact, quantitative 3-D structure is hard to compute though, and explicit assumptions about the scene (smoothness, planarity, etc.) usually have to be made in the models employed.

Considering all the time that has been spent on the computation of metric shape and that has not yet given rise to any system working in a real environment, a glimpse at nature might give us some inspiration. Maybe it is a hopeless task to aim at deriving metric shape or depth information. Psychophysical experiments indicate that binocular stereopsis in the human visual system does not produce an explicit representation of the metric depth structure of the scene. Psychophysical evidence (Collett, Schwartz & Sobel, 1991; Johnston, 1991) suggests that human performance in tasks involving metric structure from binocular disparities is very poor. Also, other cues don't seem to allow humans to extract the kind of depth information that has usually been considered. In their experiments, Todd and Reichel (1989) had subjects estimate the depths of points on a drape-like surface shown on video images. Subjects could accurately report the relative depth of two points if they were on the same surface on the same side of the "fold", but were quite poor at determining the relative depth if the points were on different "folds". This experiment leads one to conclude that humans possess relative depth judgment for points within a local area lying on a surface. However, they cannot estimate even relative depth correctly for large distances in the visual field, when depth extrema are passed.

We also know that in humans the area of the eye in which detailed (high resolution) information can be extracted covers only a small region around the fovea (about 5 deg of visual angle at normal viewing distance). The low resolution at the periphery does not allow one to derive accurate depth information. However, human eyes are nearly always in motion. The eyes are engaged in performing fixations, each lasting about 250 msec. Between fixations, saccadic movements are carried out, during which no useful information is extracted.

The biological evidence gives us good reason to argue for alternative shape models. The experiments mentioned above give rise to the following conclusions:

(a) Shape or depth should not be computed in metric form, but only relative depth measurements (ordered depth) should be computed.

(b) Shape/depth information should be computed only locally. Then, the information derived for different patches has to be integrated. This integration, however, should not take place in the usual form, leading to complete, coherent spatial descriptions. The result should not be a complete reconstructed 3-D shape model, obtained by exact combination ("gluing") of the local shape representations to form a global one. Instead, we have to look for alternative representations that suffice for accessing the shape information one needs to solve particular tasks.

These or similar arguments also find support from computational considerations. Concerning argument (b), one might ask why one should compute only local information, if from a technical standpoint there is no difference whether the devised sensors have different or the same resolution everywhere. If stereo systems are used—the most obvious for deriving shape information—and the two cameras fixate a point, the disparity measurements are small only near the fixation point, and thus can also only be computed exactly there. In particular, if continuous techniques are employed to estimate the displacement (due to stereo or also due to motion), the assumption of continuity of the spatio-temporal imagery does not have to be greatly violated. The measurements which are due to rotation increase with the distance from the image center and the translational measurements are proportional to the distance from the epipole or the point denoting the direction of translation. Another argument is that computing shape only locally gives legitimacy to the orthographic projection model for approximating the image formation. The exact perspective projection model makes the computation of distance and shape very hard, since the depth component appears inversely in the image coordinates, which in turn leads to equations that are nonlinear in the unknown parameters.

However, concerning argument (a), we don't want to prescribe the computation of ordinal as opposed to metric shape information. Why should we limit ourselves to ordinal depth and not be even less restrictive? Throughout this chapter, we have argued for task-dependent descriptions. This also applies to the shape descriptions; a variety of shape descriptions subserving different tasks can be accepted. To derive metric depth or shape means to compute exact values of the distance between the camera and the scene. In order to solve, for example, the general structure from motion problem, theoretically we require at least three views of the scene, or two views and some additional information, such as the length of the baseline for a stereo setting. From two perspective views, only scaled distance, or distance up to a so-called relief transformation, can be derived. Assume that a depth computation returned values that were related to depth by an unknown monotonic function, for example of the form $f(Z) = (-a/Z) + b$, $f(Z) = aZ + b$, and so on, where a and b are unknown parameters. If the form of f is unknown, then the results of the computation effectively provide only ordinal depth information. We argue that one could try to compute even less informative depth or shape information by aiming at deriving more involved depth functions.

Under the influence of the reconstructionists' ideas, all effort in the past has been devoted to deriving metric measurements. A new look at the old research with a different goal in mind might give us new insights. From different cues, depth and shape information of different forms might be computed and then appropriately fused. A representation less complete than an ordinal one by itself does not seem to be sufficient for 3-D scene understanding. However, by combining two or more such representations, additional information can be obtained. It seems that the study of fusion of information for the purpose of deriving form and shape description will definitely be of importance.

It should be noted that whereas shape and depth measurements are equivalent for a metric 3-D representation, they are not for ordinal representations. Dealing with metric measurements, if absolute depth is given, shape (defined as the first order derivatives of depth) can be directly computed, and vice versa. The same, however, does not hold for ordinal, or even less informative representations.

Our goal is to derive qualitative, as opposed to quantitative, representations, because the computations to be performed should be robust. This requires that we not make unreasonable assumptions and employ computations that are ill-posed. Qualitativeness, for example, does not mean performing the same computations that have been performed under the reconstruction philosophy, making the same assumptions about the 3-D world, and at the end separating the computed values by a threshold in order to end up with "qualitative" information in the form of "greater or smaller than some value." Our effort should be devoted to deriving qualitative shape descriptions directly from well-defined input. For example, it wouldn't make sense to assume exact optical flow or stereo disparity measurements—which are impossible to obtain—to derive shape descriptions less powerful than the one of scaled depth because, if we had exact 2-D image measurements, we could compute scaled shape, and there is nothing we would gain computationally from computing less.

By concentrating on simpler shape descriptions, new mathematical models and new constraints might be found. Purely mathematical considerations can reveal what kind of information could possibly be computed from a certain input allowing a defined class of operations. The study of Koenderink and van Doorn (1991) on affine structure from motion might serve as an inspiration. They investigated a hierarchy of shape descriptions based on a stratification of geometries.

9.4.7 Spatial understanding

Since in the past the actions of the observer were not considered as an integral part of perceptual investigations, computational modeling, and in particular AI research, has dealt with space only at a symbolic level. For example, some early systems (Winston, 1975) dealt with the spatial relationship of objects in a blocks world. Assuming that objects can be recognized and thus can be stored as symbols, the spatial configuration of these objects under changing conditions was studied. Also, in existing studies on spatial planning (e.g., path planning), solutions to the problems of recognizing the objects and the environment are assumed to be available for the phase of coordinating motions.

Within the framework of behavioral vision a new meaning is given to the study of space perception. The understanding of the space surrounding an observer results from the actions and perceptions the observer performs and their relationship. For a static observer that does not act in any way, space does not have much relevance. But, to interact with its environment it has to have some knowledge about the space in which it lives, which it can acquire through actions and perceptions. Of course, the knowledge of space can be of different forms at various levels of complexity depending on the sophistication of the observer/actor and the tasks it has to perform. At one end of the scale, we find a capability as simple as obstacle avoidance, which in the most parsimonious form has to capture only the distance between the observer and points in the 3-D world, and at the other end of the scale, the competence of homing, which requires the actor to maintain some kind of map of its environment.

To obtain an understanding of space by visual means requires us to identify entities of the environment and also to localize their positions; thus both basic problems, the one of "where" and the one of "what" have to be addressed.

The problem of recognizing three-dimensional objects in space is by itself very difficult, since the object's appearance varies with the pose it has relative to the observer. In the computer vision literature two extreme views are taken about how to address the 3-D recognition problem, which differ in the nature of the models to be selected for the descriptions of objects in the 3-D environment. One view calls for object-centered models and the other for descriptions of the objects by means of viewer-centered views (3-D vs 2-D models). In most of the work on object-centered descriptions the form of objects is described with simple geometric 3-D models, such as polyhedra, quadrics, or superquadrics. Such models are suited to represent a small number of man-made (e.g., industrial) parts. However, to extend 3-D modeling to a larger range of objects will require models of more complex structural description, characterizing objects as systems and parts of relations. Recently a number of studies have been performed on viewer-centered descriptions approaching the problem from various directions. Here, we review a few of them. Based on some results in the literature of structure from motion, that show that under parallel projection any view of an object can be constructed as a linear combination of a small number of views of the same object, a series of studies on recognition using orthographic and paraperspective projections have been conducted (Ullman & Basri, 1991; Jacobs, 1992). The body of projective geometry has been investigated to prove results about the computation of structure and motion from a set of views under perspective projection (Faugeras, 1992). The learning of object recognition capabilities has been studied for neuronal networks using nodes that store viewer-centered projections (Poggio, Edelman & Fahle, 1992), and geometric studies on the so-called aspect graph have investigated how different kinds of geometric properties change with the views the observer has of the geometric model (Koenderink & van Doorn, 1979).

The problem of solving both localization and recognition is exactly the antagonistic conflict at the heart of pattern recognition. From the point of signal processing, it has been proved (Gabor, 1946) that any single (linear) operator can answer only one of these questions with high accuracy. In theory, thus, a number of processes are required to solve tasks related to space perception.

Results from the brain sciences reveal that the receptive field sizes of cells are much larger in the specialized visual areas involved in later processing than in those of the early stages. Many cells with large receptive field sizes respond equally well to stimuli at different positions. For example, in V5 cells with large receptive fields respond to spots of lights moved in certain directions, no matter where the stimulus occurs in the receptive field; nevertheless, the position of the light in the visual field can be localized accurately. Neurobiologists have suggested several solutions to this problem. The following interesting results deserve special mention. In the visual cortex cells have been found which are "gaze-locked," in the sense that they only respond to a certain stimulus if the subject is gazing in a particular direction. These cells probably respond to absolute positions in ego-centric space (Zeki, 1993).

It seems that nature has invented a number of ways for perceiving space through recognition and localization of objects in the 3-D world. Also, neurophysiological studies have been conducted that give good reason to assume that the perception of space in primates is not only grounded on object-centered or ego-centered descriptions, but that some descriptions are with regard to some action. For example, in an area called TEA, cells have been reported which are involved in the coding of hand movements (Perrett, Harries, Mistlin & Chitty, 1990). These cells respond when an action is directed towards a particular goal, but they do not respond to the component actions and motions when there is no causal connection between them. Monkeys were shown a video of hand movements and object movements contiguous or separated in space or time, for example, of a hand and a cup. The hand was retracted and after a short delay the cup moved (as if by itself) along the same trajectory as the hand. As the discrepancy between hand and object movement widened the impression of causality weakened. The above-mentioned cells tuned to hand actions were found to be less responsive when the movement of the hand and the object were spatially separated and appeared not to be causally related.

Humans possess a remarkable capability in recognizing situations, scenes, and objects in the space surrounding them from actions being performed. In the computer vision literature a number of experiments (Johansson, 1973) are often cited in which it has been shown that humans can recognize specific animals and humans that move in the dark and are visible only from a set of light bulbs attached to their joints. These experiments demonstrate very well the power of motion cues. Since actions give rise to recognition, and actions are largely understood from motions, it seems worthwhile to investigate further motion models, more complicated than the rigid one, to describe actions. For example, situations occurring in manipulation tasks might be modeled through nonrigid motion fields. The change of the motion field or parts of it may be expressed in the form of space-time descriptions that can be related to the tasks to be performed. It should be mentioned that recently some effort along this line has started. A few studies have been conducted exploiting motion cues for recognition tasks. In particular, periodic movements, such as the motion of certain animal species, have been characterized in frequency space (Nelson & Polana, 1992; Shavit & Jepson, 1993). Statistical pattern recognition techniques have been applied in the time-domain to model highly structured motions occurring in nature, such as the motions of flowing water or fluttering leaves (Polana & Nelson, 1993). Attempts have been made to model walking or running humans by describing the motion of single limbs rigidly (Qian & Huang, 1992), and also various deformable spatial models like superquadrics and snakes have been utilized to model nonrigid motions of rigid bodies, for the purpose of face recognition (Pentland, Horowitz & Sclaroff, 1991).

Representations used for understanding space should be allowed to be of any of three kinds: with regard to the viewer, with regard to an object, or action-driven. An appropriate representation might allow us to solve tasks straightforwardly that would require very elaborate computations and descriptions otherwise. Perrett, Mistlin and Chitty (1988) give a good example underpinning this point of view: A choreographer could, for example, use a set of instructions centered on the different dancers (such as to dancer M. who is currently lying prostrate and oriented toward the front of the stage: "Raise head slowly" and to dancer G., currently at the rear of the stage facing stage left: "Turn head to look over left shoulder"). Alternatively the choreographer could give a single instruction to all members of the dance troupe ("Move the head slowly to face the audience"). To allow for the choice of different systems of representation will be a necessity when studying spatial descriptions. These descriptions, however, must be related in some form. After all, all measurements are taken in a frame fixed to the observer's eye. Thus, a great deal of work in spatial understanding will amount to combining different representations into an ego-centered one.

The competence of homing is considered to be the apogee of spatial behavior. The amazing homing capabilities of some animals have attracted the attention of researchers for many years. Effort has been spent on investigating the sensory basis of animals' perception. Discoveries have been made of sensory guidance by sunlight, light patterns in the sky, and moonlight, such as the use of ultraviolet light by ants (Lubbock, 1889) and polarized light by bees (Frisch, 1949). More recent research concerns how particular species organize the spatial information acquired through their motor sequences and sensors (Sandini, Gandolfo, Grosso & Tistarelli, 1993; Srinivasan, Lehrer, Zhang & Horridge, 1989).

Zoologists differentiate between two mechanisms for acquiring orientation: the use of ego-centered and geo-centered systems of reference. Simple animals, like most arthropods, represent spatial information in the form of positional information obtained by some kind of route integration relative to their homes. The route consists of path segments each of which takes the animal for a given distance in a given direction. This form of representation related to one point of reference is referred to as an ego-centered representation.[2] More complicated than relying only on information collected en route is the use of geo-centered reference systems where the animal in addition relies on information collected on site (recognition of landmarks) and where it organizes spatial information in a map-based form.

However, research from studies on arthropods (Wehner, 1992; Collett, Dillmann, Giger & Wehner, 1992; Collett, Fry & Wehner, 1993) shows that already in these simple animals, the competence of homing is realized in seemingly any possible way. A large variety of different ways employing combinations of information from action and perception have been discovered. The way the path is stored, the way landmarks are recognized, etc., is different for every species. Not many general concepts can be derived. It seems that the physical realizations are tightly linked to the animal's physiology and overall performance. This has to apply to artificial systems as well. Computations and implementations cannot be separated. Obviously, the more storage capability a system has, the more complex operations it can perform. The number of classes of landmarks that a system can differentiate and the number of actions it can perform will determine the homing capability of a system. Our suggested strategy is to address competences involving spatial representations (and in particular the homing competence) by synthesizing systems with increasing action and perception capabilities and study the performance of these systems, considering constraints on their memory.

9.5 Conclusions

The study of vision systems in a behavioral framework requires the modeling of observer and world in a synergistic way and the analysis of the interrelationship of

---

[2]In the computer vision literature the term *ego-centered* reference system is used with a different meaning than in zoology.

action and perception. The role that vision plays in a system that interacts with its environment can be considered as the extraction of representations of the space-time in which the system exists and the establishing of relations between these representations and the system's actions. We have defined a vision system as consisting of a number of representations and processes, or on a more abstract level, as a set of maps which can be classified into three categories: the visual competences that map different representations of space-time (including the retinotopic ones) to each other, the action routines which map space-time representations to motor commands or representations of various kinds residing in memory, and the learning programs that are responsible for the development of any map. To design or analyze a vision system amounts to understanding the mappings involved. In this paper we have provided a framework for developing vision systems in a synthetic manner, and have discussed a number of problems concerning the development of competences, learning routines and the integration of action and perception. We have also described some of our technical work on the development of specific motion-related competences.

To achieve an understanding of vision will require efforts from various disciplines. We have described in this study work from a number of sciences, computational as well as empirical ones. Besides these, the general area of information processing has various fields of study from which the design and analysis of vision systems can benefit. Some studies of possible interest include the realization of specific maps in hardware (VLSI chips or optical computing elements); the study of the complexity of visual tasks under the new framework; information-theoretic studies investigating the relationship between memory and task-specific perceptual information; and the study of control mechanisms for behavioral systems.

Acknowledgments: The support of ONR, NSF, ARPA, and the Austrian "Fonds zur Förderung der wissenschaftlichen Forschung" project No. S7003 is gratefully acknowledged.

9.6 References

Aloimonos, Y. (1990). Purposive and qualitative active vision. In *Proceedings of the DARPA Image Understanding Workshop* (pp. 816–828). Mountain View, California: Morgan Kaufmann.

Aloimonos, Y. (Ed.). (1993). *Active Perception*, Volume 1 of *Advances in Computer Vision*. Hillsdale, New Jersey: Lawrence Erlbaum.

Aloimonos, Y. & Brown, C. M. (1984). Direct processing of curvilinear sensor motion from a sequence of perspective images. In *Proceedings of the Workshop on Computer Vision: Representation and Control* (pp. 72–77). Annapolis, Maryland: IEEE.

Aloimonos, Y. & Shulman, D. (1989). *Integration of Visual Modules: An Extension of the Marr Paradigm*. Boston: Academic Press.

Aloimonos, Y., Weiss, I. & Bandopadhay, A. (1988). Active vision. *International Journal of Computer Vision*, *2*, 333–356.

Bajcsy, R. (1988). Active perception. *Proceedings of the IEEE, 76*, 996–1005.

Ballard, D. H. (1991). Animate vision. *Artificial Intelligence, 48*, 57–86.

Boussaud, D., Ungerleider, L. G. & Desimone, R. (1990). Pathways for motion analysis: cortical connections of the medial superior temporal fundus of the superior temporal visual areas in the macaque monkey. *The Journal of Comparative Neurology, 296*, 462–495.

Brady, M., Hollerbach, J. M., Johnson, T. L., Lozano-Perez, T. & Mason, M. T. (Eds.). (1983). *Robot Motion*. Cambridge, Massachusetts: MIT Press.

Brooks, R. A. (1986). A robust layered control system for a mobile robot. *IEEE Journal of Robotics and Automation, 2*, 14–23.

Brunswik, E. (1956). *Perception and the Representative Design of Psychological Experiments*. Berkeley, California: University of California Press.

Collett, T. S., Dillmann, E., Giger, A. & Wehner, R. (1992). Visual landmarks and route following in desert ants. *Journal of Comparative Physiology A, 170*, 435–442.

Collett, T. S., Fry, S. N. & Wehner, R. (1993). Sequence learning by honeybees. *Journal of Comparative Physiology A, 172*, 693–706.

Collett, T. S., Schwartz, U. & Sobel, E. C. (1991). The interaction of occulomotor cues and stimulus size in stereoscopic depth constancy. *Perception, 20*, 733–754.

Duffy, C. J. & Wurtz, R. H. (1991). Sensitivity of MST neurons to optical flow stimuli I: a continuum of response selectivity to large field stimuli. *Journal of Neurophysiology, 65*, 1329–1345.

Ernst, G. W. & Newell, A. (1969). *GPS: A Case Study in Generality and Problem Solving*. New York: Academic Press.

Faugeras, O. (1992). *Three Dimensional Computer Vision*. Cambridge, Massachusetts: MIT Press.

Fermüller, C. (1993). Navigational preliminaries. In Y. Aloimonos (Ed.), *Active Perception*, Volume 1 of *Advances in Computer Vision* (pp. 103–150). Hillsdale, New Jersey: Lawrence Erlbaum.

Fermüller, C. & Aloimonos, Y. (1992). Tracking facilitates 3-d motion estimation. *Biological Cybernetics, 67*, 259–268.

Fermüller, C. & Aloimonos, Y. (1993). The role of fixation in visual motion analysis. *International Journal of Computer Vision: Special issue on Active Vision, M. Swain (Ed.), 11(2)*, 165–186.

Fermüller, C. & Aloimonos, Y. (1994). On the geometry of visual correspondence. Technical Report CAR-TR, Center for Automation Research, University of Maryland.

Frisch, K. (1949). Die Polarisation des Himmelslichts als orientierender Faktor bei den Tänzen der Bienen. *Experientia, 5*, 142–148.

Gabor, D. (1946). Theory of communication. *Journal of IEE, 93, Part III*, 429–457.

Gelernter, H. (1959). Realization of a geometry theorem-proving machine. In *Information Processing: Proceedings of the International Conference on Information Processing* (pp. 273–282). Paris: UNESCO House.

Gibson, J. J. (1979). *The Ecological Approach to Visual Perception*. Boston: Houghton Mifflin.

Goodale, M. A., Milner, A. D., Jacobson, L. S. & Carey, D. P. (1991). A neurological dissociation between perceiving objects and grasping them. *Nature*, *349*, 154–156.

von Helmholtz, H. L. F. (1866). *Handbuch der Physiologischen Optik*, Volume 2. Hamburg-Leipzig: Voss.

Hervé, J. Y. (1993). *Navigational Vision*. PhD thesis, University of Maryland, Computer Vision Laboratory, Center for Automation Research.

Horn, B. K. P. (1986). *Robot Vision*. New York: McGraw Hill.

Horn, B. K. P. & Weldon, E. J. (1987). Computationally efficient methods for recovering translational motion. In *Proceedings of the 1st International Conference on Computer Vision* (pp. 2–11). London: IEEE.

Horridge, G. A. (1987). The evolution of visual processing and the construction of seeing systems. *Proceedings of the Royal Society, London B*, *230*, 279–292.

Horridge, G. A. (1991). Evolution of visual processing. In J. R. Cronly-Dillon & R. L. Gregory (Eds.), *Vision and Visual Dysfunction*, Volume 230 (pp. 279–292). New York: MacMillan.

Hubel, D. H. & Wiesel, T. N. (1968). Receptive fields and functional architecture of the monkey striate cortex. *Journal of Physiology*, *195*, 215–243.

Jacobs, D. W. (1992). Space efficient 3D model indexing. In *Proceedings of the IEEE Conference on Computer Vision and Pattern Recognition* (pp. 439–444). Annapolis, Maryland: IEEE.

Johansson, G. (1973). Visual perception of biological motion and a model for its analysis. *Perception and Psychophysics*, *14*, 201–211.

Johnston, E. B. (1991). Systematic distortions of shape from stereopsis. *Vision Research*, *31*, 1351–1360.

Kanizsa, G. (1979). *Organization in Vision: Essays on Gestalt Perception*. New York: Praeger.

Kant, I. (1786/1973). *Critique of Pure Reason*. London, England: MacMillan. Translated by N. K. Smith.

Koenderink, J. J. & van Doorn, A. J. (1979). The internal representation of solid shape with respect to vision. *Biological Cybernetics*, *32*, 211–216.

Koenderink, J. J. & van Doorn, A. J. (1991). Affine structure from motion. *Journal of the Optical Society of America A*, *8*, 377–385.

Köhler, W. (1947). *Gestalt Psychology*. New York: Liveright.

Lubbock, J. (1889). *On the Senses, Instincts, and Intelligence of Animals with Special Reference to Insects*. London: K. Paul Trench.

Marr, D. (1982). *Vision*. San Francisco: W. H. Freeman.

Maunsell, J. H. R. & van Essen, D. C. (1983). Functional properties of neurons in middle temporal visual area of the macaque monkey I. Selectivity for stimulus direction, speed and orientation. *Journal of Neurophysiology*, *49*, 1127–1147.

Movshon, J. A. (1990). Visual processing of moving images. In H. Barlow, C. Blakemore & M. Weston-Smith (Eds.), *Images and Understanding* (pp. 122–137). Cambridge, England: Cambridge University Press.

Nalwa, V. (1993). *A Guided Tour of Computer Vision*. Reading, Massachusetts: Addison-Wesley.

Nelson, R. C. (1991). Qualitative detection of motion by a moving observer. *International Journal of Computer Vision, 7*, 33–46.

Nelson, R. C. & Polana, R. (1992). Qualitative recognition of motion using temporal texture. *CVGIP: Image Understanding, 1*, 33–46. Special Issue on Purposive, Qualitative, Active Vision, Y. Aloimonos (Ed.).

Nilsson, N. J. (1980). *Principles of Artificial Intelligence*. Palo Alto, California: Tioga Publishing Co.

Orban, G. A. (1992). The analysis of motion signals and the nature of processing in the primate visual system. In G. A. Orban & H.-H. Nagel (Eds.), *Artificial and Biological Vision Systems*, ESPRIT Basic Research Series (pp. 24–57). Heidelberg, Germany: Springer-Verlag.

Pentland, A. P. (Ed.). (1986). *From Pixels to Predicates: Recent Advances in Computational and Robot Vision*. Norwood, New Jersey: Ablex.

Pentland, A. P., Horowitz, B. & Sclaroff, S. (1991). Non-rigid motion and structure from contour. In *Proceedings of the IEEE Workshop on Visual Motion* (pp. 288–293). Annapolis, Maryland: IEEE.

Perrett, D. I., Harries, M., Mistlin, A. J. & Chitty, A. J. (1990). Three stages in the classification of body movements by visual neurons. In H. Barlow, C. Blakemore & M. Weston-Smith (Eds.), *Images and Understanding* (pp. 94–107). Cambridge, England: Cambridge University Press.

Perrett, D. I., Mistlin, A. J., Harries, M. & Chitty, A. J. (1988). *Vision and Action: The Control of Grasping*. Norwood, New Jersey: Ablex Pub.

Poggio, T., Edelman, S. & Fahle, M. (1992). Learning of visual modules from examples: A framework for understanding adaptive visual performance. *CVGIP: Image Understanding, 56*, 22–30. Special Issue on Purposive, Qualitative, Active Vision, Y. Aloimonos (Ed.).

Polana, R. & Nelson, R. C. (1993). Detecting activities. In *Proceedings of the IEEE Image Understanding Workshop* (pp. 569–574). Mountain View, California: Morgan Kaufmann.

Qian, R. J. & Huang, T. S. (1992). Motion analysis of articulated objects. In *Proc. International Conference on Pattern Recognition* (pp. 220–223). Annapolis, Maryland: IEEE.

Sandini, G., Gandolfo, F., Grosso, E. & Tistarelli, M. (1993). Vision during action. In Y. Aloimonos (Ed.), *Active Perception* (pp. 151–190). Hillsdale, New Jersey: Lawrence Erlbaum.

Shavit, E. & Jepson, A. (1993). Motion using qualitative dynamics. In *Proceedings of the IEEE Workshop on Qualitative Vision* (pp. 82–88). Annapolis, Maryland: IEEE.

Sommer, G. (1994). Architektur und Funktion visueller Systeme. Künstliche Intelligenz, 12. Frühjahrsschule.

Srinivasan, M. V., Lehrer, M., Zhang, S. W. & Horridge, G. A. (1989). How honeybees measure their distance from objects of unknown size. *Journal of Comparative Physiology A, 165*, 605–613.

Tanaka, K. & Saito, H. A. (1989). Analysis of motion of the visual field by direction, expansion/contraction, and rotation cells illustrated in the dorsal part of the Medial Superior Temporal area of the macaque monkey. *Journal of Neurophysiology, 62*, 626–641.

Thompson, W. B. & Pong, T.-C. (1990). Detecting moving objects. *International Journal of Computer Vision, 4*, 39–57.

Todd, J. T. & Reichel, F. D. (1989). Ordinal structure in the visual perception and cognition of smoothly curved surfaces. *Psychological Review, 96*, 643–657.

Ullman, S. (1979). *The Interpretation of Visual Motion*. Cambridge, Massachusetts: MIT Press.

Ullman, S. & Basri, R. (1991). Recognition by linear combination of models. *IEEE Transactions on Pattern Analysis and Machine Intelligence, 13*, 992–1006.

Ungerleider, L. G. & Desimone, R. (1986). Cortical connections of visual area MT in the macaque. *The Journal of Comparative Neurology, 248*, 190–222.

Ungerleider, L. G. & Mishkin, M. (1982). Two cortical visual systems. In D. J. Ingle, M. A. Goodale & R. J. W. Mansfield (Eds.), *Analysis of Visual Behavior* (pp. 549–586). Cambridge, Massachusetts: MIT Press.

Warrington, E. & Shallice, T. (1984). Category specific semantic impairments. *Brain, 107*, 829–854.

Wehner, R. (1992). Homing in arthropods. In F. Papi (Ed.), *Animal Homing* (pp. 45–144). London: Chapman and Hall.

Winston, P. H. (1975). Learning structural descriptions from examples. In P. H. Winston (Ed.), *The Psychology of Computer Vision* (pp. 157–205). New York: McGraw-Hill.

Zeki, S. (1993). *A Vision of the Brain*. Boston, Massachusetts: Blackwell Scientific Publications.

10

A Framework for Information Assimilation

Arun Katkere[1]
Ramesh Jain[1]

ABSTRACT Most complex intelligent systems deal with information from different sources which are "disparate". A powerful paradigm of assimilation of information from disparate sources is needed in these systems. We posit that to do this seamlessly and efficiently, we should represent and use the "state" of the system (the Environment Model). Also, to handle differences in resolution and data rates, and to provide flexibility and dynamic reconfiguration, we should assimilate these different information sources independently into the model. Interaction between the information streams, if any, is indirectly achieved using the Environment Model. In this paper we will develop a formal framework (and design methodology) for developing and evaluating systems dealing with multiple information sources. Design of such a system will involve building an Environment Model that matches the task with the input information, independent assimilation tracks for these sources, methods of evaluating "usefulness" of input information and methods of updating the Environment Model. We will discuss this framework using robot navigation as an example.

10.1 Introduction

Many complex systems deal with information from multiple sources. An autonomous robot, like its human counterpart, needs to interact with its environment using multiple sensors and actuators. A system collecting multi-modal information from information servers for presentation to the user should combine multiple sources of information. A system guiding a pilot should be able to dynamically assimilate information from on board sensors and communication. One common feature in all these systems, varying in degrees of autonomy, is the inherent *disparateness* of information sources. For example, in mobile robot applications input information can include point range data from sonar, $2\frac{1}{2}$−D depth maps from stereo and laser range-finders, motion and object location data from monocular cameras, self position information from GPS and landmark recognition, symbolic scene information from a human operator and *a priori* information about objects in the scene from a map. All this information, although unregistered and unsynchro-

[1]Visual Computing Laboratory, University of California, San Diego

nized, is complementary and describes a single environment and therefore is all necessary for navigation tasks. The disparate nature of this information, however, makes it difficult to "combine" different information streams at the data level. In this paper we propose a novel scheme to assimilate "relevant" information into an Environment Model and make available correct information at the correct time.

Assimilation of information from disparate sources has not received enough attention. Much of the previous work in perception for complex systems has focused on single or similar sensors. Our framework takes advantage of the strengths of these perception systems and handles disparate information. Therefore, it overcomes the inherent weaknesses of such systems. In our approach, we focus on relevant information content in each information stream rather than on the raw data. We use this to assimilate each stream directly into a single representation: the Environment Model. The synergy between disparate sources occurs through their common information content.

We will discuss the process of assimilation and present a formal framework in Section 10.2. This formal framework is based on a layered *Environment Model* to represent the state of an agent's perception, cognition and action. The assimilation of information into the Environment Model in this framework is done in an exploratory fashion using expectations generated using the existing state information. An autonomous robotics application provides us with enough mechanisms and an experimental environment to implement and test our framework. In Section 10.3 we will describe an autonomous *gofer* application and discuss our framework in that context.

10.2 Information Assimilation: Formal Framework

A significant body of work in robotic perception has concentrated on iterative interpretation of sensory information. This has produced some excellent formulations for assimilation of individual sensory sources into the model. But for solving most natural problems, we need to process and assimilate information from different information sources which may be disparate.

The advantages of using multiple senses in perception are well known. Often, cues provided by one sensor help in the interpretation of information from another sensor.

Redundancy and overlap between sensors increase the robustness of perception. Deficiencies of one sensory source may be avoided by use of another sensor which does not have such deficiency. An illustration of the naturalness of using multiple senses in literature is the use of aural metaphors to describe visual events. Marks (1978), in his treatise on the unity of senses, notes the interplay between senses (synesthesia) in human perception and cognition, and the importance of using multiple senses in perception[2]. The answer to the problem of assimilating

[2]For example, Kipling's use of the simile "the dawn comes up like thunder" (1898).

information from disparate sources lies in understanding how this apparent interchangeability of the senses occurs. We believe that this is due to the presence of a model containing data-independent representations reflecting the "information content".

Research in sensor fusion has focused on *fusion* of information from multiple sensors (Aggarwal & Nandhakumar, 1990; Aloimonos, Weiss & Bandopadhay, 1988; Landy, Maloney, Johnston & Young, 1995; Maloney & Landy, 1989, etc.). While these systems provide excellent formulations for fusion of information from *similar*, *registered* sensors, they do not address assimilation of information from "disparate" sources. We believe that there is a strong need for perception systems that can handle disparate information sources easily. In the next section, we will formally compare sensor fusion based approaches with our approach (Roth & Jain, 1994).

10.2.1 Perceptual cycle

Neisser argues that perception is guided by explorations of the environment guided by anticipatory schemata (Neisser, 1976). In the perceptual cycle, proposed by Neisser and shown in Fig. 10.1, an agent continuously modifies the model (schema) of the scene or the environment (real world). In this formulation, perception occurs exclusively via verifications of expectations generated by our beliefs in the state of our environment. Schemata in Neisser's theory are active elements that seek information to confirm their model of the world. They create goals for sensor activation and direction.

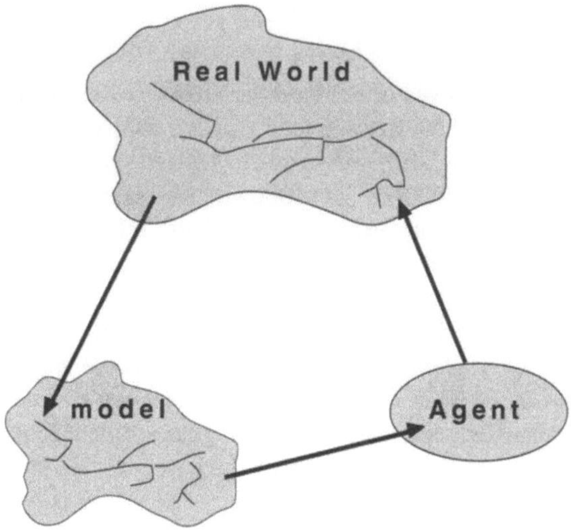

FIGURE 10.1. Neisser's perceptual cycle. Internal schemata are used to generate expectations about what is perceived. This guides exploration, the outcome of which modifies the original schema.

The above formulation emphasizes two important ideas: use of a *model* (i.e., schema) as a representation of state and for guiding exploratory perception, and use of a predict-verify model of assimilation. These ideas are central to our paradigm of information assimilation. The role of predictions of expected sensory input has been suggested by researchers in robotics (Albus, McGain & Lumia, 1987). Similarly, researchers in active vision emphasized the use of active exploration in simplifying perceptual analysis through the known control of a sensing device (Aloimonos et al., 1988; Andress & Kak, 1987; Bajcsy, 1988; Ballard, 1987).

10.2.2 Sensor fusion and information assimilation

In information assimilation, the focus is on the physical world being modeled, and the sensory information is just a means to this end. Traditional sensor integration/fusion approaches (for example, see Durrant-Whyte, 1988; Henderson, Weitz, Hansen & Mitiche, 1988; Mitiche & Aggarwal, 1986; Richardson & Marsh, 1988; Shafer, Stentz & Thorpe, 1986) combine sensory data based on general probabilistic weights associated with each sensing device or mode, rather than use the existing state of the system to estimate the instantaneous model features. To clarify, let us characterize the problem of sensor integration in the following traditional way: let x represent an m-dimensional state vector, and let y be an n-dimensional vector of measured quantities (the sensed data). The relation between x and y can be defined as

$$y = p(x) + \nu$$

where ν is a random n-D vector representing errors, noise, etc. The function $p()$ represents a projection of the state vector x onto the observation space y. For multiple observations y_i (e.g., different sensors, different geometry, etc), there will be different projection functions $p_i()$ following the same type of equation. The integration/fusion approaches explore probabilistic relations that usually relate to the reliability associated with a particular sensor and a particular observable feature. Regardless of the techniques employed (Bayesian probabilities, belief functions and evidence theory, fuzzy sets and possibility theory), the existing (current) state of the system is completely ignored. These approaches place the sensory analysis as the main goal of the system, ignoring the role for which this interpretation is intended, and therefore ignoring the instantaneous state of the system. Under the integration/fusion approach the goal is to find an inverse function $g(y_1, y_2, \ldots, y_i) = x$. In essence the function $g()$ is oblivious to the current state of the system (operates entirely in the *discovery* mode). Ignoring the current state of the system inhibits the system's continuity.

Fig. 10.2 compares the sensor integration approach with our assimilation approach, and shows how disparate information may be assimilated into one model. Note the ease of adding and removing sources in Fig. 10.2(B).

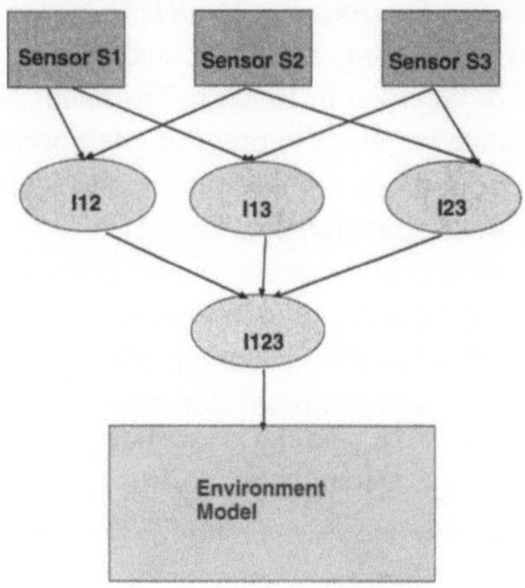

A. Integration

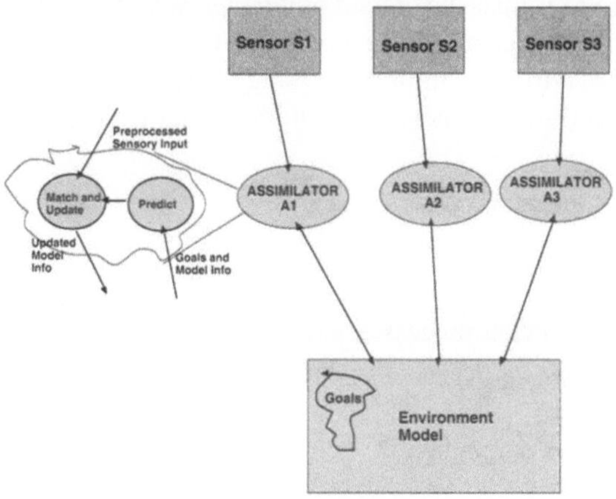

B. Assimilation

FIGURE 10.2. Sensor integration versus information assimilation. (A) In sensor integration, the emphasis is on combining features in 2-D images to refine the 2-D description of the real (3-D) world. (B) In information assimilation, the Environment Model(EM) plays a central role in combining different kinds of information coming from different sources. Only features that are important, based on previous processing and current goals of the agent, are predicted in each source's information space, and the "error signal" produced by verification is used for assimilation into the EM.

10.2.3 Environment Model

The concept of state plays a central role in the analysis and design of dynamic systems and digital computers. The information in the state of the system is determined by the task of the system. In the context of mobile robot applications, we have proposed and used *Environment Models* (EMs) to represent the state of an agent that combines perception, cognition, and action (Jain, 1989; Khalili & Jain, 1991; Roth-Tabak & Weymouth, 1990a,b; Tirumalai, Schunck & Jain, 1990). The EM is like a set of state variables in providing temporal continuity to the system, but is very different in its nature and representation due to the presence of information at both symbolic and signal levels.

To assimilate information from disparate sources of information and present this information to the controlling entity, it is important to have an EM that stores information at multiple levels of abstraction and multiple resolutions. Such a model requires indexing in spatial and temporal dimensions, and facilities to manage uncertainties and ambiguities. This model is the heart of our assimilation and synchronization system; it is responsible for interaction among different components, for providing temporal coherence, for combining information from multiple information sources, and for purposive behavior of the system (Jain, 1989). The organization of environment and world models for assimilation of information, which we have developed in the context of mobile robot systems, addresses many of these requirements and features.

Formally, an EM is made up of a set of *interdependent* objects $\mathcal{O}_i(t)$. The set of values of these objects at any instant comprises the state of the system $\mathcal{S}(t)$. Each object has several attributes, most basic being the *confidence* in its existence. The value of an object $\mathcal{O}_i(t)$, and hence the state $\mathcal{S}(t)$, may change due to the following factors:

- New input information (assimilation),
- Change in related model information (propagation),
- Advice from higher processes, and
- Decay (due to aging).

The set of object types is finite, the rules for transformation (formalized in subsequent sections) are finite. This can be characterized by a finite nondeterministic automaton.

An EM consists of objects at different levels of abstraction. There is a strong correlation between objects at different abstractions. To ensure consistency, we should propagate any changes that occur at one level to other levels (higher and lower) or tag this apparent inconsistency for future updating. In general, propagation from higher to lower levels of abstractions is easier than vice versa. Conversion from higher to lower levels of abstraction is usually done using techniques like rendering which are usually well defined. For example, conversion from an environmental object (defined using a CAD structure) to a set of filled voxels is simple. The opposite operation in this case, determining whether a set of voxels belong to an "object", is difficult and cannot be done without some form of expectations and domain specific assumptions. Hence, we need to attempt to assimilate at as high a level of abstraction as possible.

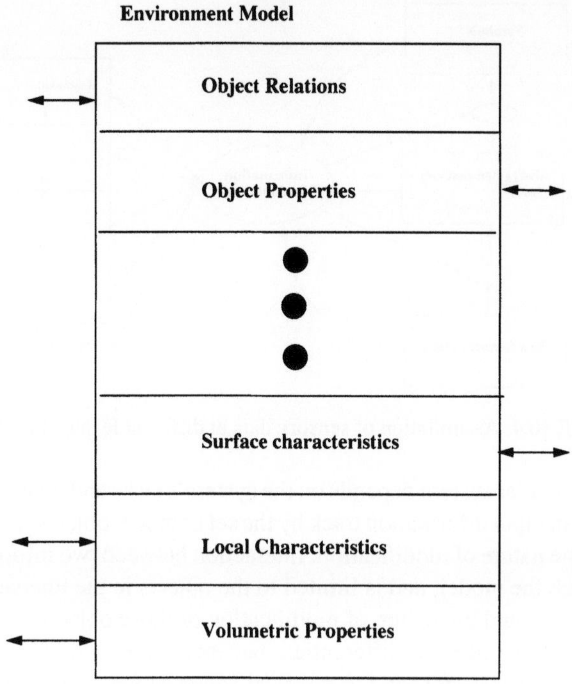

FIGURE 10.3. Different layers of the Environment Model.

The lowest level of the Environment Model is a discretized grid/voxel representation. Each of these *grid objects* has a membership set: confidence that it belongs to one of the composite objects at a higher level of abstraction (e.g., road edge vs. road). *A priori* data at higher levels (like map data) are converted to information at this level during the initialization of the system. Direct information acquisition at higher levels must be followed by conversion of that information to the information at the densest level, so that information at all levels is consistent. It is important to come up with efficient access (and update) strategies at this level since this could potentially be the bottleneck of the entire representation/assimilation module. Fig. 10.3 shows some of the possible layers of the Environment Model, and how each layer communicates independently with other modules. It should be noted that this approach is similar to the blackboard model of information processing (Nii, 1986). However, there is more structure to conversion between levels of abstraction and input/output than the usually unstructured blackboard systems.

Fig. 10.4 shows how information from sensors is assimilated into different layers of the Environment Model.

10.2.4 *Input information tracks*

In our paradigm, we handle each input information track independently. Each track is characterized by the *information content* of the track. Information content of a

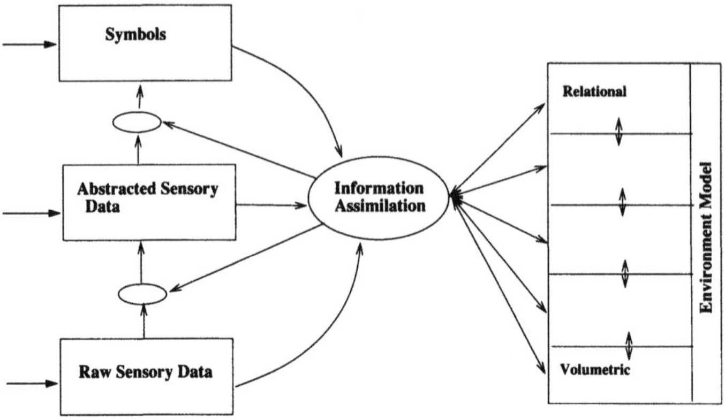

FIGURE 10.4. Assimilation of sensory data at different levels of abstraction.

track for a particular system depends on the system's tasks and goals. Thus, we can characterize an input information track by the set of model objects it could possibly modify and the nature of modification. Interaction between two information tracks occurs through the model, and is limited to the objects in the intersection of their modification sets and the nature of modification of those objects.

At this point, we need to differentiate between data *sources* and information *tracks*. Data *sources* correspond to physical sources of data like a camera which generates image data. These image data may be used in different information tracks like *roadfollowing*, *obstacle avoidance*, and so on, which independently assimilate different types of model information. In this paper, we concentrate on modeling these information tracks and processing in these tracks. The correspondence between a track and corresponding physical data source is not very important for the assimilation process.

10.2.5 Task modeling

The relative importance of the content of the input streams and model components should be determined in light of the tasks at hand. We need to model the tasks and goals of the system and use these to assimilate only relevant information into the model. The task model should include mechanisms for evaluation of relevance. Goals which are a means of accomplishing the tasks are useful in evaluating the relevance of model components and input information.

10.2.6 Information assimilation

Information integration or data fusion approaches, used extensively to combine data from multiple "sensors" in robotic systems, are strong in providing a solid mathematical framework for combining information, but are very weak in combining information from disparate sources (like symbolic sources, signal based sources, etc.). We believe that the information assimilation approach allows combination of information from disparate sources because different levels of abstractions are easily developed in this approach. Two important characteristics of this approach are independent assimilation of different tracks of media information into a common, coherent model, and the use of a prediction-verification mode of processing instead of a strictly bottom-up data-driven approach. The former ensures that the assimilation system is scalable, modifiable and degrades gracefully, since each assimilation track can be plugged in or unplugged without much effort. The latter eliminates redundant processing by making use of relationships between features in successive frames/samples. An important assumption made during assimilation is that an approximate model is available at the start. This is a reasonable assumption in most real systems.

10.2.7 Knowledge caching for assimilation

One of the greatest problems in most large knowledge-based systems is that the large size of the knowledge base causes a crippling effect on the understanding system. In our case, however, at any point in time only a small set of these assumptions will be relevant to immediate assimilation. We can add additional assumptions about the structure of environments to the knowledge base that media information deal with. These assumptions allow us to dynamically hide large portions of knowledge that are irrelevant at a given time. We call this approach *Knowledge Caching*. We have implemented this approach (*Context-based Caching*) in indoor mobile environments (Roth & Jain, 1991, 1992; Roth, Zhang & Jain, 1992) in which knowledge items are swapped based on precompiled relations between knowledge items. Roth and Jain (1994) discuss knowledge caching in a formal framework, and describe a *context-based caching* implementation.

10.3 Example Application: Autonomous Outdoor Navigation

Autonomous navigation in an outdoor environment using 3-D perception data from on-board sensors has been actively pursued by a number of research groups (Arkin, Riseman & Hanson, 1987; Thorpe, 1990). The application we concentrate on involves multiple outdoor mobile robot platforms that serve as Gofers[3] on the UCSD Campus. The goal is to demonstrate the information assimilation and control architectures in a navigation oriented setting. The scenario we envision is that given a destination point, these navigators use the on-board sensor and all other available information sources to arrive safely at their destination. The information sources we are considering include sensory data, map-based data and communication data from a central station.

[3]go.fer ['*goE-fer*, alter. of go for, 1970] n.: an employee whose duties include running errands.

Information derived from these disparate sources is assimilated into an Environment Model and used by the system to guide the vehicle along its path. Sensory data are obtained using both active and passive sensors. The sensor suite on-board the vehicle includes sonar, a laser range finder, several color CCD cameras (for short and intermediate range sensing with two arranged as a binocular stereo pair), vehicle state sensors (like dead-reckoning), GPS, and so on. The laser range finder, sonar array and stereo-based depth data will provide the necessary information about local static and dynamic obstacles. Color CCD cameras supply motion information as well as other relevant visual data.

Another important source of information is *a priori* information derived from maps describing the environment. Map-based data are comprised of relief maps, land-category maps and landmark locations. Digital relief maps provide elevation data about the navigation area. Other information such as location of roads, buildings and vegetation are provided by the land-category maps.

Autonomous navigation requires dynamic path planning, obstacle avoidance and vehicle control. Path planning involves determining an initial general path which is modified dynamically when moving or previously unknown stationary objects appear in this path. These obstacles must be detected by the on-board sensors early enough for the system to be able to react in time to avoid collision. Given an obstacle-free path, the system must control the vehicle's speed and trajectory to stay as close to this path as possible. All these issues should be addressed when building a working autonomous mobile robot.

10.3.1 Design

In going from a formal framework to an actual working system we have to address the following issues: exactly how many layers are needed for autonomous navigation tasks and what is the exact nature of these layers; what scheme is used to represent information and associated uncertainty at that layer; how does each layer interact with other layers and the assimilation process; how are moving objects handled; and what is the best way to incorporate decay into this scheme.

10.3.2 System architecture

A high level block diagram of the components of each agent is shown in Fig. 10.5. Basic components of each agent are the following.

Information Assimilation. Our main emphasis for this proposal is on this component. The *Information Assimilation* component has two basic functionalities: to build and iteratively refine the Environment Model from input information sources using the goals of the system and to answer EM queries (especially by the planner). These two tasks are performed by two modules: the *Assimilator* and the *Query Interface*.

Sensor Preprocessing. The *Assimilator* assimilates information from the sensors at mostly a symbolic level. The *Sensor Preprocessing* component contains

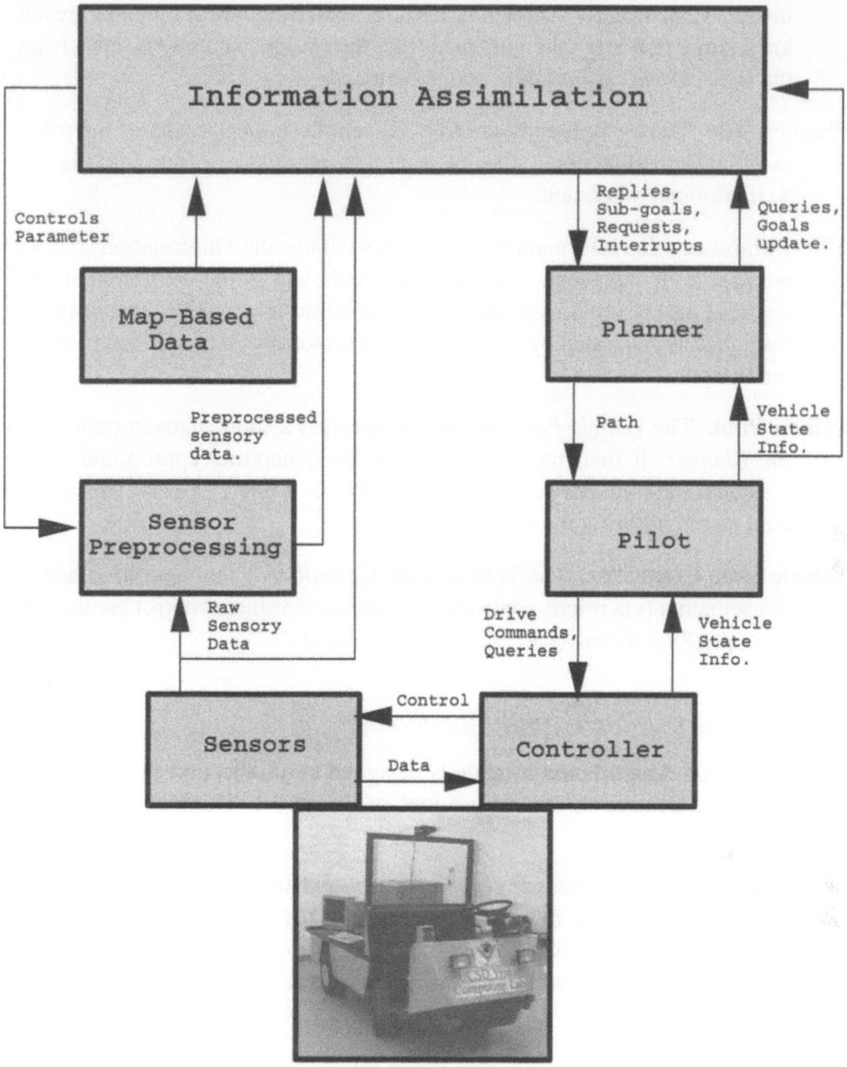

FIGURE 10.5. A high level block diagram of the autonomous agent. Basic components of each agent are *Information Assimilation*, *Sensor Preprocessing*, *Planner*, *Map Database*, *Vehicle Pilot* and *Vehicle State Controller*. Planning for the agent is done by the *Planner*. The *Vehicle State Controller* maintains a database of the vehicle state and changes this state by issuing actuator commands based on direct or standing instructions from the *Pilot*. The *Pilot* translates high-level instructions from the *Planner* to low level control commands. The *Information Assimilation* module cannot directly control the vehicle state. If certain information requested by the *Planner* is unavailable, the *Information Assimilation* unit creates a sub-plan for acquiring that information, and the *Planner* should decide whether it is feasible to execute the plan.

modules that do early vision to facilitate assimilation. These modules are not knowledge-rich and take guidance (like thresholds, parameters, etc.) from the *Information Assimilation* component.

Planner. The *Planner* indirectly controls all actions and behaviors of the robot using information about the environment provided by the *Information Assimilation* component.

Map Database. Map information is unlike most of the other information sources we handle. It represents a slightly outdated state of the environment and contains highly organized and mostly symbolic information. Our map information is generated by digitization and vectorization of scanned maps of the campus.

Vehicle Pilot. The *Vehicle Pilot* component receives a short segment path from the *Planner*. It then traverses that path by generating appropriate vehicle control commands (e.g., `set_speed(.2 mps)`, `turn(.2 deg)` etc.) and monitoring the vehicle state.

Vehicle State Controller. The *Vehicle State Controller* is the main interface to the actuators (via micro-controllers). It converts vehicle control commands from the *Pilot* to low level controller commands.

10.3.3 Information assimilation module

The Information Assimilation module is designed as a collection of coroutines, communicating through their modifications of the model. These coroutines fall into three types.

Assimilation. These routines use a predict-verify-update cycle to assimilate useful information from the input sources. Some of the sensor processing coroutines expect some bottom-up sensor preprocessing to be performed already. This bottom-up processing is usually done on dedicated image processing hardware.

Model Refinement. Some coroutines refine information already existing in the model using reasoning. These routines also convert information at one level of abstraction into another.

Query Processing. Query management is done by a set of coroutines which "plan" how information needed to satisfy the query may be accessed or obtained.

As an example of a set of assimilation streams, we will consider how different types of features extracted from a color CCD camera will be assimilated into the model. A diagram of information flow in this case is shown in Fig. 10.6. The figure shows how different types of information extracted from the CCD camera are assimilated independently using separate predict-verify-update cycles.

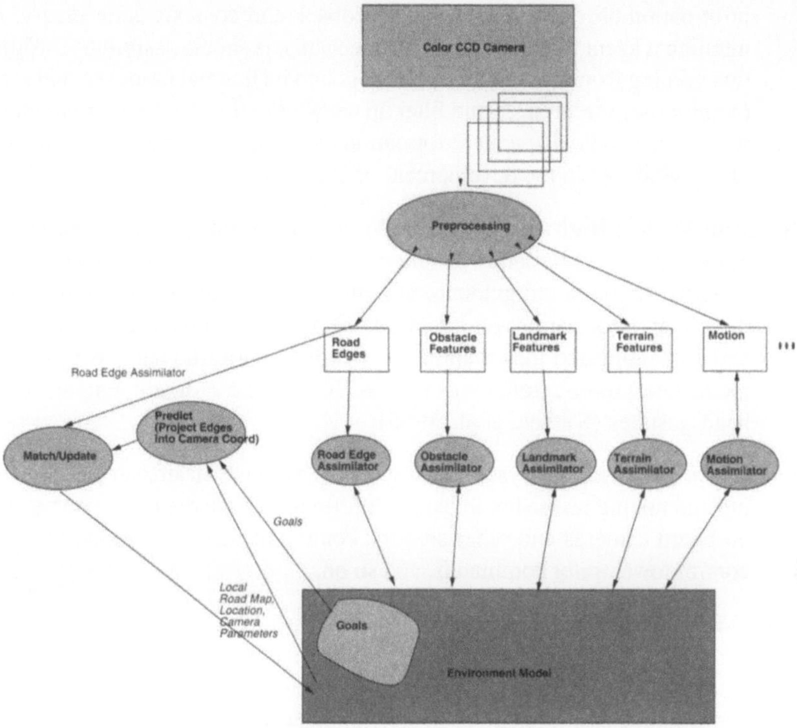

FIGURE 10.6. Assimilation of information from a color CCD camera. Each assimilation stream generates predictions based on current goals of the assimilation module.

10.4 Applications of Information Assimilation

Our information assimilation formulation as presented in the previous sections is very general and is applicable not only to autonomous navigation but to other to diverse applications as well. Essential characteristics of systems we address are: *a priori* information, disparate information sources and quantizable system goals. A few of the systems we are addressing are the following.

Multiple Perspective Interactive Video. The Modeling and Interaction components of our system are ideally suited for providing interactivity to traditional TV (Chatterjee et al., 1995; Jain & Wakimoto, 1995). Our implementation of the *MPI-Video* system shares several components (at the software level) with the autonomous gofer system.

Intelligent Access Agents. In distributed information systems containing multimedia information like the proposed *digital libraries*, end users access disparate information such as audio, video, images, text, and so on, from multiple sites. Since the amount of information accessible from these distributed server sites is large, mechanisms to intelligently access relevant information user profiles, relevant feedback and contexts is necessary. An intelligent agent that provides these mechanisms should assimilate information coming from the distributed servers into its internal model so that it can monitor user preferences and filter unnecessary information. Our paradigm, which can handle disparate information like video, sound and symbols, will play a vital role in the development of such systems.

Intelligent Vehicle Highway Systems. While our current application strives for autonomy, our assimilation paradigm can also be useful in semi-autonomous systems. An on-board guidance system in IVHS applications requires handling map information, on-board sensor information like camera and GPS and communicated information like weather and traffic data. We have implemented a semi-autonomous system $\mathcal{ROBOGEST}$ driven remotely using hand-gestures (Katkere et al., 1995).

Pilot Landing Assistance Systems. A system that assists an aircraft pilot in landing and taxiing tasks should handle information from disparate sources like on-board cameras and other sensors, communicated information from the control tower, pilot commands, and so on.

Smart Manufacturing Systems. Flexible manufacturing systems require strong coupling among specification, design, inspection, marketing and customer feedback phases. Information obtained at each of these phases is disparate and could be assimilated via an Environment Model similar to ours.

10.5 Conclusion

Autonomous behavior using perception involves acquisition and processing of information from disparate sensory inputs and representing this information in such a way that it can be accessed and updated for certain specified tasks. In this paper, we presented a novel approach towards assimilation of information from disparate sensors. The information assimilation framework put forth here has the capability of handling multimodal, disparate information. Also, it can deal with incomplete information. Robustness is embedded in the framework through the assumptions. This alleviates certain common low-level problems such as sensor registration, calibration and noise. These characteristics will make our framework important and useful in many diverse applications. Currently, most of the system is on paper and in the early stages of implementation. Efforts are underway to use this framework on some of the applications mentioned in Section 10.4, specifically the MPI-Video and Pilot Assistance systems.

Acknowledgments: We thank Saied Moezzi, Shankar Chatterjee and Don Kuramura for their involvement and comments. These three and several others in our

PLAS group have helped with ideas and implementation issues. Many of the ideas for this paper came from our earlier work at Michigan. We thank Terry Weymouth, Yuval Roth and all others involved with the project at Michigan. Finally, thanks to everyone at the Visual Computing Lab for making this an interesting place to work. This work is partially supported under grant NCC 2-792 from NASA.

10.6 References

Aggarwal, J. K. & Nandhakumar, N. (1990). Multisensor fusion for automatic scene interpretation. In R. C. Jain & A. K. Jain (Eds.), *Analysis and interpretation of range images* (pp. 339–361). New York: Springler-Verlag.

Albus, J. S., McGain, H. G. & Lumia, R. (1987). NASA/NBS standard reference model for telerobot control system architecture (NASREM). Technical Report NBS Technical Note 1235, National Bureau of Standards, Robotics Systems Division.

Aloimonos, Y., Weiss, I. & Bandopadhay, A. (1988). Active vision. *International Journal of Computer Vision, 1*, 333–356.

Andress, K. & Kak, A. C. (1987). A production system environment for integrating knowledge with vision data. In Kak, A. C. & shing Chen, S. (Eds.), *AAAI Workshop on Spatial Reasoning and Multisensor Fusion* (pp. 1–11). Los Altos, California: Morgan Kaufmann.

Arkin, R. C., Riseman, E. M. & Hanson, A. R. (1987). AuRA: An architecture for vision-based robot navigation. In *DARPA Image Understanding Workshop* (pp. 417–431). Los Altos, California: Morgan Kaufmann.

Bajcsy, R. (1988). Active perception. *IEEE Proceedings, 76*, 996–1005.

Ballard, D. H. (1987). Eye movement and spatial cognition. Technical Report TR-218, Computer Science Department, The University of Rochester.

Chatterjee, S., Jain, R. C., Katkere, A., Kelly, P., Kuramura, D. Y. & Moezzi, S. (1995). Modeling and interactivity in MPI-Video. Submitted to the 1995 International Conference on Computer Vision.

Durrant-Whyte, H. F. (1988). *Integration, coordination and control of multi-sensor robot systems*. Norwell, Massachusetts: Kluwer Academic Publishers.

Henderson, T., Weitz, E., Hansen, C. & Mitiche, A. (1988). Multisensor knowledge systems: interpreting 3d structure. *International Journal of Robotics Research, 7(6)*, 114–137.

Jain, R. C. (1989). Environment models and information assimilation. Technical Report RJ 6866 (65692), IBM Almaden Research Center, San Jose, California.

Jain, R. C. & Wakimoto, K. (1995). Multiple perspective interactive video. Submitted to IEEE Multimedia.

Katkere, A., Hunter, E., Kuramura, D. Y., Schlenzig, J., Moezzi, S. & Jain, R. C. (1995). *ROBOGEST*: Telepresence using hand gestures. Submitted to IROS '95.

Khalili, P. & Jain, R. C. (1991). Forming a three dimensional environment model using multiple observations. In *Proceedings of the IEEE Workshop on Visual Motion* (pp. 262–267). Princeton, New Jersey: IEEE Computer Society Press.

Kipling, R. (1898). *Mandalay*. New York: M. F. Mansfield.

Landy, M. S., Maloney, L. T., Johnston, E. B. & Young, M. J. (1995). Measurement and modeling of depth cue combination: In defense of weak fusion. *Vision Research, 35*, 389–412.

Maloney, L. T. & Landy, M. S. (1989). A statistical framework for robust fusion of depth information. In Pearlman, W. A. (Ed.), *Visual Communications and Image Processing IV: Proceedings of the SPIE*, Volume 1199 (pp. 1154–1163).

Marks, L. E. (1978). *The Unity of the Senses*. New York: Academic Press.

Mitiche, A. & Aggarwal, J. K. (1986). Multiple sensor integration/fusion through image processing: a review. *Optical Engineering, 25(3)*, 380–386.

Neisser, U. (1976). *Cognition and Reality: Principles and Implications of Cognitive Psychology*. San Fransisco: W. H. Freeman and Company.

Nii, H. P. (1986). Blackboard systems: The blackboard model of problem solving and the evolution of blackboard architectures. *AI Magazine* (pp. 38–53).

Richardson, J. M. & Marsh, K. A. (1988). Fusion of multisensor data. *International Journal of Robotics Research, 7(6)*, 78–96.

Roth, Y. & Jain, R. C. (1991). Context-based caching for control and simulation in mobile platforms. In *Proceedings 1991 International Simulation Technology Conference* (pp. 559–564). Orlando, FL.

Roth, Y. & Jain, R. C. (1992). Integrating control, simulation, and planning in mosim. In *SPIE Conference 1708 Applications of Artificial Intelligence X: Machine Vision and Robotics, Simulation and Visualization Environments for Autonomous Robots*. Orlando, Florida.

Roth, Y. & Jain, R. C. (1994). Knowledge caching for sensor-based systems. *Artificial Intelligence, 71*, 257–280.

Roth, Y., Zhang, G.-Q. & Jain, R. C. (1992). Situation caching. Technical Report CSE-TR-124-92, Computer Science and Engin Division, The University of Michigan, Ann Arbor, Michigan.

Roth-Tabak, Y. & Weymouth, T. E. (1990a). Environment model for mobile robots indoor navigation. In *SPIE Vol. 1388 Mobile Robots V* (pp. 453–463). Boston, Massachusetts.

Roth-Tabak, Y. & Weymouth, T. E. (1990b). Using and generating environment models for indoor mobile robots. In *International Association for Pattern Recognition Workshop on Machine Vision Applications* (pp. 343–346). Tokyo.

Shafer, S. A., Stentz, A. & Thorpe, C. E. (1986). An architecture for sensor fusion in a mobile robot (autonomous vehicle). In *Proceedings 1986 IEEE International Conference on Robotics and Automation*, Volume 3 (pp. 2002–2011). Washington, DC: IEEE Computer Society Press.

Thorpe, C. E. (1990). *Vision and navigation: the Carnegie Mellon Navlab*. Boston, Massachusetts: Kluwer Academic Publishers.

Tirumalai, A. P., Schunck, B. G. & Jain, R. C. (1990). Robust self calibration and evidential reasoning for building environment maps. In *SPIE Symp. on Advances in Intelligent Systems*.

11

Task-Oriented Vision

Katsushi Ikeuchi[1]
Martial Hebert[1]

ABSTRACT
In this paper, we introduce a systematic approach for tailoring perceptual modules to specific tasks. In this approach, modules for image segmentation, object representation, and manipulation are selected based on the constraints of the target task and on the environment. The goal is to generate the most efficient set of perceptual modules based on the constraints. This approach is a high-level equivalent to the more traditional low-level active control of sensing strategies. We illustrate this approach through two example systems. In the first system, the task is to manipulate natural objects. In the second system, the task is to pick industrial parts out of a bin. We show how the two tasks were decomposed and analyzed to yield the best match between perceptual capabilities and task requirements. Then, we introduce a general framework for task-based development of vision systems.

11.1 Introduction

Intelligent control of perception for robotic systems involves not only active control of the early vision modules, such as sensor parameters and image processing algorithms, but also dynamic selection of the higher-level components of the perceptual system. Such components include object representations, features, and segmentation methods. The conventional practice of using a fixed sequence of algorithms, even with a sophisticated sensor control scheme, will lead to severe limitations on the applicability of the perceptual system. Moreover, it is impossible to design a sensing strategy that is optimal for all the robotic tasks that may be undertaken. Rather, a methodology that automatically selects the appropriate perceptual components must be developed.

In this paper, we investigate such a methodology for organizing the components of a perception system based on the requirements of the target task. This approach can be seen as the equivalent for the higher-level components of perception to active sensor control.

There are two approaches to the problem of organizing the components: general-purpose and task-oriented. The general-purpose school claims that we should build

[1]School of Computer Science, Carnegie Mellon University

vision systems to solve all vision tasks using a single architecture, that is, an architecture in which a fixed selection of components is executed in a fixed order. They also claim that we should avoid using any task-specific constraints in building vision systems.

We, researchers in the task-oriented school, claim that we should prepare different architectures for vision systems and that, depending on the task (i.e., the final goal of the system and the environment in which such a task is achieved), a robot vision system should change its architecture so that it uses the optimal selection of components to achieve a given task.

In the task-oriented approach, however, few attempts have been made to clarify and establish theories for task-oriented vision and methodologies to implement theories for building a vision system of optimal architecture beyond several ad hoc attempts.

This paper proposes a task-oriented vision approach and investigates the design of vision systems in a systematic way. We focus on the design of robotic systems that involve the localization and grasping of objects because these two tasks illustrate the advantages of the task-oriented approach. To illustrate this approach, this paper will overview two vision systems recently completed: rock sampling vision for planetary rover robots and bin picking vision for industrial robots. We first describe the current implementation of the two systems. We then show how the two systems were designed using a systematic systems analysis approach. Finally, we show how the analysis can be generalized to generic perception systems for robotic tasks. We conclude by showing how this task-oriented approach differs from the traditional bottom-up approach to vision.

11.2 Systems Description

11.2.1 Rock sampling system

One of the most important goals of a planetary exploration mission is to collect and analyze terrain samples. As part of the CMU Ambler Project (Bares et al., 1989), we have investigated techniques for retrieving small rocks from sand. This section gives an overview of the architecture of the rock sampling system.

11.2.1.1 Image acquisition sensor and segmentation

The upper-left panel of Fig. 11.1 shows a typical scene for the system. A range image is acquired using a range finder (Sato, Yamamoto & Inokuchi, 1987). Each pixel in the range image stores the three-dimensional coordinates of a point measured with respect to the sensor.

Three types of features are extracted from the range image:

- shadows,
- orientation discontinuities, and

- range discontinuities.

These features give an indication of where the boundaries of the rocks may be located in the scene. Unfortunately, because these features are fragmented, they are not sufficient for reliably extracting rocks from the scene; and therefore, we cannot use a simple region extraction technique that would assume that the features are grouped into closed boundaries.

Given that the scene contains a cluster of features which is assumed to be due to the presence of a rock, we can use a geometric sensor model to derive the approximate location of the center of the rock from the distribution of the corresponding shadow regions and depth discontinuities.

We implemented an iterative segmentation algorithm, similar to the "snake" algorithm (Kass, Witkin & Terzopoulos, 1988). A snake has a closed deformable boundary, somewhat similar to that of a rubber band, which is attracted by features and tries to shrink to its original size. The algorithm enlarges the boundary of the snake until it connects features.

In order to implement the attractive and shrinking forces, we use two kinds of energy fields: external (attracting) and internal (shrinking).

The external energy is supplied by the following three forces:

- shadow attractor,
- orientation discontinuity attractor, and
- range discontinuity attractor.

Drawn by the attractive forces, the boundary moves toward the surrounding features. At the same time, over-expansion is avoided by using the internal energy of the contour. We implemented the following internal energy fields:

- center attractor and
- region attractor.

Attracted by its central position, and also by itself, a snake tries to form a compact shape.

A small snake is initially located at the hypothesized center of a rock region. It grows iteratively while deforming its shape until it sits along the features where the external forces from the features and its own internal forces are in equilibrium. The lower-left panel of Fig. 11.1 shows a schematic diagram of our segmentation algorithm.

The segmentation result in Fig. 11.1 shows the rock region, shown in black, which has been extracted by this algorithm. Because this approach interpolates the missing gaps between features, it allows us to locate rocks in the scene even when only a very small number of visual features are extracted from the image. This is a departure from other vision systems which implicitly assume that strong and reliable features can always be extracted, and therefore would not perform well in this unstructured environment.

11.2.1.2 Representation

To grasp a rock, we need parameters based on three-dimensional information, such as its center of mass and its axis of inertia. A snake segmentation provides the two-dimensional outline of the rock. The range data within the extracted contour provides the three-dimensional shape. However, it describes only the shape of the visible side of the rock. We have to infer the whole shape of a rock from the shape of the visible side to obtain the 3-D parameters necessary for grasping.

We use a superquadric surface to approximate the whole rock shape from the range data. A superquadric is a generalization of an ellipsoid that can represent a wide variety of shapes using a small number of parameters (Bajcsy & Solina, 1987; Pentland, 1986). The generic implicit equation of a superquadric is given by

$$\left(\left(\frac{x}{a}\right)^{\frac{2}{e_2}} + \left(\frac{y}{b}\right)^{\frac{2}{e_2}}\right)^{\frac{e_2}{e_1}} + \left(\frac{z}{c}\right)^{\frac{2}{e_1}} = 1, \qquad (11.1)$$

where a, b, c are the size parameters. By changing the parameters, e_1 and e_2, we can represent several shapes as shown in the lower-middle panel of Fig. 11.1.

We chose superquadrics as our representation for two reasons.

- Superquadrics are appropriate for blob-like shapes.
- Fitting superquadrics to a set of points from a partially visible object gives an estimate of the whole shape of the object, whereas local surface representations would only provide a representation of the visible part of the object.

We implemented a standard gradient descent method to fit a superquadric surface to the range data (Solina, 1987). The upper-middle panel of Fig. 11.1 shows a superquadric representation of the rock.

11.2.1.3 Grasping strategy

The superquadric fitting module provides the following shape parameters:

- mass center position,
- size, and
- axis direction.

FIGURE 11.1. Overview of the rock sampling system. The image of a scene is acquired by a range finder. Features such as shadows and discontinuities are extracted from the image. A deformable contour is initialized at the center of each cluster of features and is deformed until it fits the fragmented set of features detected on the outline of the rock by the segmentation algorithm. A superquadric is fit to the 3-D data inside the final contour. A grasp position and orientation is planned for a clam-shell gripper by aligning the axis of the gripper with the axis of the superquadric. Finally, the gripper grasps the object according to the plan generated by the vision system.

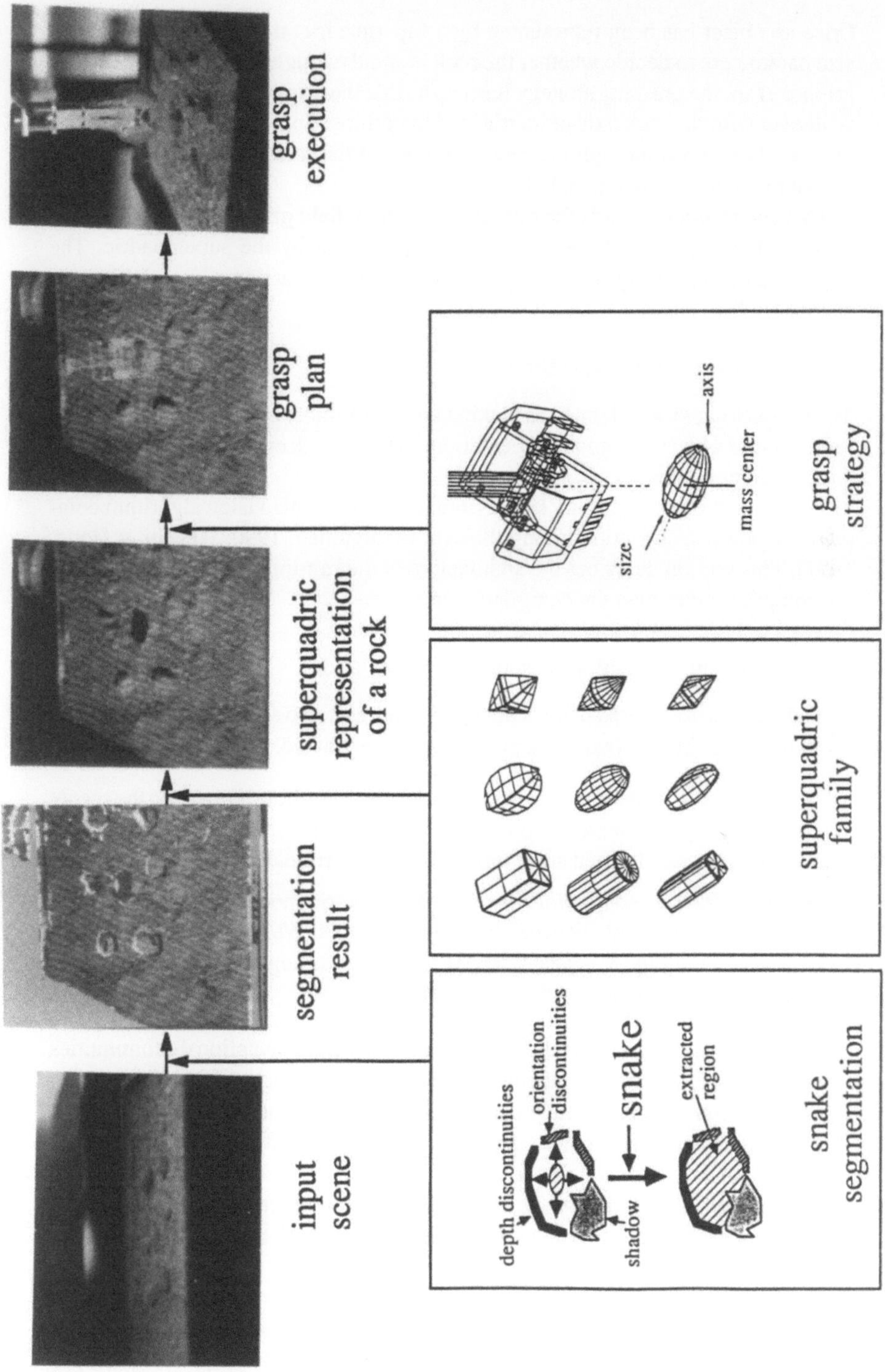
input
scene
segmentation
result
superquadric
representation
of a rock
grasp
plan
grasp
execution
depth discontinuities
orientation
discontinuities
shadow
snake
extracted
region
snake
segmentation
superquadric
family
size
mass center
axis
grasp
strategy

Once an object has been represented by a superquadric, the system examines its size parameters to decide whether the rock is small enough to be picked up by the gripper. If so, the grasping strategy is set up so that the gripper orientation direction is aligned with the rock axis of inertia, and the gripper approach direction is along the z-axis and goes through the center of mass of the rock. The grasp strategy is shown in the lower-right panel of Fig. 11.1.

This configuration yields the minimum potential field given by the relationship between the gripper and the rock that is represented by the superquadric. The upper-right panel of Fig. 11.1 shows the actual grasping action.

11.2.2 Bin picking system

The bin-picking task is defined as picking up the top-most object from a pile of the same kind of objects in random orientations. The input scene in Fig. 11.2 shows a typical example of a bin.

We have developed a bin-picking system using the CMU vision algorithm compiler for object localization (Hong, Ikeuchi & Gremban, 1990; Ikeuchi & Hong, 1991). This section describes the architecture of the run-time system generated by the compiler rather than the compilation techniques.

11.2.2.1 Image acquisition sensors and segmentation

The range data is acquired using a dual photometric stereo system (Ikeuchi, 1987). From the range image, two types of features are extracted.

- *Shadows* — A photometric stereo system projects three lights onto the scene; each light generates a shadow region. By distributing three lights in a triangle shape, the part located at the top of the bin is surrounded by shadow regions.
- *Orientation discontinuities* — From a geometric model of the part, we can determine the angle between two adjacent faces. We can find the minimum angle among these adjacent angles, and use this angle as the threshold for determining orientation discontinuities.

A simple segmentation method based on shadows and orientation discontinuities works quite well in the bin-picking case due to the characteristics of the scene. In a bin of industrial parts, each part is enclosed by a clear occluding boundary. It is not necessary to use a more detailed segmentation method such as the snake segmentation method used in the previous system. The segmentation result in Fig. 11.2 shows that given by the simple segmentation method; the parts (dotted regions) are separated from each other by shadows and orientation discontinuities (white regions).

The highest region is determined among the regions extracted by the segmentation program. From this highest region, the object localization process begins. This is because the highest region usually corresponds to the top-most part, and the top-most part is usually the easiest one to pick up.

11.2.2.2 Representation

To perform the object localization, the vision algorithm compiler automatically generates a localization program from the object and sensor model. This program can perform object localization in the least amount of computational time among several possible localization programs (the lower-left panel of Fig. 11.2).

Several geometric features are extracted by the localization program in the predetermined order by the compiler, such as area, inertia and distance between two adjacent regions. The program compares the extracted features with those from the model and determines the attitude and position of the part.

Using the resulting position and attitude, the program generates a part representation using the geometric model as shown in the recognition results of Fig. 11.2. The neighboring regions are represented by dodecahedral prisms. These dodecahedral prism representations are used for collision checking while constructing a grasping strategy.

11.2.2.3 Grasping strategy

The grasp configuration should satisfy two conditions (Ikeuchi et al., 1986).

- It should produce a mechanically stable grasp, given the gripper's shape and the part's shape. Such configurations will be called *legal grasp configurations*.
- The configuration must be achieved without collisions with other parts. Grasp configurations are limited by the relationship between the shape of the gripper and the shapes of neighboring obstacles. Such configurations will be called *collision-free grasp configurations*.

In compile mode, possible legal grasp configurations are compiled and stored at each representative attitude of the part in a grasp catalogue as shown in the lower-middle panel of Fig. 11.2.

In execution mode, the system determines to which representative attitude the current attitude of the top-most part belongs. Then, the legal grasp configurations corresponding to the representative attitude are retrieved from the grasp catalogue. These configurations are then converted into the world coordinate system based on the observed configuration of the industrial part.

The system has to find a collision-free grasp configuration from among these configurations. It generates a cube representation corresponding to the work-space of each legal grasp configuration in the geometric representation as shown in the collision check panel of Fig. 11.2. Then, the system examines whether an intersection exists between the gripper work space cube and the obstacle prisms in the geometric representation.

Among the possible collision-free configurations found by the system, the optimal configuration (currently, the one nearest to the part's center of mass) is chosen. The system picks up the part using the configuration shown in the upper-right panel of Fig. 11.2.

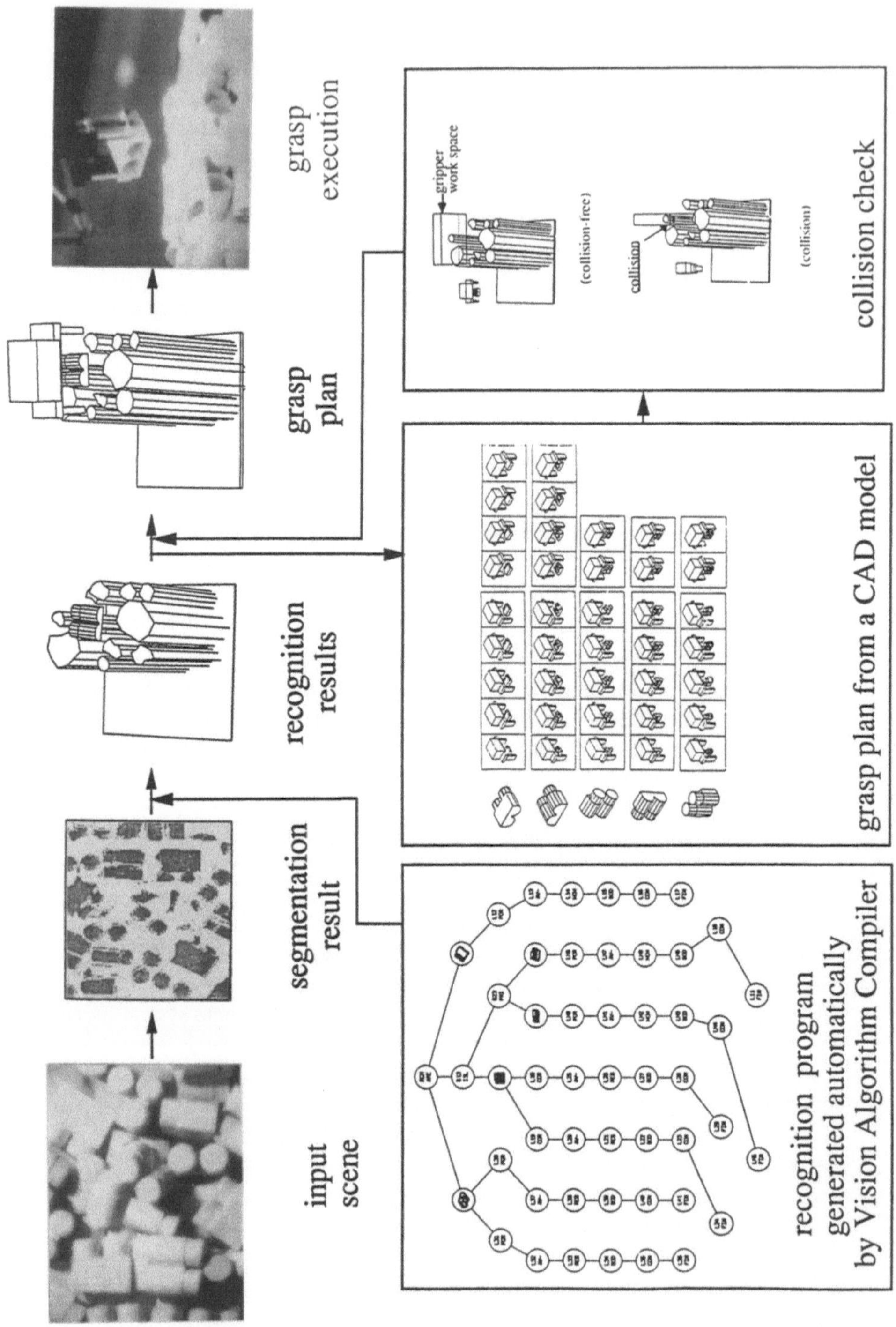
input scene
segmentation result
recognition results
grasp plan
grasp execution
recognition program generated automatically by Vision Algorithm Compiler
grasp plan from a CAD model
gripper work space
(collision-free)
collision
(collision)
collision check

11.3 System Analysis

The tasks that the rock-sampling and bin-picking systems aim to achieve are roughly same: observe a scene, determine a grasp strategy and, based on the result of the scene analysis, pick up an object. However, the two systems have completely different architectures. This section will examine the reason why such different architectures are necessary.

11.3.1 Rock-sampling system

Fig. 11.3 shows the design flow of the rock sampling system. The design proceeds from the task specification of the rock sampling through the image acquisition method.

11.3.1.1 Task specification

The task of this system is to grasp a rock in the sand under the following conditions.

- The rocks are far enough away from each other so that it is not necessary to consider collisions between the gripper and the neighboring rocks, when picking up a rock.
- We can allow collisions between the gripper and the neighboring sand because:
  - damaging the neighboring sand grains is not important,
  - collisions between the gripper and neighboring sand do not cause a configurational change of the rock.
- We do not know the exact shape of a rock beforehand.

11.3.1.2 Grasping

Under this task specification, it is appropriate to use a spherical grasp as illustrated by the lower-left panel of Fig. 11.3. This grasp has the following characteristics.

- It requires a large empty volume around an object to be grasped, because all the fingers approach the object from all directions.
- It may grasp the material neighboring the object, if any, as well as the object.
- It does not require the precise attitude and position of the object, because it grasps an object as if it wrapped the object.

FIGURE 11.2. Overview of the bin picking system. An intensity image of a bin of parts is first acquired by a video camera. A needle map of the surface normals on the objects seen in the image is computed using photometric stereo. Based on the needle map, the scene is segmented into planar regions. A grasp of a two-finger gripper is planned based on the position and orientation of the top-most object.

To realize this spherical grasp, we built a clam-shell gripper. See the lower-right panel of Fig. 11.3.

11.3.1.3 Representation

Using a clam-shell gripper imposes two constraints on the representation:

- the expected center of mass of a rock should be inside the gripper, and
- the expected size of a rock should be smaller than the inner hull of the gripper.

While working from only a partial observation of a rock and without any prior knowledge of the rock shape, we still need to recover the above information. We do not need to recover a precise shape representation of a rock, however. Based on these observations, the superquadric representation was chosen because it is described by a small set of parameters which can be recovered by using a fitting method such as gradient descent.

11.3.1.4 Segmentation

For a rock partially buried in the sand, orientation discontinuities and depth discontinuities are small. Thus, it is usually difficult to detect these discontinuities reliably and to extract a closed boundary based on them.

Because there is no a priori rock model available, the segmentation cannot be guided by a model as in the bin-picking system. The only available information is that a rock forms a closed boundary. Along this closed boundary, the following three boundary elements exist:

- shadow boundaries,
- orientation discontinuities, and
- depth discontinuities.

Thus, it is necessary to use a segmentation method that connects these boundary elements and extracts a closed boundary. For this purpose, a model-based segmentation method based on the snake algorithm described in Section 11.2.1.1 was employed.

11.3.2 Bin-picking system

Fig. 11.4 shows the design flow of the bin-picking system. The design flow begins with the task specification of the bin-picking and specifies all successive components down to the image acquisition method.

11.3.2.0.1 Task specification

The task of this system is to grasp the top-most industrial part in a bin of parts under the following conditions.

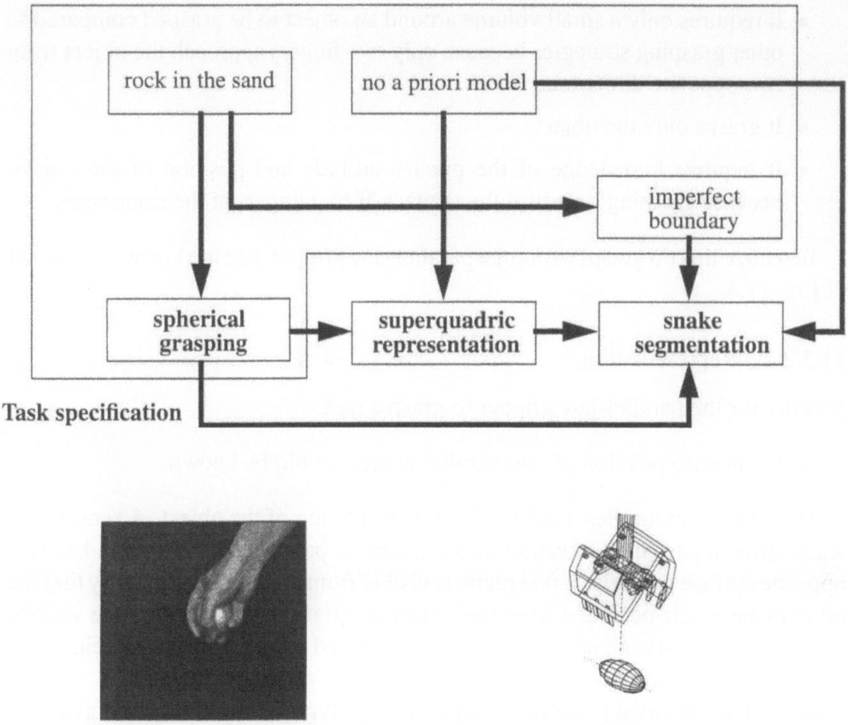

FIGURE 11.3. Design flow of the rock sampling system. The components of the system, snake segmentation, superquadric representation, and spherical grasp, are selected from the initial task description (picking up rocks in sand). The constraints are shown in boldface.

- The parts are close to each other. Some collisions may occur between the gripper and the neighboring parts if a random grasping strategy is chosen.
- Collisions between the gripper and the neighboring parts must be avoided because:
  - collisions may cause damage to the neighboring parts, and
  - collisions may cause the configuration of the part to be grasped to change, because the part is supported by the neighboring parts, and thus, the gripper may fail to grasp the part.
- The exact shape of a part is known beforehand.

11.3.2.1 Grasping

Under these task-specifications, it is appropriate to use a tip grasp. See the lower-left panel of Fig. 11.4. This grasp has the following characteristics.

- It requires only a small volume around an object to be grasped compared to other grasping strategies because only two fingers approach the object from two opposite directions.
- It grasps only the object.
- It requires knowledge of the precise attitude and position of the object, because grasping occurs at the contact of two fingers at the same time.

To realize this tip grasp, we built a parallel-jaw gripper. See the lower-right panel of Fig. 11.4.

11.3.2.2 Representation

In order for the parallel-jaw gripper to grasp a part,

- the precise position of two parallel planes should be known.

This constraint implies a polyhedral representation of the object. A sensor typically gives a partial observation of an object. A pair of parallel planes has two opposite surface normals. If one plane is visible from the sensor, it is likely that the other plane is self-occluded from the sensor. Even if the two planes were visible, it is necessary to use an n^2 search out of n observed planes. Thus, we decided not to find such parallel plane pairs at run time.

Instead, we decided to represent a part by a polygonal approximation given by a geometric model, to search for plane pairs in the representation at compile time, and to establish the relationship between pairs and observed part attitude. At run time, we concentrate on using this relationship to recover the orientation of the part, as well as the orientation of the pairs.

11.3.2.3 Segmentation

Distinct depth discontinuities can be observed around an object, because an industrial part sits on other parts, as opposed to the rock-sampling case in which a rock may be partially buried in the sand. Also, from the geometric model of the object, the threshold value used to find surface discontinuities can be found from the minimum angle between adjacent faces.

The following facts are utilized for segmentation.

- An object boundary is surrounded by a shadow. Since the current implementation of the photometric stereo system employs three light sources, the top-most object in the bin is always surrounded by a shadow.
- The threshold value that defines the surface discontinuities can be defined by computing angle differences of every face pair in the model.

Since these two classes of boundaries are distinct and connected, we do not need an additional step for connecting boundaries, such as the snake-based segmentation used in the rock sampling.

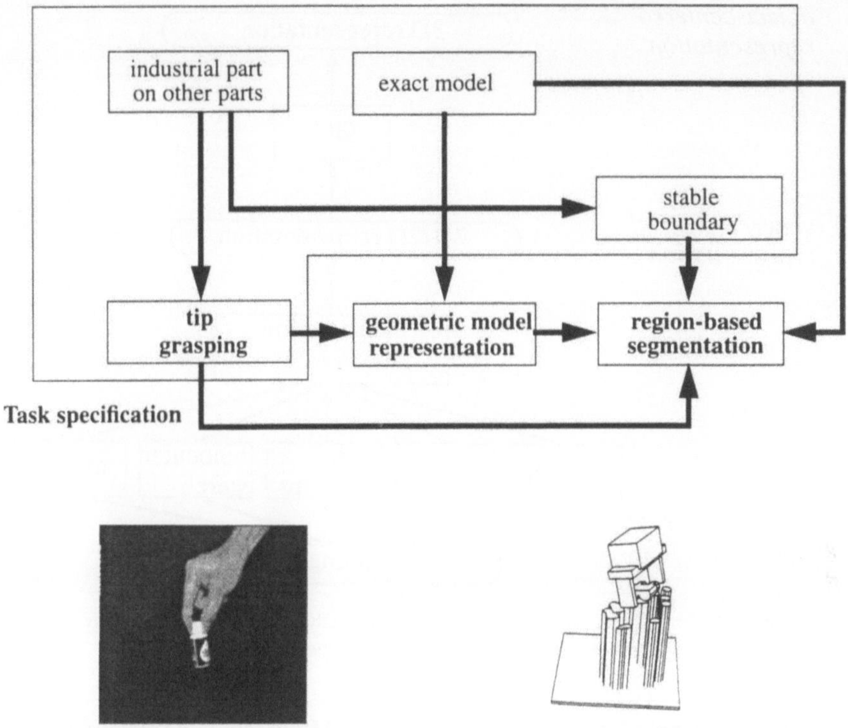

FIGURE 11.4. Design flow of the bin-picking system. The components of the system, region-based segmentation, planar representation, and tip grasping, are selected from the initial task description, picking parts from a bin. The constraints are shown in boldface text.

As shown in this section, to construct a vision system, it is not enough to investigate algorithms for each vision module, such as representation methods or segmentation methods, individually. It is also necessary to investigate the constraints and interactions among vision modules. Such constraints and interactions provide valid assumptions from which each vision module should be developed and the expected performance which each vision module should generate.

11.4 Task-Oriented Approach

An important goal for computer vision research is the development of a vision system which can serve as a complete and self-contained artificial vision unit. Currently, the majority of the vision community adheres to an approach which emphasizes general-purpose vision machines constructed using a single architecture — an approach epitomized by Marr's paradigm.

Fig. 11.5 shows the architecture in Marr's paradigm (Marr, 1982). An intermediate representation ($2\frac{1}{2}$-D sketch) is generated from several 2-D image cues such as shading, texture and motion. Then, a final 3-D representation, based on

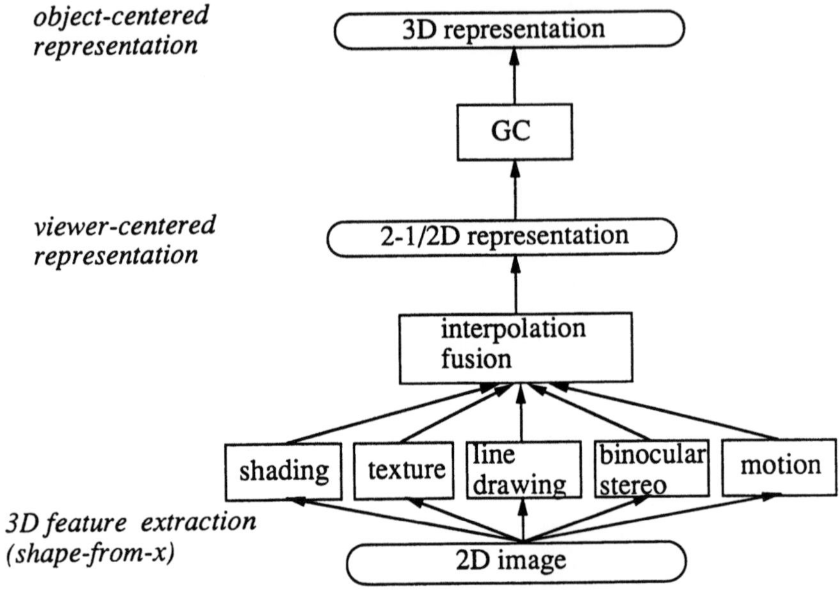

FIGURE 11.5. Marr's paradigm. Intermediate representations, from 2-D image to 3-D representation, are computed from the input image independent of the task.

the object-centered coordinate system, is generated from this $2\frac{1}{2}$-D representation. Independent of the nature of the tasks, the visual information is processed in a bottom-up fashion. Research focuses on each module in the system rather than on the overall system. Accordingly, intermodule interactions and the system's connections to specific tasks are deemphasized.

We propose to investigate task-oriented vision systems. We assume that without aiming to see a target object (without having some specific task), we cannot see it (we cannot achieve the task). Under this assumption, we claim that one particular visual task should govern the choice of representations, vision modules, and image acquisition sensors. Thus, a task determines the optimal architecture for the vision system.

Fig. 11.6 shows the paradigm of the system we are proposing. The basic collection of modules is the same as Marr's with the exception of a box labeled "Task". The interaction between modules will change as a result of the "decisions" from the Task box and will compose the optimal architecture for the vision task at hand. We refer to this paradigm as a task-oriented approach.

Our approach concentrates on developing not only intramodule algorithms within vision modules, which is addressed by the traditional approach, but also intermodule interactions which depend on tasks. In other words, we consider a vision system as a whole, and for a particular task, we investigate how vision modules interact. The key element is the logical order in which the vision components are selected and built. From the previous examples, we can see that this task-oriented approach is critical in building a working system. For example, the bin-picking

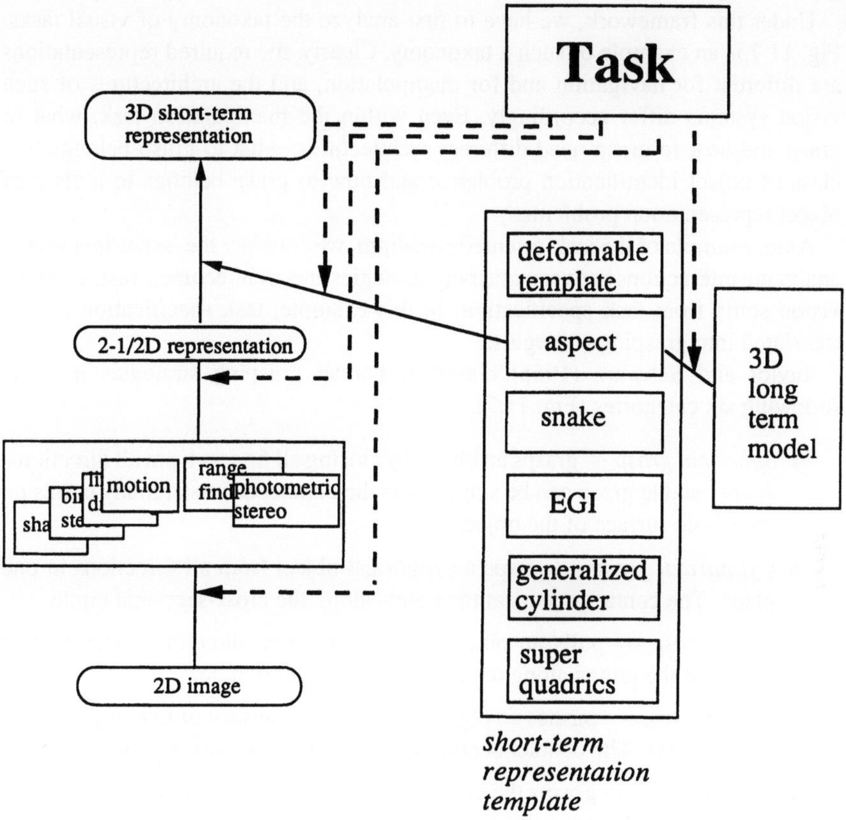

FIGURE 11.6. Task-oriented paradigm. Every component, from segmentation to grasping strategy, is selected based on the constraints imposed by the task.

system would not work if a superquadric representation were used: The difference between the superquadric surface and the actual surface may cause the contact of one finger to occur before the other, thus causing the object to move, and even possibly to fall from the top of the bin.

To investigate architectures of vision systems using the task-oriented paradigm, we have to consider the following:

- task specifications,
- functional capabilities of a representation required by the task,
- representations having such functional capabilities,
- features appropriate for extracting such representations,
- segmentation methods appropriate for extracting such features and representations, and
- image sensors and their strategies appropriate for obtaining such segmentation methods and features.

Under this framework, we have to first analyze the taxonomy of visual tasks. Fig. 11.7 is an example of such a taxonomy. Clearly, the required representations are different for navigation and for manipulation, and the architectures of such vision systems differ accordingly. Even within the manipulation task, what to grasp and how to grasp need different architectures; what to grasp belongs to a class of object identification problems, and how to grasp belongs to a class of object representation problems.

As an example of the task-oriented paradigm, we consider the issues inherent in analyzing interactions between grasping strategies and architectures. Task-oriented vision starts from task specification. In this example, task specification can be translated into grasping strategies.

Taylor and Schwarz (1955) classified human grasping strategies into the following six categories (Fig. 11.7).

- *Spherical grasp* — grasps an object by closing all fingers from all directions. A very stable grasp can be achieved as the contact occurs at many points on the whole surface of the object.
- *Cylindrical grasp* — grasps a cylindrical object from all directions in one plane. The contact occurs at the points along the cross-sectional circle.
- *Hook grasp* — pulls an object toward particular directions. The contact occurs at the points along the cross-sectional semicircle.
- *Lateral grasp* — pushes an object on a soft side surface of one finger by the other fingers. The contact occurs at a point and points on a plane.
- *Palmar grasp* — grasps the end of bar by closing three fingers. The contact occurs at the three points.
- *Tip grasp* — grasps an object by closing two fingers from two opposite directions. We can achieve very fine grasping. The contact occurs at the two opposite points.

Spherical, cylindrical and hook grasps are grouped as *power* grasps, while lateral, palmar and tip grasps are grouped as *precision* grasps.

Once a grasping strategy is given, we have to choose one particular representation suitable to the strategy. Here, the issue is to investigate the relationship between required functional capabilities, representations, and grasping strategies.

Fig. 11.8 summarizes the required functional capabilities for representations by these six grasping strategies. In the figure, the sign " $\sim$ " indicates that a quantity needs to be known only approximately. For example, the spherical grasp requires the approximate radius and the approximate center of the object.

We can summarize that the three power grasping strategies — spherical, cylindrical, and hook — require only approximate values of the parameters. For these grasping strategies, weak models such as superquadric representations are suitable for representing the grasped object.

The three precision grasping strategies require a detailed (e.g., geometric) model. The lateral grasp requires the exact position of the two planes. The palmar grasp

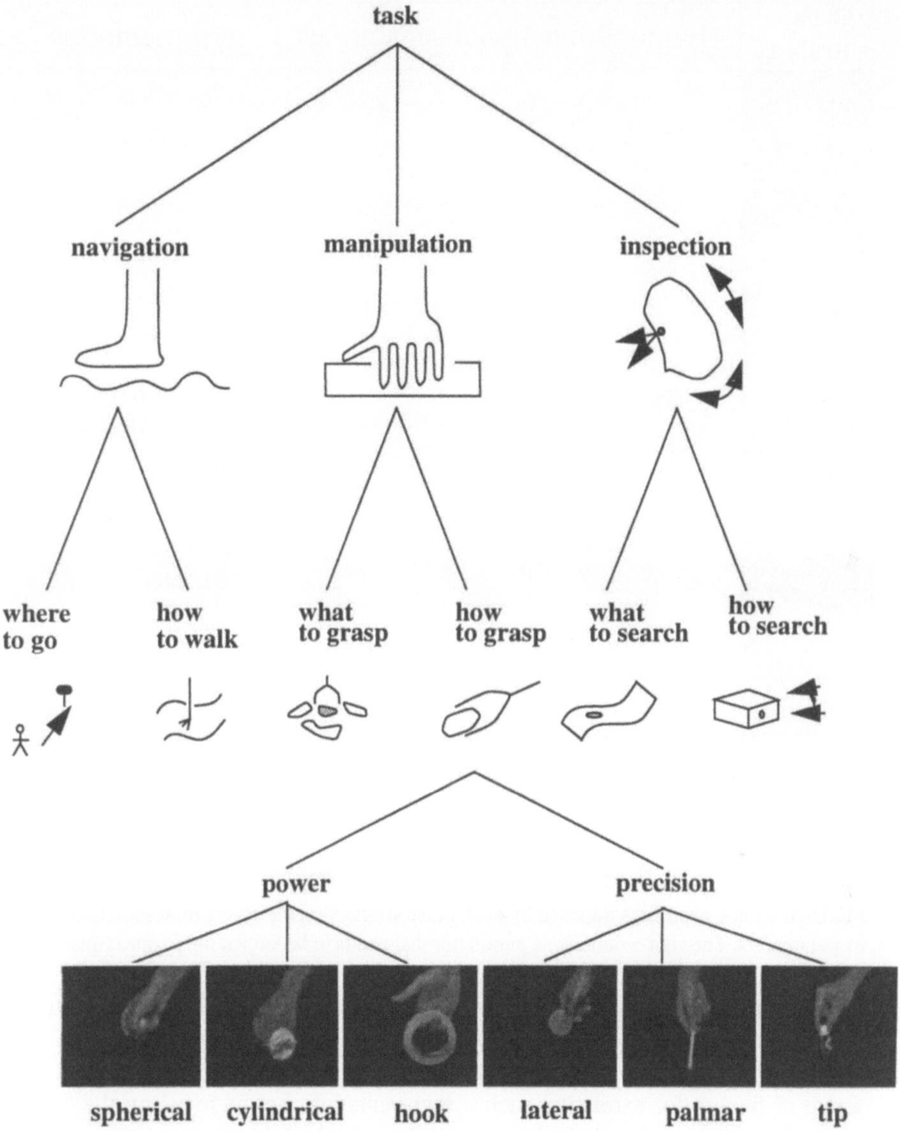

FIGURE 11.7. Taxomony of tasks. Potential tasks for a perception system can be divided into navigation, manipulation and inspection. The three types of tasks differ in the type of sensory information that is relevant and in the way the information is used. We concentrate here on manipulation tasks for which six basic manipulation strategies are illustrated.

requires knowledge of the exact position of the cross-section, while the tip grasp requires knowledge of the exact position of two contact points. To extract such exact information, we need an exact object model which is represented using polyhedra. Since it is difficult to extract such information precisely in real time, we need a

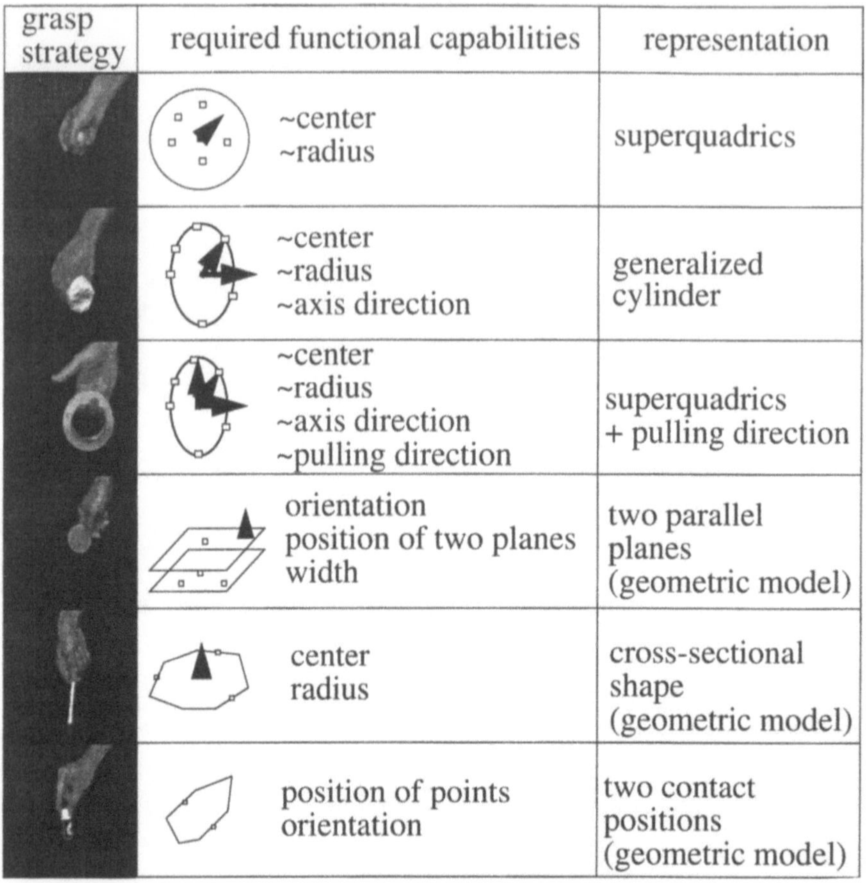

grasp strategy
required functional capabilities
representation
~center
~radius
superquadrics
~center
~radius
~axis direction
generalized cylinder
~center
~radius
~axis direction
~pulling direction
superquadrics + pulling direction
orientation
position of two planes
width
two parallel planes (geometric model)
center
radius
cross-sectional shape (geometric model)
position of points
orientation
two contact positions (geometric model)

FIGURE 11.8. Constraints imposed by each grasp strategy on the object representation and its parameters. The sign ~ indicate a parameter that needs to be known only approximately.

polyhedral approximation such as the one provided by a geometric model of the object.

The next important intermodule constraint is how to determine an appropriate group of feature for extracting such a representation. Let us focus on the case of precision grasps. For these grasps, a geometric model is required. Then, the issue can be translated into how to select optimal features automatically from a geometric model. Goad (1983) proposes a method for selecting appropriate edges for object localization. Bolles and Cain (1982) propose a focus feature method that selects important features and less important secondary features. We have been developing a CMU vision algorithm compiler which chooses the optimal set of features for object localization (Hong et al., 1990).

For further segmentation and image acquisition, such issues as "What kind of sensor should be used?", "Where should it be located to detect the necessary features?", and "What kind of features should be extracted?" have been investigated under active sensing strategy generations (Chen & Kak, 1989; Cowan & Bergman, 1989; Hansen & Henderson, 1987; Ikeuchi & Kanade, 1989; Yi, Haralick & Shapiro, 1990).

By using the task-oriented vision paradigm as the design philosophy for vision system architecture, we can unify recent active vision (Aloimonos, 1990; Bajcsy, Paul, Yun & Kumar, 1991; Ballard, 1989; Hutchinson & Kak, 1989) and vision algorithm compiler (Bolles & Cain, 1982; Ikeuchi & Hong, 1991; Yi et al., 1990) accomplishments toward the ultimate goal of efficient perceptual systems for robotic tasks.

11.5 Conclusion

In this paper, we have presented a task-oriented approach to the design of vision systems. The task-oriented vision approach involves changing the architecture of a vision system in a systematic fashion that depends on each task specification. We have presented a task-oriented approach for systems that involve the localization and grasping of 3-D objects. In this case, the general methodology involves analyzing the task specification to derive the constraints on and requirements of the vision components. This starts with the type of representation, derived from the type of grasp selected, and continues down to the type of sensor. The task-oriented approach is applicable to a wide range of vision systems. The task-oriented approach will not only build more robust systems but will also give a new direction to vision research.

Acknowledgments: The authors wish to thank H. Delingette, T. Choi and Y. Yen for writing some of the rock-sampling software, and K. S. Hong and K. D. Gremban for writing some of the bin-picking software. Takeo Kanade and Raj Reddy provided many useful comments and encouragement. Kathryn Porsche and Fredric Solomon proofread drafts of this paper and provided many useful comments which have made it more readable. The authors also thank Shree Nayar, Hideichi Sato, Yoshimasa Fujiwara and the members of the Task-Oriented Vision Laboratory, the Robotics Institute, Carnegie Mellon University, for their valuable comments and suggestions.

This research was sponsored in part by the National Science Foundation under CDC-9121797, in part by the Advanced Research Projects Agency under the Avionics Laboratory, U.S. Air Force, under Contract F33615-90-C-1465, and in part by the Advanced Research Projects Agency under the Army Research Office, the Department of the Army, under grant DAAH04-94-G-0006.

The views and conclusions contained in this document are those of the authors and should not be interpreted as representing the official policies, either expressed or implied, of NSF, ARPA, or the U.S. government.

11.6 References

Aloimonos, Y. (1990). Purposive and qualitative active vision. In *Proceedings of DARPA Image Understanding Workshop* (pp. 816–828). San Mateo, California: Morgan Kaufmann.

Bajcsy, R., Paul, R., Yun, X. & Kumar, V. (1991). A multiagent system for intelligent material handling. In *Proceedings of the 1991 International Conference on Advanced Robotics* (pp. 18–23). Los Alamitos, California: IEEE Computer Society.

Bajcsy, R. & Solina, F. (1987). Three-dimensional object representation revisited. Technical Report MS-CIS-87-19, University of Pennsylvania, Department of Computer and Information Science.

Ballard, D. H. (1989). Animate vision. In *Proceedings of the International Conference on Artificial Intelligence* (pp. 1635–1641). San Mateo, California: Morgan Kaufman.

Bares, J., Hebert, M., Kanade, T., Krotkov, E. P., Mitchell, T., Simmons, R. & Whittaker, W. (1989). An autonomous rover for exploring mars. *IEEE Computer, 22(6)*, 18–26.

Bolles, R. C. & Cain, R. A. (1982). Recognizing and locating partially visible objects: the local-feature-focus method. *The International Journal of Robotics Research, 1(3)*, 57–82.

Chen, C. H. & Kak, A. C. (1989). A robot vision system for recognizing 3-D objects in low-order polynomial time. *IEEE Transactions on Systems, Man, and Cybernetics, 19*, 1535–1563.

Cowan, C. K. & Bergman, A. (1989). Determining the camera and light source location for a visual task. In *IEEE International Conference on Robotics and Automation* (pp. 509–514). Los Alamitos, California: IEEE Computer Society Press.

Goad, C. (1983). Special purpose automatic programming for 3D model-based vision. In *Proceedings of the DARPA Image Understanding Workshop* (pp. 94–104). McLean, Virginia: Science Applications International.

Hansen, C. & Henderson, T. (1987). CAGD-based computer vision. In *Proceeding of the IEEE Computer Society Workshop on Computer Vision* (pp. 100–105). Los Alamitos, California: IEEE Computer Society Press.

Hong, K. S., Ikeuchi, K. & Gremban, K. D. (1990). Minimum cost aspect classification: a module of a vision algorithm compiler. In *10th International Conference on Pattern Recognition* (pp. 100–105). Los Alamitos, California: IEEE Computer Society Press.

Hutchinson, S. A. & Kak, A. C. (1989). Planning sensing strategies in robot work cell with multi-sensor capabilities. *IEEE Transactions on Robotics and Automation, 5*, 765–783.

Ikeuchi, K. (1987). Determining a depth map using a dual photometric stereo. *The International Journal of Robotics Research, 6(1)*, 15–31.

Ikeuchi, K. & Hong, K. S. (1991). Determining linear shape change: Toward automatic generation of object recognition programs. *Computer Vision, Graphics, and Image Processing: Image Understanding, 53*, 154–170.

Ikeuchi, K. & Kanade, T. (1989). Modeling sensors: Toward automatic generation of object recognition program. *Computer Vision, Graphics, and Image Processing, 48*, 50–79.

Ikeuchi, K., Nishihara, H. K., Horn, B. K. P., Sobalvarro, P. & Nagata, S. (1986). Determining grasp points using photometric stereo and the prism binocular stereo system. *The International Journal of Robotics Research, 5(1)*, 46–65.

Kass, M., Witkin, A. & Terzopoulos, D. (1988). Snakes: Active contour models. *International Journal of Computer Vision, 2(1)*, 321–331.

Marr, D. (1982). *Vision*. San Francisco: Freeman.

Pentland, A. P. (1986). Perceptual organization and the representation of natural form. *Artificial Intelligence, 28*, 293–331.

Sato, K., Yamamoto, H. & Inokuchi, S. (1987). Range imaging system utilizing nematic liquid crystal mask. In *Proceedings of the International Conference on Computer Vision* (pp. 657–661). Los Alamitos, California: IEEE Computer Society Press.

Solina, F. (1987). Shape recovery and segmentation with deformable part model. Technical Report MS-CIS-87-111, University of Pennsylvania, Department of Computer and Information Science.

Taylor, C. L. & Schwarz, R. J. (1955). The anatomy and mechanics of the human hand. *Artificial Limbs, 2*, 22–35.

Yi, S., Haralick, R. M. & Shapiro, L. G. (1990). Automatic sensor and light source positioning for machine vision. In *Proceedings of the 10th International Conference on Pattern Recognition* (pp. 55–59). Los Alamitos, California: IEEE Computer Society.

Part IV

Human and Machine: Telepresence and Virtual Reality

This section describes two applications that attempt to match the active, exploratory behavior of the human eye to virtual reality environments.

12

Active Vision and Virtual Reality

Brian C. Madden[1]
Hany Farid[1]

ABSTRACT The experienced quality of virtual reality is currently limited by available computational resources, and will be for some time to come. It is essential that these limited resources not be wasted on the acquisition and processing of data that do not contribute significantly to the final percept. Efficient construction of the intermediate views that are the basis of some forms of virtual reality (telepresence) depends on the proper selection of the acquired views. A fixed array of sensors that afford adequate resolution over the entire scene can present a prohibitive cost in bandwidth and computation while complete sensor mobility is technically difficult to achieve without becoming unacceptably intrusive. Active vision provides a mechanism of effective and efficient resource allocation in the transformation of real scenes into virtual ones. The use of electronic cameras and lenses mounted on positionable platforms (pan and tilt) that can track objects of interest, maintain sharp focus on salient surfaces and zoom in on regions of interest can provide the functional equivalence of a much larger array of fixed sensors.

12.1 Introduction

Scientific advancement is often limited by technology, whether it is the need for a better microelectrode or a faster microchip. Virtual reality is currently in such an elastic region of advancement. Qualitative improvements in performance mirror quantitative improvements in the hardware. While it is certainly true that computational resources will increase and pixel size will decrease, improvements can also be obtained through better utilization of current resources.

In this chapter, we describe a system for acquiring virtual views of a remote real environment using an array of active cameras (see Fig. 12.1). Depending on the position of the participant, the content of the scene and the task at hand, the configuration of the cameras can be changed to improve the utility of the provided views. In the first section, we survey the range of systems that have been applied to this problem and discuss the compromises they incorporate to accommodate current hardware limitations. We then briefly discuss the advantages of the active vision paradigm and apply it to the telepresence view acquisition problem. In the next section, we describe the collection of photometric and range data, its

[1]GRASP Lab, Department of Computer and Information Science, The University of Pennsylvania

FIGURE 12.1. Active Telepresence Example. Using this system, a clinician wearing a head mounted display could view a patient thousands of miles away through the appropriate deployment of an array of positionable cameras. The combination of photometric and range information from the different cameras in the array in a world coordinate system would provide a representation that could be sampled from an appropriate perspective to provide the desired view.

integration into a single representation and the creation of an arbitrary intermediate view. In the following section, we apply the proposed techniques to both graphical simulations and digitized images. In the final section, we discuss some of the strengths and weaknesses of this approach and the direction of future work.

12.1.1 Virtual reality and telepresence

> "Imagine a wraparound television with three-dimensional programs, including three-dimensional sound, and solid objects that you can pick up and manipulate, even feel with your fingers and hands. Imagine immersing yourself in an artificial world and actively exploring it, rather than peering in at it from a fixed perspective through a flat screen in a movie theater, on a television set, or on a computer display. Imagine that you are the creator as well as the consumer of your artificial experience, with the power to use a gesture or word to remold the world you see and hear and feel. That part is not fiction. The head-mounted displays (HMDs) and three-dimensional computer graphics, input/output devices and computer models that constitute a VR (virtual reality) system make it possible, today, to immerse yourself in an artificial world and to reach in and reshape it (Rheingold, 1991)."

The basic principle of such a virtual reality system was introduced by Ivan Sutherland (1965). He proposed a system that supported real time spatial and temporal interactions with created worlds instead of numeric or symbolic simulations.

Since that time, the contributions of numerous researchers have made "virtual worlds" increasingly life-like. There is, however, a considerable distance to go before rendering and reality become indiscriminable. By approaching the problem in a different way, a new vision of telepresence has emerged (Fuchs & Neumann, 1993; Minsky, 1979), where *graphical worlds* are replaced with *real worlds*. The telepresence worlds are real in the sense that physical relations present in an actual environment serve to constrain the displayed scene whereas virtual reality scenes in general can be made up out of whole cloth and need not suffer from any external constraints. This distinction becomes especially important when the real environment being emulated is dynamic.

Telepresence is a simulation system where participants wear a head-mounted display to look around an actual, but physically remote, environment. The surface geometries of objects in the participant's line of sight are continuously sensed and updated as they maneuver through the remote environment. The challenge is to create for the participant an arbitrary 2-D image of the 3-D scene from the collection of available 2-D images currently being acquired. This creation must be done in real time (and potentially separately for each eye) to provide the views appropriate for the observer's current orientation in the scene and maintain the illusion of immersion within the remote environment.

A further distinction can be made within the domain of telepresence between systems that represent real worlds by manipulating digital images of the environment or by graphically rendering simulacra. The limited number of polygons that can be created in real time by graphics engines results in an assembly of subsampled surfaces that discards most of the minute variations that make up real objects. While it is true that the visual system itself discards most of these variations in the process of extracting and estimating parameters of the environment, their absence disrupts the feel of realism. Ultimately, as graphics *verité* improves, these distinctions will disappear. Until that time, however, digital images of dynamic complex scenes will be the best source of lifelike surface patterns and, at the very least, the source of the dynamic updating of the position of objects in the scene used for rendering.

The question is how to efficiently acquire these digitized views. Consider a camera that can be instantaneously repositioned with six degrees of translational (left/right, up/down and toward/away) and rotational (pitch, roll and yaw) freedom and can transmit images with no delay. With such a universal sensor, a telepresence system could be realized easily by simply displaying to each of the participant's eyes the appropriate camera views. Lacking such a device, researchers have applied various compromises to bridge the gaps in acquisition technology. Most approaches to telepresence are variations or combinations of the following methods.

Impersonal viewpoint: In the 1970s, Artificial Reality was the term applied by Myron Krueger (1991) to allocentric[2] systems that used nonintrusive tracking of participants based on pattern recognition techniques. The recovered position is

---

[2] Viewpoints are usually either first person (yours) or third person (someone else's). The application here uses the unusual perspective of viewing yourself remotely, hence the term allocentric, or *other-centered.*

used to update the patterns on a projection display that shows a representation of the participant's body in the virtual environment. This *second person* approach avoids many technical difficulties by restricting itself to capturing only the relation between the participant's body and the environment and not the associated personal point of view. Personalizing the view would require tracking of the six degrees of freedom of head position plus the direction of gaze and even the state of accommodation.

Flat view: Another technology that falls in the periphery of virtual reality is the large field of view displays used in some remote meeting applications. These displays afford percepts that are flat, with no stereo or motion parallax cues. Even for fields of view exceeding 60°, the participant gets no feeling of immersion at the remote site due to the lack of depth. Virtual gaze and movement are restricted and the participant becomes an observer. These devices fall more within the domain of multimedia and teleconferencing than in that of virtual reality.

Stereo view: Large field of view displays can take advantage of devices such as liquid crystal shutters mounted on glasses that alternate in synchrony with slightly disparate scenes on the display to induce stereopsis. The screen width acts as an aperture into a remote site. Ohya, Kitamura, Takemura, Kishino and Terashima (1993) used a 70-inch stereoscopic display as the vehicle to present wire-frame 3-D models of remote participants. Markers on the participants' faces were monitored by two cameras mounted on a helmet and magnetic markers were placed on the head and body to facilitate tracking of position. Articulation of the hands and fingers was measured by data gloves. The paucity of presentation detail allowed the wire-frame renderings to be presented in near real time (e.g., a 1000 node head model could be displayed at 6 frames per second). This rate is more than a factor of 3 away from that required to generate the illusion of smooth motion. In addition, latencies integral to tracking caused delays in the updating of the display. Initially distracting, these latencies can ultimately induce severe nausea. Although not incorporated into many systems yet, predictive techniques such as Kalman filtering can be used to compensate for much of the tracking lag (Azarbayejani, Starner, Horowitz & Pentland, 1993). Stereo systems bring with them another burden. By presenting images on a screen with a common objective depth, the participant is forced to dissociate accommodation from accommodative convergence. While human observers have some plasticity in this domain, the dissociation brings with it both fatigue and adaptation. Long term immersion could result in lowered performance both in the virtual world and subsequently in the real world.

Enveloping views: The CAVE[3] system (Cruz-Neira, Sandin, DeFanti, Kenyon & Hart, 1992) extends the large screen stereo approach to envelop the participant in a cube of imaging surfaces. The advantage of this configuration over head-mounted displays is that only translation of the head need be tracked. Rotations and eye movements are unrestricted due to the simultaneous presentation of much

---

[3]CAVE is a recursive acronym for CAVE Automatic Virtual Environment.

of the remote scene on the facets of the cube. The apparent advantage of the panoramic presentation may be muted by the limited aperture of some stereo glasses. If the virtual objects are not too proximal, the tracking/display lags are not very intrusive since the large majority of perspective changes are small for minor head translations and mostly vanish for virtual distance exceeding 2 m. There can be, however, some distortion when objects overlap facet boundaries. It is difficult to compensate disparities in more than one plane for tangent distortion for all points of view. Other difficulties occur with this type of display when real objects such as hands improperly occlude virtual objects. Currently, the only way to avoid this conflict is to keep the virtual world behind the screen (*fish tank* VR).

Model-based: A different approach takes the form of a type of predictive coding wherein a model at the receiver is animated by key parameters extracted at the remote site. The savings in bandwidth are paid for with the increased computational load of reconstruction. Caudell, Janin and Johnson (1993) trained a neural net to reconstruct the silhouette of a face during speech given the location of a set of features (selected manually). They suggested that this technique can be extended to 2-D using polygons in the place of line segments. Terzopoulos and Waters (1993) applied a multilayer anatomical model that used the physical properties of the tissue, muscle and bone to generate deformations of the skin. The collection of these facial deformations structured expressions in a realistic manner. Complete range and photometric maps were obtained for each individual and were used to customize an epidermal mesh that overlaid the physical-anatomical model. The photometric information was texture-mapped onto the mesh to enhance realism. Extraction of key physical features (artificially enhanced) was used to control the deformation process from frame to frame. Extraction of facial features can also be used to track positional changes and allow the virtual head to be embedded in the remote environment (Azarbayejani et al., 1993). While the commercial potential of teleconferencing has resulted in much of the work being concentrated in the presentation of talking heads, these ideas may be extended to other classes of objects.

The approaches that incorporate high level knowledge of the remote environment overcome bandwidth limitations that often hobble systems that depend on real time transmission of digital images. However, the cost is great in terms of flexibility. The same *a priori* knowledge that allows efficient and detailed rendering also precludes the introduction of arbitrary (nonmodeled) participants, an implicit presumption of static world composition. In addition, while the local generation of images is quite advanced in some cases, the remote extraction of features in real time continues to be a stumbling block. Many of the actual facial deformations result in only the slightest of luminance gradients. In addition, the very extraction of the facial features of participants who themselves are wearing head-mounted displays is problematic. The advancement of hardware will both increase the advantage of encoding by improving the quality of rendered images, and decrease it as the bandwidth available to transmit images from remote sites increases.

Mobile sensor: The ability to move a boom-mounted stereo camera pair that is controlled by the head movements of a remote observer through a remote environment (or a mockup of one) is a solution at the other extreme in the trade between remote sensor positioning and local image manipulation. Some of the earliest flight simulators that incorporated visual feedback did so by passing a camera over a scaled landscape model. As the trainee altered the controls of the simulated aircraft and viewed the *out-the-window* displays, the path of the camera would be altered proportionately. The advantage of boom-mounted cameras is that the dynamic changes in the remote environment are independent of its complexity. For many applications the use of remote sensors *in loco ocularis* can provide a relatively inexpensive solution while providing a good deal of functionality in a dynamic environment.

Pan and tilt as well as boom-mounted sensors exhibit positioning latency problems. The acquisition latencies caused by delays inherent in moving the platforms are often far below the formidable rates that the participants exhibit when changing their position and especially their gaze. The use of predictive systems here as well may reduce errors due to tracking latency to an acceptable level. Some systems that don't track the participant's position require skilled operation of the cameras or, at the very least, occupy the hands of the participant to move the sensors (Cruz-Neira et al., 1992). This control requirement reduces the ability of the participant to interact with the environment, both locally and virtually. Furthermore, the booms can be intrusive and sometimes even dangerous for certain applications (e.g. tele-surgery, Adam, 1994; Fuchs & Neumann, 1993).

Local dome: An intermediate approach that employs a remote camera on a pan and tilt platform to acquire a series of images from the remote site can be used to texture-map images onto a virtual dome. Hirose, Yokoyama and Sato (1993) presented observers with a series of views on a head-mounted display that reflected the sensed position of their heads. The local updating eliminates acquisition latencies (transmission limitations and delays in camera positioning) that could disrupt the illusion of remote presence and eventually induce nausea. Additional preprocessing that yields the relative parallax of objects at different depths by rotation about a point other than the nodal point of the lens was used to deform the virtual dome and to provide enhanced realism through parallax distortion of near (within 2 m) objects. Although new photometric information could be acquired for each new viewing direction, the system as proposed was designed to allow the observer only rotational degrees of freedom in a static world. As with other systems that rely on local computation based on models, this approach is limited in its ability to represent novel events in the remote world.

Camera array: One way to improve the speed of image acquisition over that available with a mobile system is to preposition a fixed array of cameras about the perimeter of the remote environment. Fuchs et al. (1994) proposed using a *sea of cameras* to acquire both photometric and depth information. They used a multibaseline stereo algorithm to compute the disparity of local image patches and then used the disparities to construct a polygonal model for a given camera position. The photometric information was texture-mapped onto the model and the result was used as the basis of display while the participant was in a viewing position near that of the selected camera. As the participant moved about the remote environment, the process was repeated for the cameras in the array most appropriate for the current point of view.

An array of cameras mounted along the ceiling and walls eliminates the positioning problem but incurs a host of problems inherent with static sensors. It is difficult to adjust the camera parameters to be appropriate for image acquisition over a large volume of a remote environment. Very often the most appropriate camera will be obstructed. In addition, calculating accurate, dense, high resolution depth maps from a pair of images offset in space is far from being a solved problem. In particular, specularities disrupt accurate depth estimation even when problems in obtaining correlations due to occlusions do not occur. Beyond the issue of depth acquisition, the computational cost of constructing a model from stereo data of the remote environment at each instance in time is formidable.

Retinal projection: The virtual retinal display (Kollin, 1994) proposes to project a high resolution, panoramic, chromatic image directly on the retina of the eye. Using Maxwellian view optics and by tracking the pupil of the eye, they intend to pass a 140° unoccluded image through the entrance pupil of each eye. They also propose to control the divergence of the beam thereby inducing a stimulus to accommodation. If successful, this technological leap would go a long way toward eliminating many flaws currently present in virtual displays. Even this device, however, will not solve the problem of how to obtain the views that are projected.

Although some of these methods exhibit a considerable amount of telepresence, each also inflicts considerable limitations on the illusion of remote presence. These limitations are the consequence of the formidable task of determining the viewing state of the observer, obtaining the appropriate views (or the information necessary to recreate the views) and presenting the views. To match the available resources, compromises are made in the degree to which the participant can interact with the remote environment, the level of detail available, and the flexibility of the approaches in coping with dynamic or novel events. As the presentation technology improves, some of the compromises will be relaxed; however, many of the difficulties arise from a need to acquire remote views in a versatile, timely and economical fashion. Perhaps, then, in a manner similar to the way that a different approach (telepresence) was needed to improve upon the original concept of virtual reality to overcome hardware limitations, an extension of the concept of telepresence itself would now be useful.

12.1.2 Active vision

As long as bandwidth limitations exist in the real time presentation of remote scenes, there will be a need for strategies that maximize the percentage of transmitted information that is relevant to the requirements of the observer. The essential canon of active vision[4] (Aloimonos, Weiss & Bandopadhay, 1987; Bajcsy, 1988; Ballard, 1991) is based on feedback from the current image altering future processing and acquisition. Active vision is a collection of sensor control strategies wherein the cost of mobility and calibration is more than offset by the increase in the density of information appropriate for the task, information that is selected from the staggering amount available. Passive (static) sensing of the environment often results in attempts to reconstruct the world that are underconstrained. There are limits as to how detailed a representation of a given scene can be obtained by simply pounding on a set of pixels, and even if theoretically soluble, limits exist on the stability of solutions in the face of acquisition noise. Beyond these considerations, a complete and detailed recreation of the entire surroundings is not required for most tasks (recognition does not require total reconstruction, Aloimonos, 1990).

Active vision techniques use the control of the sensors' pose relative to the objects' surfaces both to select manageable portions of the environment relevant to the current task as well as to simplify the extraction of the required parameters from those selected views. Active vision is not just a change in viewing parameters, but is an adaptive response to the demands of the task given the environment. The question arises, at what level are sensors active (e.g., at the level of pixels, algorithms, positioning mechanisms, filter shapes, goals or priorities) and what is the difference between being active and merely adaptive? Clearly, circumnavigating an obstacle to obtain information previously obscured on a remote object is active; but is stopping down an aperture active as well? Yes, if the manipulation extracts a needed new dimension from the scene not previously available (e.g., depth), and no, if only the signal to noise ratio is being improved. As Pahlavan, Uhlin and Eklundh (1993) proposed, to be more than just trivially active, a system must be making a choice between at least two alternatives. These choices can be implemented in many ways, on many levels.

Perhaps the most basic operation of active vision is tracking (the ability to acquire and to maintain acquisition of a region of interest). Stabilization of a moving target within a sensor window offers several advantages (Madden & Cahn von Seelen, 1995). A constant target size facilitates the use of object-centered coordinate systems. Multiple estimates of minimally varying targets can be obtained. The motion blur of a moving target can be reduced (or eliminated) while the induced motion blur of the background can be used to help segment objects of interest. Since the dynamic range of target motion need not fit into a static window, the resolution of the region of interest can be higher. All of these consequences of actively tracking targets serve to reduce the complexity of higher-level processing.

In cases where the Lambertian reflectance component can be determined, object motion can be derived simply by servoing on the location of a fixed intensity pattern (Aloimonos et al., 1987). The motion of the object is evident from the

[4]"Active" is used in the sense of exploring the environment, not to distinguish sensors that emit energy (laser rangefinders) versus passive collectors (ccd cameras).

camera movement required to maintain fixation rather than from some optic flow calculation. In cases where optic flow is useful, the flow patterns can be optimized by controlling the motion of the sensor to extract the desired parameters. Even a static scene can yield flow information relevant to depth if the sensor moves through it.

If the responses of two sensors are combined to derive stereo information, active cameras can be verged to improve disparity resolution for a fixed baseline or, by moving the sensor, larger effective baselines can be created by incrementally combining the ease of correspondence in a short baseline to obtain the disparity resolution available in the larger. The technique of continually improving pose to obtain more independent estimates or to eliminate obstructions to aid under-constrained computations yields more accurate estimates of the target because no regularization is required (Aloimonos et al., 1987). By actively varying the focus of the camera lens, depth estimates can be obtained with a single sensor, thereby avoiding the correspondence problem altogether (Krotkov, 1989).

The advantages of active vision go beyond these (considerable) aids to parameter extraction. Some researchers have expanded the concept that began as a sensory acquisition strategy to cover virtually all cognitive tasks that rely on visual input (Aloimonos, 1990; Ballard, 1991). Active vision can be shown to facilitate the execution of complex tasks. Many problems cannot be solved from a static perspective. Visual problem solving often requires a sequence of views and the scanning patterns that are optimal will vary with the task. Conversely, visual properties are often dependent on the behaviors (and views) that generated them.

The proper design of such techniques and tools is an open question, but one thing is clear: The observer is more than just a recorder. Variables at many levels will contribute to optimal performance. Interaction with the environment depends on the task at hand. The active pursuit of information that will allow determination of one set of parameters also uncovers future questions that need to be asked. Active visual processing is inextricably part of cognitive processing and problem solving.

12.1.3 Active telepresence

What problems need to be solved in a telepresence application? The essential service that must be provided is the delivery of views appropriate for the participant's state relative to that of the remote environment. The nature of the supplied views and the weighting of errors in timing, geometry and photometry will vary with the application. Certainly the veridicality of views in a clinical diagnostic setting will have a higher weight than in an entertainment application where accuracy may be sacrificed to smooth over deficits and preserve the suspension of disbelief. Lacking the universal sensor described above, arbitrary viewpoints must be interpolated from available images. We propose to do this by combining the photometric information with a depth map obtained for the same view. Both maps are then supersampled using bicubic interpolation and the result is projected into a world coordinate system. Forward ray tracing is used to determine the projection of the points onto the virtual sensor appropriate for the desired point of view. Points distal to the sensor along a common ray are eliminated as obstructed and the remaining projections are mapped to the nearest pixel locations. This procedure is repeated for each remote sensor and the virtual view is a weighted composite using a registration algorithm and coarse camera calibration to align the components. (The method of interpolating views will be described in detail in Section 12.2.) While this technique provides a required service, it is not active in the sense developed above.

In what ways can the image acquisition techniques of active vision enhance the performance of telepresence systems? While the physical sensor support for active telepresence could take many forms, we have developed a scenario wherein an array of active sensors (pan/tilt/focus/zoom) arranged about the periphery of an environment can unobtrusively acquire information from that environment. Our approach builds on ideas contained in the *sea of cameras* configuration (Fuchs et al., 1994) as well as some of the dynamic aspects of the *local dome* system (Hirose et al., 1993). This configuration is a balance between intrusion (interference with the activities of the remote environment) and functionality (the ability to provide intermediate views of dynamic and novel scenes containing concavities and interpositions). The active component comes from using sensor input to reconfigure the array so as to improve the view presented to the participant.

Whereas a fixed array of cameras must distribute its coverage evenly over the environment (or cope with blind spots), positionable sensors can dynamically alter the sampling density and bring resources to bear where they are currently needed. In addition, differences in the location of the camera platforms allow a redundancy in the coverage of any given area to also contain a range of available pose. For example, high resolution zoom images of critical regions could be merged with data from coarser wide field of view images. In contrast to fixed sensor systems, targets of interest would not be restricted to a particular sweet spot but could be mobile. This concentration of resources could follow targets of interest around the environment. Alternatively, some sensors could be dedicated to the acquisition of fine-grained depth information without incurring the penalty of the associated photometric distortion. This specialization can be carried further through the use of different types of sensors in the array. Cameras with megapixel sensors could be mixed with color cameras or, as they become available, sensors designed for real time focus ranging (e.g., Krishnan & Ahuja, 1993). This sensor diversity does not necessarily require multiple arrays since most sensors can provide multiple functionality. It is just the ordering of their performance on these tasks that varies.

The information that can be extracted from a scene will vary with the scene dynamics. While we propose that the information required for the construction of arbitrary views can be extracted from the acquired images and does not need *a priori* models, this does not mean that all information must be extracted anew from each frame. The bottom-up models are indeed *disposable* in that new models can be constructed from the projection of the photometric data of subsequent images into world coordinates. However, it is also true that in many applications not much displacement occurs in 1/30th of a second. As is often true, in this case the faster (more frequently) a process is executed, the simpler it becomes. In scenes where not much is changing, small deformations detected in the new data can be accommodated in the world projection of the old data at a lower cost than recomputation from scratch. At the very least, the previous results can be used as the seeds for calculations based on the new data. The savings in time accrued by this reuse will allow the representation to be augmented in terms of both resolution and dimensionality. If the changes are occurring apace, then the acquisition parameters must be ready to handle a much larger range. To accommodate variations in demand, we propose a coarse-to-fine presentation of photometric and depth values. In cases where this is unacceptable in terms of perceived quality, an alternative would be to display the closest acquisition view and then transform it into the desired view as sufficient data are acquired. While sensor lags in positioning do occur and could potentially limit performance, the consequences are less intrusive since the interpolated views only have to be positioned somewhere in the space spanned by the available basis views[5]. There need be no delay, therefore, as would be the case with a one-to-one acquisition/presentation methodology.

Another dynamic that must be accommodated is the often severe spatial gradients brought on by perspective, interposition and differences in viewpoint. The quality of surface representation depends on the pose of the virtual view relative to that of the acquisition views. If the projection from the virtual view is much closer to the surface normal than that of an acquisition view, the sampling will be coarse. Estimates of the surface normals can be obtained from the depth gradients and coverage by the array could be altered to improve the sampling. The closest camera does not always afford the best view. Cameras more lateral in the array could be panned to supplement the sampling of a particular surface. Sensors with optical axes 90° or more away from that of the desired view, depending on the orientation of the surfaces in the scene, can contribute to the final composite. Determining the optimal allocation of pose among the cameras in an array is an open question in active vision which we hope to address with this configuration. The actual coverage will depend on the array spacing which, in turn, will depend on the application. However, two or three rows of positionable sensors spaced about 1 m apart should provide adequate coverage for simple viewing tasks over a working volume 2 m high and several meters in depth for the length of the array.

The distribution of coverage by the array can also be altered to minimize the extent of hidden surfaces (Fig. 12.2). There are at least two conditions where boundaries of missing data can be detected. If an object (a continuous region of depth more proximal than the background) lies along a boundary of an acquisition image and that boundary does not project inside of one of the other acquisition images, it is likely that that edge marks a region of missing data. In addition, if the pixels on the edge of a continuous region of similar depth values (bounded by depth discontinuities) do not project into a continuous region of continuous

[5]The acquisition views supplied by the camera array do not necessarily afford a complete representation of all possible surfaces within their field of view. The determination of the set of interpolated views completely covered by a set of acquisition views given a particular scene is not a solved problem.

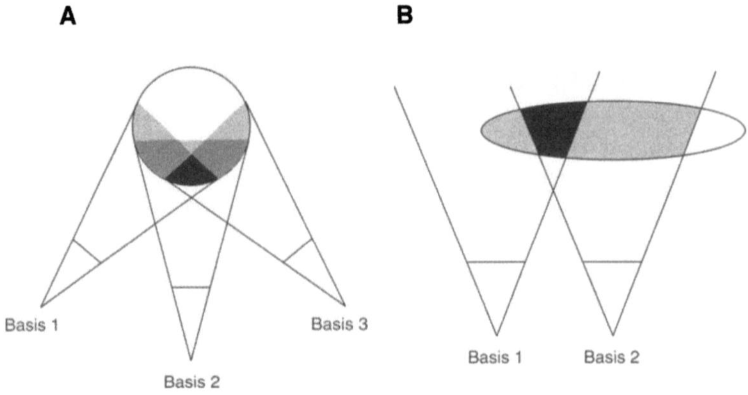

FIGURE 12.2. Hidden surfaces. In a given scene, not every surface will necessarily be completely within the span of the sensors; however, an occlusion boundary does not always indicate the edge of missing surface information. In A, each basis view is seen to cover a region of the sphere's surface that is slightly less than a hemisphere. The overlapping coverage often results in an occlusion boundary projecting to the middle of another view. It is only the boundaries that are not visible in other views that delimit the regions of missing data (unshaded). Similarly in B, while the ellipsoid is truncated by both sides of basis 2, only the right limit is a boundary of missing data. The minimization of such boundaries is one measure of the goodness of a given camera configuration.

depth values in another image, then they likely mark regions of missing data as well. Again, cameras in the array could be positioned to minimize the missing data boundaries or to eliminate them in a particular critical location.

In a manner analogous to the way the notions of purposive and animate vision extended the concept of active vision to more complex cognitive activities, it is possible to envision similar enhancements augmenting telepresence. A recent National Academy of Sciences report on virtual reality concluded that there are many applications where the enhancement of special features may be more important than realism (Durlach, 1994). Using data extracted from previous views, a participant could be guided to an appropriate pose for a given task. Work on using virtual fixtures has been shown to be a valuable tool for performing tasks remotely (Sayers & Paul, 1993, 1995). Fixtures don't just react to the state of the participants, they strive to actively assist their actions. The fixtures are always present in the environment but are activated only under the appropriate conditions. They channel the activity in a particular direction, but do not compel compliance. For example, a fixture could be used to maintain the position of a clinician relative to a freely moving remote patient but would still allow sufficiently large or rapid deviations in position or gaze to break off the attachment. Over time, use of such a *visual toolbox* would become second nature.

In summary, then, the extension we propose to the concept of telepresence is the assimilation of the techniques of active vision into the acquisition of images for remote viewing — an active telepresence. The essence of this proposal is that a universal sensor can be approximated and even improved upon by the application of sensor strategies that interact with the remote environment.

In the following section, we will detail how the continuum of intermediate views that are necessary for an observer to experience a remote presence can be generated from a limited number of sensors. For images obtained from an array of active cameras that are adjusted appropriately in focus, zoom and direction, we will demonstrate that, barring any narrow concavities, acquired photometric data can be brought into registration, merged and warped into a pattern that approximates any viewpoint in the span of the camera positions (and, at increasing risk, may extend beyond that span).

12.2 Generating Views

The remote environment is a collection of point sources. Each point on an object acts as a radiator, throwing off a near continuum of spherical waves. With the exception of some special circumstances where these wavefronts exhibit phase coherence and produce interference patterns, the waves simply expand and superimpose on one another. A manifestation of the particle nature of light occurs at low intensity levels in the statistical distribution of photons. An active sensor acts as a sampler of this quantized superposition of diverging light. At a basic level, vision is the adjustment of the position and optical parameters of a sensor so as to select the appropriate portions of the wavefronts necessary to disambiguate the superposition with a criterion level of reliability. Beyond the contributions of the degrees of freedom inherent in the scene geometry, this disambiguation is rendered somewhat more complex by the superposition of specularities that not only retain varying degrees of the spectral composition of the illuminants but the associated wavefront curvature as well.

The fundamental resource of active telepresence is a sensor or sensors that can alter their properties to obtain the necessary estimates of surface reflectance in response to the state of the participant. A small array of active cameras can provide dense coverage of an environment by adjusting the fields of view to abut or overlap in a manner that supports the interpolation of intermediate views. With control of zoom, abutting fields of even frontoparallel cameras can be formed at arbitrary depth planes — an approximation to translation in depth. With control of pan and tilt, cameras can be brought to bear on regions throughout a wide range and, therefore, subjects need not be restricted to a special focal area. Each acquired view can be thought of as a *basis* vector in the multidimensional space that is the remote environment. The task is to obtain a sensor configuration that forms a basis set that spans that portion of the space containing the desired views. A major advantage of using such a basis set to construct virtual views is the decoupling of the acquisition frustum (the 3-D pyramidal view of the camera), from the participant's field of view, thereby easing restrictions on both.

There are as many potential sensor configurations as there are applications. The scenario developed in this paper is based on an array of cameras arranged about the periphery of a scene. The proposed array is composed of three rows of sensors spaced 1 m apart horizontally and vertically. The center row is approximately at head height. This configuration provides good coverage (a good basis set) over a large volume (a variation is also being examined that shifts the center row laterally by half the sensor spacing to improve the acquisition of concave surfaces). The cameras are mounted on motorized platforms that allow a range of pan and tilt on an axis through the nodal point of the lens. The lenses are motorized and allow control of zoom and focus. The following sections detail the physical and algorithmic components that generate the interpolated views.

12.2.1 Camera calibration

A standard camera calibration method based on the method of Tsai (1987) is implemented in our system. This method calibrates both the extrinsic (rigid-body transformation between camera and world coordinate systems) and intrinsic camera parameters (focal length, lens distortion, and scaling parameters). The calibration technique requires a *calibration fixture* with at least seven, noncoplanar, calibration points with precisely known positions relative to an arbitrary world coordinate system. Each basis view may be calibrated by first determining the position of each of the calibration points in image coordinates. With calibration points in hand, the extrinsic camera parameters may be recovered by solving a set of, possibly overconstrained (i.e., more than seven calibration points), linear equations. Intrinsic parameters are recovered through a gradient descent, or comparable minimization (see Tsai, 1987, for more details). Although the mathematical principles underlying this calibration technique are sound and well understood, highly accurate camera calibration has proven to be quite elusive in practice. Problems exist in determining the *ground truth* of targets as well as locating these calibrated features in the digitized image with sufficient precision. In addition, nonlinear minimization is susceptible to being trapped by local minima during the selection of solutions for the camera parameters. As such, we propose to use the results of this camera calibration as the initial conditions for the image registration stage described below. To model the response of the lenses over the available range of focus and zoom, a number of static calibrations need to be performed (Willson, 1994). These results will form a lookup table that can be interpolated for any arbitrary condition of a lens.

12.2.2 Digitization

For the purposes of experimentation and validation, the array of active sensors will be approximated by positionable electronic cameras (a TRC BiSight 2-axis camera platform) mounted on a robotic arm (PUMA 560). Additional static cameras may be added to provide a wider range of simultaneous views. The combined eight degrees of positioning freedom can be used both to simulate the pan and tilt of the active sensor and the position of the sensors in the array. Two SONY XC77-RR CCD cameras can be asymmetrically verged by the TRC platform with an inter-pan axis baseline of 25 cm. The pan axes are capable of a peak velocity of 1000 deg/sec and a peak acceleration of 12,000 deg/sec^2 while maintaining a precision of 2′.

Offering three additional degrees of freedom per camera are the Fujinon motorized lenses (H10x11E-MPX31). The lenses have motorized focus and zoom (focal length: 11 to 110 mm). The 10 to 1 change in magnification has a corresponding effect on the field of view (approximately 4 to 40°). Light levels can be controlled either by a motorized iris in the lens or by an adjustable flux integration time (electronic shutter) in the camera. The monochromatic NTSC camera output is digitized (Data Translation DT1451; 512 by 480 pixels) and read into a SPARCStation (Sun Microsystems) for analysis.

12.2.3 Active estimation of surface depth

As pointed out by Pahlavan et al. (1993), the projection of the world onto a camera sensor is a 3-D to 3-D mapping. Although the sensor records the impinging wavefronts in a 2-D array of pixels, the distribution of the light from each point in the environment is distributed in a manner characteristic of its distance and the optics of the lens (Hopkins, 1955). This curvature of the optical wavefronts is commonly neglected in the formulation of the plenoptic function of the elemental sources of visual information (e.g., Adelson & Bergen, 1991). Along each ray from the nodal point of the lens, there is a distance at which a point radiator will produce the smallest blur circle in the sensor plane. The collection of these points form a (sometimes complicated) surface that will have the sharpest image. As a point being imaged moves along the ray between it and the nodal point in either direction, the distribution of the light from that point is enlarged. For incoherent light, a good approximation to the changing point image is a disk of varying diameter. As the point moves further from the surface that is conjugate with the sensor, the increasing disk diameters act as omnidirectional lowpass filters with ever lower cutoff frequencies.

Krotkov (1987) developed a technique to use these changes to assign depth values to local patches of the image. His focus ranging algorithm was based on determining the focus setting of a calibrated variable focus lens that produced the image with the greatest amplitudes in the high spatial frequency range. The Tennengrad metric of local image content was computed by convolving each small (10 to 20 pixels square) region of the image with 3 by 3 vertical and horizontal Sobel edge operators and combining them to form the oriented edge gradient at each pixel position. The sum of these high frequency bandpass filters over the local patch was tracked. Then, by knowing the relation between the focus setting and the distance of the surface that was conjugate with the sensor, a depth value could be assigned to a local image patch that corresponded to the peak of the Tennengrad metric. Depth resolution was found to be 1% of distance over a range of 1 to 3 m.

It is somewhat surprising that more effort has not gone into developing this technique given the advantage of not having to solve the correspondence problem. One reason focus ranging has been relegated to the backwaters of computer vision may be the lack, until recently, of a range of commercially available motorized lenses. Advances in technique should increase its application. Xiong and Shafer (1993) improved depth resolution by an order of magnitude by replacing the peaks found by a Fibonacci search method with the mean of a Gaussian fit to a range of focus settings about the maximum found by the Fibonacci search. Equally large advances could be made with better hardware. Much of the optics used today is designed to have a large depth of field. It is considered advantageous to have as much of the field of view in focus as the light level and sensor sensitivity will allow. Focus ranging, on the other hand, would benefit from a very small depth of field. In addition, many of the commercial motorized lenses are designed for surveillance applications where precision positioning of the optical components is not a requirement.

Advantage can be taken of redundancies and parallelism in the current application. Portions of the scene are viewed by more than one camera. The overlap can not only be used to consistently merge the different views into a world coordinate system, but can provide multiple opportunities for focus ranging. These overlapping views may have very different optical parameters, allowing very precise regions of depth measurements to be combined with larger regions of coarse values. Parallelism exists in the form of the hundreds of small patches in each acquisition image that are used to determine the Tennengrad metric. While it is possible that each patch is independently displaced in depth between frames, it is far more likely that in the 30th of a second frame interval the motion of the objects in the scene relative to the camera can be well modeled by a rigid (or at least a piecewise rigid) transformation. Depending on the reliability of the depth measures, significant deviations from rigidity can be incorporated into the projection of the data into the world view. The parallelism of focus metric computation is itself redundant in the regions of view overlap. In addition to range from focus (Krotkov, 1987) and range from defocus (Pentland, 1987), which requires detailed knowledge of the optics and two distinct focus settings, the method proposed here forms a third way to use focus information to constrain depth estimates — slight perturbations of hundreds or even thousands of depth computations that in the aggregate capture the approximately rigid change in the camera to object transformation that occurred between the acquisition of the previous and current frames.

12.2.4 Registration

Knowledge of the position of the basis cameras in a common world coordinate system is essential to successful interpolation. As was stated in Section 12.2.1, recovering accurate camera position (extrinsic parameters) from standard camera calibration techniques (Tsai, 1987) has proven to be a formidable task. As such, we propose to use the results of camera calibration as initial conditions for an image registration algorithm. The formulation of image registration developed by Besl and McKay (1992) starts with a model shape in a model coordinate system and aligns 3-D data taken from a sensor coordinate system by determining the rotation and translation that minimizes the distance between the shapes using a mean-square distance metric. In our application, there is no model. We are interested in bringing two or more data sets, whose location in the world coordinate system is known only approximately, into mutual registration by aligning their overlapping regions. The data from the camera location nearest to the point of view being interpolated are taken as the basis to which the other shapes are registered. Using the results of camera calibration as initial conditions, we apply the the iterative closest point (ICP) algorithm of Besl and McKay (1992) to compute the accurate position of the basis views relative to one another. For views that have no overlap, registration into the world coordinate system is based solely on the camera calibration of the acquiring sensor.

12.2.5 Warping

A standard *forward warping* technique (Wolberg, 1990) is used to *warp* or *project* each of the basis views to the desired virtual view. Both the intensity image and depth map, as discussed above, are required for this procedure. The warping is performed by first projecting each sensor element in the basis view into a common world coordinate system, as specified by the associated depth map. This point is then reprojected into the virtual sensor plane (Fig. 12.3(A)). The intensity of the basis sensor element is assigned to the virtual sensor element whose region was intersected by the projection. Although this warping technique is straightforward, several technical difficulties arise. It is very common that a single sensor element in the basis view will not project into a single sensor element in the virtual view (Fig. 12.3(B)). As such, we currently perform a nearest neighbor interpolation and assign the intensity to the nearest sensor element. This interpolation is necessary for both the generation of the final digital image and for the hidden surface computation. That is, if two points project into the same sensor position in the virtual view after nearest neighbor interpolation, their depth values are used to determine which point should "shadow" the other. To avoid aliasing artifacts due to the nearest neighbor interpolation, the original basis views are upsampled by a factor of two before warping. The virtual view is subsequently subsampled accordingly.

Although the upsampling eliminates some of the aliasing, several artifacts still remain. We are currently investigating several approaches for eliminating these artifacts. With each of the basis views warped to the virtual position, a complete view is generated by integrating across each of these views.

12.2.6 Integration of views

The integration of warped basis views is currently based on a somewhat *ad hoc* weighting scheme (Fig. 12.4). In particular, for each pixel in the virtual view, the contribution of each warped basis view is given by $e^{-0.05\theta}$, where θ is the angle

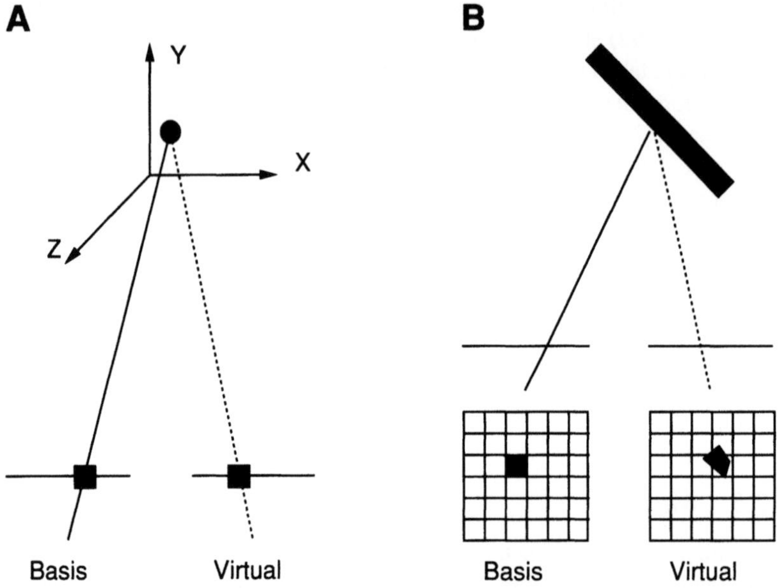

FIGURE 12.3. Warping of a pixel. A basis view is three dimensional. Each photometric map that is acquired must have a corresponding depth map. Each depth value is associated with a frustum delimited by the square or rectangular region that is the area of a pixel as seen from the nodal point. When projected into the world coordinate system and then back to the virtual view, the projection of the pixel area undergoes a perspective distortion. The resulting region that corresponds to a given basis pixel may cover (all or part of) several virtual pixels, or it may only fill a small portion of a single pixel.

between the rays connecting the object in 3-space and the virtual and basis views. This weighting function was chosen so as to have a value of 1 when $\theta = 0$ (i.e., the virtual view is coincident with a basis view) and fall off gradually as θ increases (i.e., the virtual view moves away from the basis view). More sophisticated methods are currently being investigated which weight the contribution of each basis view by its position relative to the surface normal. In particular, the finest sampling of a surface (and hence the richest representation) is not necessarily obtained from the closest view, rather it is obtained when the imaging sensor is parallel to the surface being imaged (i.e., perpendicular to the surface normal). Estimates of the surface normals can be obtained from the depth gradients and the contribution of each basis view can then be specified as a function of the surface normal.

12.2.7 Distortions

The interpolated views contain two sources of error, intensive and positional. It is helpful to separate out the two kinds of error when analyzing the quality of the generated views. Comparisons are ideally made between the actual view that would be obtained at an intermediate position and the interpolated view for corre-

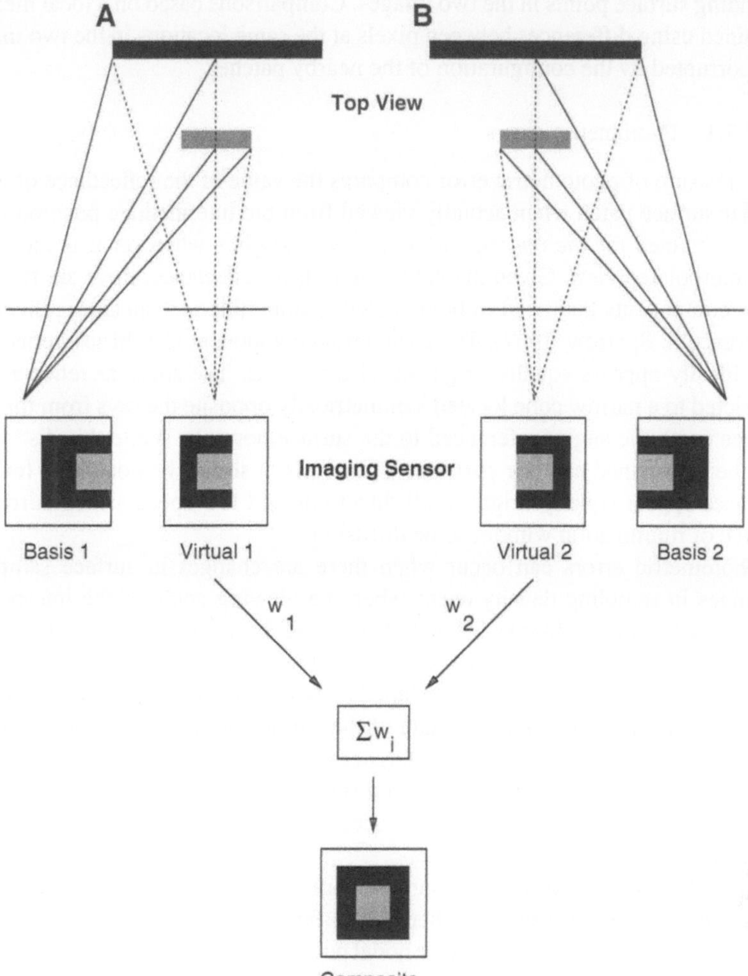

FIGURE 12.4. Integration of views. In general, due to a limited field of view or by interposition, a surface needed for a virtual view may not be completely visible from any position of any sensor in the array. By combining the multiple acquisition views in a single world coordinate system, it is possible to form composite surfaces. A ray is computed between each point projected from a given basis view to the nodal point of the virtual view. The virtual projection of the point occurs where the ray intersects the virtual sensor. Each projection is labeled with its distance from the virtual nodal point. Multiple projections to the same sensor region with different depth values indicate an occlusion (e.g., the dotted vertical line in (A) and (B)). The distal point is removed. The remaining projections are combined to form the regular sampling grid of the virtual image. Each partial virtual image is then weighted by the angular difference between the basis and virtual views. The weighted sum forms the composite image.

sponding surface points in the two images. Comparisons based on a local measure obtained using differences between pixels at the same locations in the two images are corrupted by the configuration of the nearby patches.

12.2.7.1 Photometric errors

The measure of photometric error compares the value of the reflectance of a particular surface patch when actually viewed from the intermediate position to the value obtained *for the corresponding patch of surface* wherever it is located in the interpolated view. Given the nature of surface reflectance, there are two separate components that need to be modeled, Lambertian and specular reflectance (Torrance & Sparrow, 1967). The Lambertian component is a diffuse reflectance that ideally appears equally bright in all directions. The specular reflectance is restricted to a narrow cone located symmetrically opposite the rays from the light source (with the angles referenced to the surface normal). While this distinction will be maintained here for purposes of analysis, it should be noted that few real surfaces appear equally bright in all directions and few specularities mirror the source of illumination without some diffusion.

Photometric errors can occur when there are changes in surface sampling. Changes in sampling density occur when the viewing angle of the interpolated image relative to a surface is different than that of the acquisition view. If the interpolated view is closer to the surface normal, the surface will be dimmer than the actual view. It is possible (but not included in the current system) to use the gradient of depth values to obtain an estimate of the surface normal. This estimate could be used to correct brightness for differences in surface slant. Pattern information on a relative slanted surface suffers from such subsampling. This distortion is not improved by simply scaling by differences in slant and can only be corrected by acquiring an image with better pose.

Specular error is linked to the surface normal in a different way. Specularities appear at positions on a surface where the surface normal bisects the angle formed between the point, the source and the nodal point of the view. In general, as the view changes, so will the distribution of specularities. Several methods exist to remove the specular component of an image (Klinker, Shafer & Kanade, 1987; Lee & Bajcsy, 1992; Madden, 1993). It is possible to actively acquire the location of the light source(s) and to obtain multiple views of the same surfaces from different orientations. Together with an estimate of the distribution of surface normals, this information could be used to locate, extract and reposition specularities so that the specular component of the interpolated view is correct. These techniques are also not incorporated in the current system. In practice, an uncorrected interpolation will give the incorrect impression that the light source has shifted or that there are multiple sources (a specular diplopia).

12.2.7.2 Geometric errors

The positional error measure compares the location of a surface patch when actually viewed from the intermediate position to the position of the same surface patch in the interpolated view. Very few of the incorrect positional displacements are due to sensor distortion. With the quality lenses available today, what little distortion there is can be quantified by calibration and corrected by a radial lens model. By far the largest component of the geometrical error is in the depth estimates. Impact of these errors increases with the angle between the acquisition and interpolation views. When the desired view matches one of the basis views, there is no need at all for depth measures. The distortion caused either by absolute errors in the projection of points into the world coordinate system or by their quantization reaches a maximum when the interpolated view is 90° away from the acquisition view. The depth errors are brought about in part by the presence of steep gradients. The Tennengrad metric is computed over a region of several hundred pixels. If the region has point images that form at very different depths, the metric will be degraded. One of the measures of a good sensor configuration is how close the pose of each sensor is to the surface normals of the region of interest.

12.3 Results

By increasing the density of an array of active sensors, the errors in interpolated views can be kept arbitrarily small. The cost in hardware and computation cannot be said to do the same. We need to determine how much of the look and feel of actual camera motion between two positions can be obtained simply through appropriate merging of available information in the displayed scene. In this section we will present images corresponding to interpolated views created using both graphically generated and digitized basis views. Examples are presented that show the quality of images within the span of the basis views as well as what happens when the desired views extend beyond this. It should be noted that these static images give only a partial indication of the appropriateness of the interpolated views. Full judgment should be reserved for assessment of the images presented at frame rates in phase with movements of the participant.

The simulations are 3-D arrangements of spheres and ellipsoids rendered using varying surface reflectance models (both Lambertian and specular) and a single light source. The rendered images are created to correspond to simulated camera positions. Each rendered image is matched with a depth map. Together with the location of the nodal point and image plane of the sensor, the depth map can be used to locate each pixel in the simulation at a position in the world coordinate system. Specification of the nodal point and image plane of the view to be interpolated is all that is required to map (warp) the pixels in the world coordinate system into the desired view. Relatively simple antialiasing and defocus routines are applied to the warped image. The individual warped virtual views are blended to form the final image.

12.3.1 Simulations

We built a graphics-based test-bed of our proposed telepresence system. This system allowed us to generate arbitrary views of a graphics world through standard ray tracing techniques. Virtual views from arbitrary positions were interpolated from a small set of ray-traced basis views. The errors in the virtual view were analyzed by comparing the virtual view to a ray-traced image of the scene from the same position as the virtual camera.

Our graphics world consisted of an arbitrary number of quadric surfaces. The general form of a quadric surface, $f(x, y, z)$, is given by the following equation:

$$ax^2 + by^2 + cz^2 + 2dxy + 2eyz + 2fxz + 2gx + 2hy + 2jz + k = 0. \quad (12.1)$$

This quadric equation is a 3-dimensional extension of the general conic equation. Spheres, cones, cylinders, ellipsoids, and paraboloids and hyperboloids of revolution are among the shapes that may be generated from a quadric equation. This representation is desirable since quadric objects have surface normals that can be computed analytically. Objects in the world had a Lambertian surface property with specular highlights. Lighting consisted of a point light source plus ambient light. When objects were positioned between the point light source and a surface, shadows were cast accordingly. Arbitrary views of the world could be generated by specifying the position and viewing direction of a camera, along with its focal length and aperture.

A standard ray tracing algorithm (Foley, van Dam, Feiner & Hughes, 1992) was employed to generate views of the graphics world. A ray was projected from the camera's nodal point (a pin-hole camera model was assumed) through each sensor element. Calculations were done to determine whether the ray struck any of the quadric surfaces in the world. Since all objects in the world were analytic, it was possible to compute the intersection between a line in the world coordinate system (the ray) and any quadric surface. If the ray intersected more than one surface, the closest surface was assumed to obstruct the other surfaces (i.e., no transparent surfaces were allowed). At obstruction boundaries where depth discontinuities occurred, the quantization of depth in the hidden surface calculation led to a certain amount of aliasing. For the rays that were not determined to be obstructed, the intensity (I) reflected by a point on the surface was given by:

$$I \;=\; \alpha(I_{\text{background}} + I_{\text{point}}(\beta\cos(\theta_1) + \gamma\cos^n(\theta_2))/\sqrt{d}), \qquad (12.2)$$

where the surface reflectance (α), the diffuse-reflection coefficient (β), the specular-reflection coefficient (γ), and the specular-reflection exponent (n) are user-defined parameters that specify both the Lambertian reflectance (the intrinsic absorptance of the surface) and the specularity (an image of the source of illumination warped by both the macroscopic and microscopic characteristics of the surface geometry). $I_{\text{background}}$ and I_{point} are the intensity of the background ambient and point light source, respectively. θ_1 is the angle between the surface normal at the point of intersection and the point light source and θ_2 is the angle between the viewing direction and the point light source mirrored about the surface normal. d is the distance from the nodal point of the camera to the object.

Shadows from the point light source were determined using the same principles as the ray tracing algorithm. A point, P, on a quadric surface, S_1, was in the shadow of another surface, S_2, if the ray from the point light source to P intersected S_2. The intensity of each point determined to be in the shadow of another object was given the value $\alpha I_{\text{background}}$. That is, the contribution from the point light source in Eq. 12.2 was set to zero.

Until now, a pin-hole camera model has been adopted. Although chosen for its simplicity, this camera model has the undesirable property that all objects in its field of view are in focus. Thus, an otherwise strong depth cue, blurring due to depth of field, is lost. As such, we simulated a finite aperture by blurring each pixel in the rendered view by a Gaussian with a standard deviation that increased as a function of the distance from a user-specified depth of field plane.

Since they were analytic and synthetic creations, precise knowledge of the depth values existed for each of the basis views. A variety of depth recovery algorithms may be simulated by perturbing the exact depth values with noise followed by quantization of the depth values. In the current system, both the noise and quantization level were scaled linearly with viewing distance. For example, at a distance of 1 m, a standard deviation of the noise distribution of 0.1 cm and 1.0 cm quantization resolution corresponded at 2 m to 0.2 and 2.0 cm, respectively. As more detailed data on the depth acquisition processes applied to digitized images are acquired, this model will be improved. With intensity and depth estimates in hand, arbitrary (i.e., virtual) views of the world can be generated by warping each of the basis views to the virtual camera position and then integrating.

12.3.1.1 Interpolation with full depth

To demonstrate view interpolation, we selected the simple interposition of a sphere and an ellipsoid (Fig. 12.5). A sphere 13 cm in diameter was located at the center of the world coordinate system with a horizontally elongated ellipsoid (13 by 40 cm) centered about a point 25 cm behind, 75 cm to the left and 5 cm above the origin. The surface reflectance properties of the sphere (ellipsoid) were: the specular-reflection exponent n, 10 (4); the specular-reflection coefficient γ, 0.25 (0.25); the diffuse-reflection coefficient β, 1.0 (1.0); and the surface reflectance α, 0.3 (0.3). The point light source was located 150 cm in front of the sphere, 10 cm to the right of the Z-axis in the XZ-plane. Since only the quadric surfaces were actually ray-traced, the background was approximated as a uniformly radiant dome surrounding the objects. The background ambient illumination was set to 1/10th of the level of the point light source. As a computational short cut, the ambient illumination was applied uniformly to all surface points. Shadows due to the point light source can be distinguished from the background by the attenuation of the surface reflectance coefficient (α). The sensor array was in a plane parallel to the XY-plane, 1 m in front of the center of the sphere. The cameras were located in three rows in a Cartesian grid 1 m apart. The camera sensor array was a 2.54 cm square with 0.2 mm/pixel resolution (128 by 128). The simulated array was smaller by a factor of 16 than that commonly found in CCD cameras. This reduction was maintained during system development to speed up the time required for rendering. The focal length of the lens was 64 mm and the

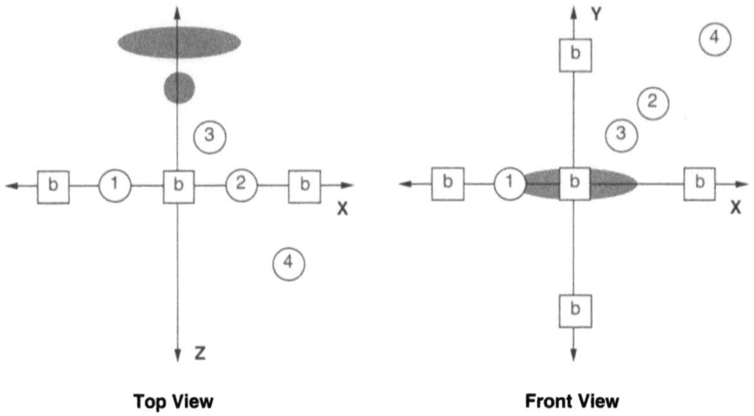

FIGURE 12.5. Configuration of simulated objects and sensors. Two quadric surfaces were positioned within the span of an array of five sensors. A sphere was centered at the world coordinate system with an ellipsoid positioned behind, above and to the left. The center camera was along the Z-axis, 1 m in front of the sphere. The other four cameras were arranged in a diamond about the center camera, each 1 m distant. Virtual views were computed for four positions: (1) midway between the center and left cameras; (2) midway between the top and right; (3) the same view as (2) but translated 0.5 m toward the sphere; and (4) the same view as (2) but translated 1.0 m away from the sphere.

equivalent aperture was approximately 50 mm. Object points 1 m from the nodal point were at best focus for each of the cameras and the blur circle diameter was increased linearly with distance from best focus. Correction of tangent errors will be incorporated in future versions so that the conjugate surface is indeed a plane. For the simulations, basis views were generated for a camera in the middle row and its four nearest neighbors. The selected views demonstrated an interpolation well within the span (view 1) as well as others on the edge of the range of views covered by the basis set (views 2-4). Fig. 12.6 illustrates the five basis views.

The virtual, actual and difference maps from a virtual camera position well within the span of the basis set (midway between the center and left acquisition images) are illustrated in Fig. 12.7. The virtual view is very close to the actual view in the position, the shape and the shading of the objects. The differences that do appear are largely due to aliasing effects in the rendering and quantization in the warping. These distortions follow from simplifications incorporated in these stages that can readily be improved upon. Another source of error comes from the camera parameters selected for this simulation. Although all cameras were verged on the center of the sphere, they were all focused 1 m in front of their nodal points. While the sphere was in sharp focus in the center image, the other four images were somewhat blurred. The differential defocus is a corruption of the photometric data which when combined with a hidden surface algorithm that tried to assign a single depth value to each location in the virtual view, resulted in the inappropriate

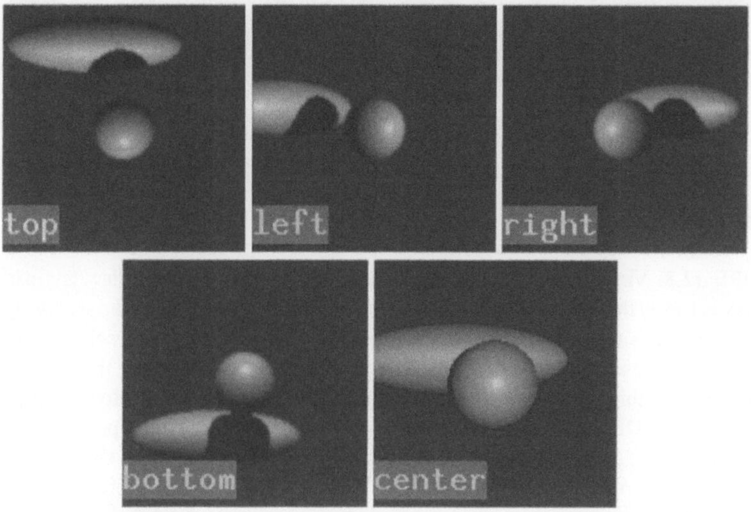

FIGURE 12.6. Simulation basis views. Views are presented that are (along the top row) above, to the left and to the right; and (along the bottom row) from below, and directly down the Z-axis toward the sphere. The center image is slightly closer (by $1/\sqrt{2}$) than the others due to the layout of the planar sensor array. The light source is slightly behind and to the right of the center camera. The effects of the light source can be seen in the movement of the specularity on the sphere as the viewpoint changes as well as the shadow of the sphere cast on the ellipsoid.

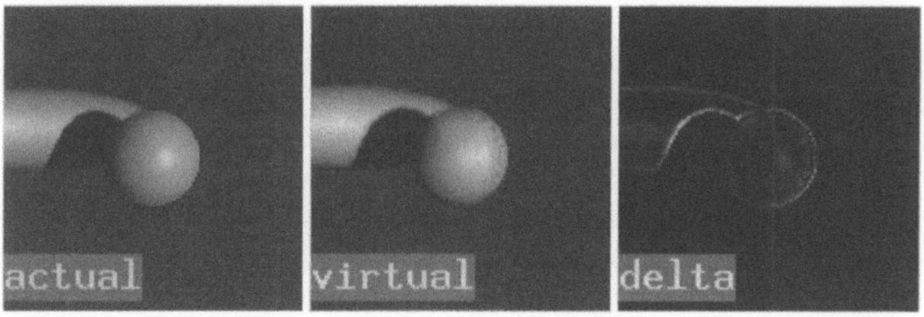

FIGURE 12.7. Virtual view within the span. A virtual view positioned midway between the center and left basis images is well within the span of the basis set. In this figure (and the series of figures that follow) both the rendering of the desired view (left) and the view interpolated from the basis set (center) are presented. On the right is the absolute value of difference between the two images scaled to the range 0 to 255. The combined errors from rendering, warping and defocus are generally small and are largest at interposition boundaries. Problems are concentrated at depth discontinuities because of aliasing in the rendering, relatively large changes in the surface sampling of steep gradients as seen from the virtual view, and the spread of blur circles across object boundaries even though the surfaces lie at different depths.

FIGURE 12.8. Virtual view on the edge of the span. When a virtual view is positioned midway between the top and right basis views, it is at one of the limits of the span supported by the five acquisition views. The weighting of the contributions from each basis view is different than those used in Fig. 12.7. Slight mismatches brought on by the simplicity of the current blending scheme can be seen in the relative brightness of the lower left portion of the ellipsoid. Diffuse specularities occur over large regions of the ellipsoid due to both low curvature and a somewhat more matte surface (than the sphere). The sharp boundaries of the different virtual view components adds to the visibility of the differences.

removal of some projections near depth discontinuities. A distortion that will be somewhat more difficult to correct is the smearing of the specularity in the virtual image due to the summation of the different specular components in the basis set as they shifted with the point of view. The interpolated surface appears much more matte but is not geometrically distorted.

Another set of virtual, actual and difference maps is presented in Fig. 12.8. These images represent a view right on the edge of the viewing space spanned by the basis set. While the span of the basis views is not a simple Cartesian metric (the slant of the surfaces relative to the acquisition views as well as the spacing of multiple surfaces can both introduce gaps within and extend coverage beyond the convex hull of the sensors), the quadric surfaces insure that positioning beyond the outer boundary defined by the sensors will incur distortions due to missing data. This view in the periphery of the coverage gives a good example of the distortions that can occur because the partial virtual views are warped separately and then combined. Variations in the weighting of the different virtual components occur because the relative contribution from each basis view depends on its proximity to the virtual view. The diffuse specularity of the ellipsoid can cause large regions to change in apparent brightness with changes in viewpoint. Even though the pixel differences involved are small, their spatial configuration is such that a human observer is very sensitive to them. Currently there is no smooth blending at the boundaries of the different virtual patches. This source of error should be eliminated when all the views are projected to the virtual sensor plane before the application of the nonlinear operations that are needed to combine them into the regular spacing of the virtual sensor grid.

The span of virtual views covered by a given basis set is, in general, three-dimensional. Even though, in our example, all of the basis views lie in a plane, it is possible to construct views that are closer or further away from the objects being

FIGURE 12.9. Virtual view in front of the sensor plane. As the point of view is shifted from position (2) (Fig. 12.5) 0.5 m toward the center of the world coordinate system (position (3)), the magnification, the perspective distortion and the field of view change appropriately. In consequence, as the point of view moves toward a set of objects, it is possible to leave the span of the basis set because of the visibility of previously obscured surfaces.

viewed. Both the change of scale and the perspective distortion are accommodated by projecting the acquisition data into a common world coordinate system and then projecting it appropriately onto the virtual sensor array. Adjustment of this sort must be done in any case for views within the plane of cameras since an arbitrary object will be at a nonuniform distance from the different nodal points. One factor that is not currently adjusted for is blur due to defocus. As is the case with the basis views of the sphere, overlapping views of a given surface patch may not all be at best focus or even have the same degree of defocus. It is possible (but also not done) to artificially add varying amounts of blur to points as they are projected into the virtual image, and it is even possible (though much more difficult) to selectively deblur portions of the surfaces through knowledge of their positions relative to the acquisition and virtual nodal points and the parameters of the two cameras. Nonetheless, moderate excursions both in front of (Fig. 12.9) and behind (Fig. 12.10) the sensor plane produce interpolated views with smooth contours that have appropriate shape, shading and geometry. The combination of appropriate scale and perspective changes dwarfs the effect of defocus errors under a wide range of conditions. Boundaries on the span of the set of basis views

FIGURE 12.10. Virtual view behind the sensor plane. When the shift in view point is 1.0 m away from the sensor plane (position (4)), changes in perspective, magnification and field of view are appropriate for the increased distance. Motion away from objects can render previously obscured surfaces visible as well.

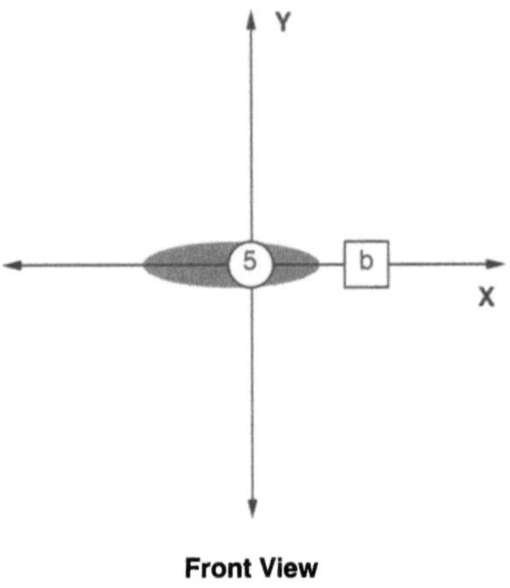

FIGURE 12.11. Inadequate sensor coverage configuration. To demonstrate the errors incurred by not having sufficient basis views to span the position of the desired view, a single acquisition image replaces the array of basis views used in previous examples (cf. Fig. 12.5). The single camera is positioned 0.5 m to the right of the Z-axis in the XZ-plane. The virtual view (5) is located where the center sensor used to be, 1 m in front of the sphere.

exist for these types of view changes as well. It is possible to see behind objects insofar as at least one oblique camera view allows you to see those surfaces (and even somewhat beyond with objects such as these quadrics, since it is difficult for participants to detect missing data errors on severe obliques).

The previous simulations demonstrated the quality of interpolated views possible within the span of the basis set. What kind of failures occur when the basis set does not provide adequate coverage of the environment? To demonstrate inadequate coverage, we have replaced the five basis views used in the previous simulations with a single view offset to the right (Fig. 12.11). As might be expected, surface data is missing from the back of the sphere (Fig. 12.12). With a single camera, it is not possible to see more than a hemisphere and the 26° shift in viewing

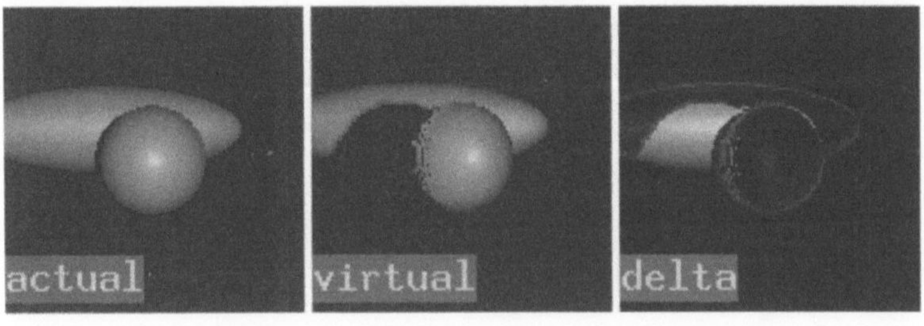

FIGURE 12.12. Consequences of inadequate sensor coverage. When there are not enough cameras, or they are not appropriately positioned, some surfaces will not be available for view. The loss of information can be due to self-occlusion (the back of the sphere) or by interposition of another object (the notch missing from the ellipsoid).

position about the center of the sphere is more than can be masked by the steep gradient at the edge of the sphere. In addition, the acquisition "camera shadow" of the sphere leaves a notch in the ellipsoid for which there are no data. With no positive data to the contrary, the warping in the region of the notch extends clear to the background. The actual shadow from the point light source (at the left edge of the sphere-ellipsoid boundary of the ideal view in Fig. 12.12) is also missing. Note that the specularity is not blurred (there is only one basis view) but that it is in the wrong position. The warping acts as if the specularity is anchored to the surface.

12.3.1.2 Interpolation with corrupted depth

In the simulations presented above, the depth map associated with a given acquisition image was known exactly (within floating point precision). In practice, however, the available depth map will have limited resolution, perhaps severely so. The focusranging results of Krotkov (1987) achieved a resolution of 1% over a distance of 1 to 3 m. Xiong and Shafer (1993) were able to improve resolution by an order of magnitude by fitting a curve to closely spaced responses about the peak of the distance metric. In Fig. 12.13, depth is quantized to 1 cm (at the 1 m viewing distance to the center of the sphere). As can be seen by comparing this figure with Fig. 12.8, the interpolated view shows very little increase in degradation due to the quantization. However, when the quantization error is increased fourfold, the distortions become apparent (Fig. 12.14). Distorted by the larger depth errors, the smooth surface takes on a mottled appearance and flanges appear about the perimeter of the objects. Note that the mottled appearance of the surface is due to very small luminance differences, generally less than 10%. Depth errors of this magnitude will require a more sophisticated combination algorithm than the simple warping now employed. When the views are degraded further by noise, any new warping algorithm will likely involve a smoothing that will result in the

FIGURE 12.13. Virtual view with 1% depth quantization. When the depth map was quantized to the approximate resolution achieved by Krotkov (1987), additional distortions brought about by the quantization were not perceptually salient. The approximately 1 cm steps in depth did not add any noticeable photometric or geometric errors. The reduction in the specular highlights resulting in a relatively matte finish was due to the movement of those contributions in the basis views and could be seen even with full depth resolution.

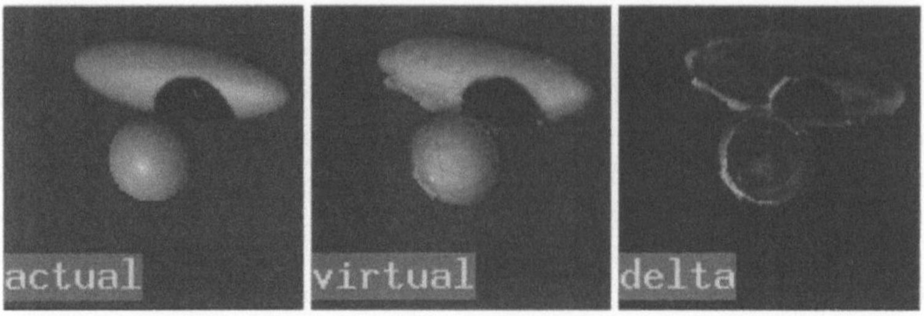

FIGURE 12.14. Virtual view with 4% depth quantization. The large 4 cm gaps between successive depth planes in the different basis views now form noticeable striations in the composite. In addition, when the depth quantization cuts through a steep gradient and the result is viewed at an angle appreciably away from that acquisition view, flanges appear extending out from the body of the object.

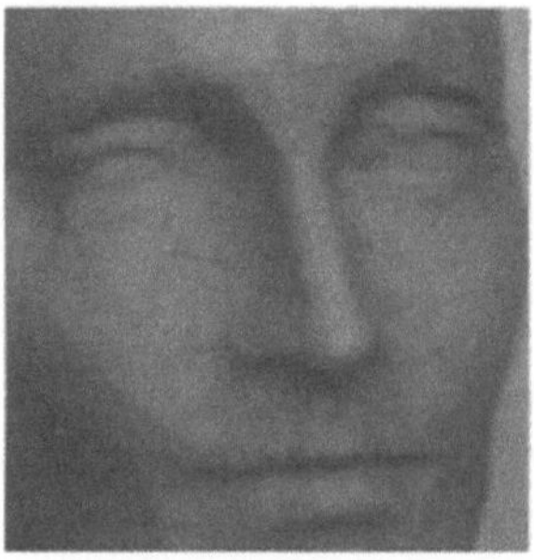

FIGURE 12.15. Mannequin face. This is an oblique view of the mannequin used in the digital image example.

loss of spatial detail — small contours, edge sharpness and spaces between nearby objects.

12.3.2 Digitized Image Interpolation

To test our active telepresence method on actual images, we digitized views of the face of a mannequin (Fig. 12.15). Due to the subtle interpretations humans make of facial expressions, this is a good test example. Since we do not yet have a real time implementation, it is also a good test example in that she doesn't move. The basis views were selected to bracket the mannequin's face (Fig. 12.16). Closeups of the eyes and mouth were obtained with an overlapping region in the middle of the nose (Fig. 12.17). Depth maps were obtained for each of the basis views (Fig. 12.18) and were brought into registration (Fig. 12.19). Six virtual views were then computed (Fig. 12.20) in two rows of three views. Three of the views were at the same vertical height as the bottom basis view (the center view coincided with the basis view). The other three virtual views were displaced vertically. All virtual views were oriented to the mannequin's mouth. This configuration resulted in virtual views that were oriented between 20° and 27° with respect

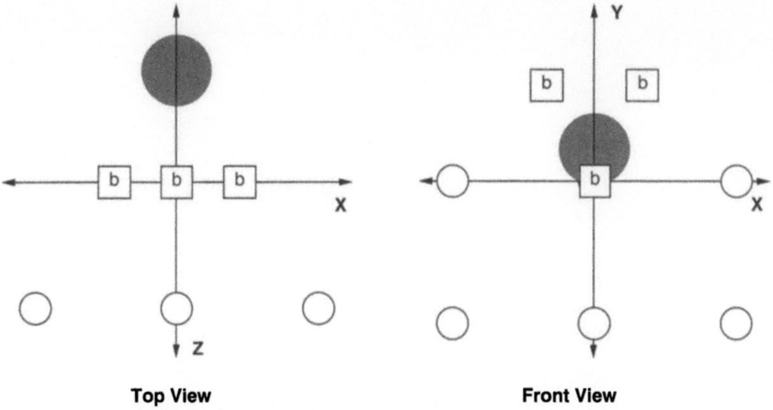

FIGURE 12.16. Layout of basis and virtual views. The origin of the world coordinate system is at the mannequin's mouth. The plane containing the basis and virtual views is parallel to the XY-plane and is translated 35 cm in the Z direction. The basis views of the eyes are 8 cm above and 3 cm to the side of the basis view of the mouth. The virtual views are in a grid, each separated by 20 cm.

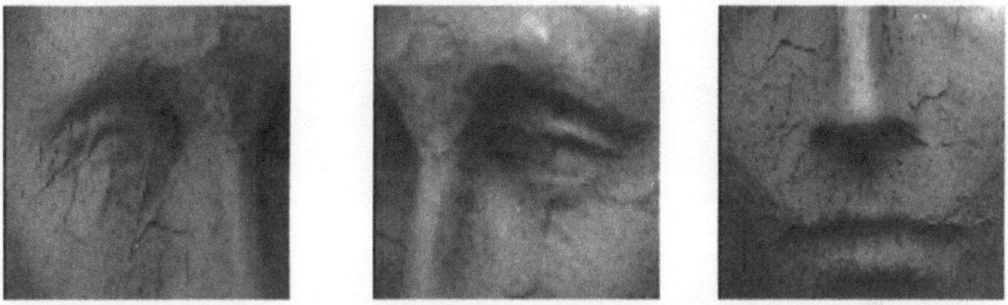

FIGURE 12.17. Photometric basis images. Closeups of the two eyes and the mouth were obtained of the mannequin's face to serve as the basis for the construction of interpolated views. The images of the eyes were displaced vertically by 8 cm and laterally by 3 cm from the view of the mouth.

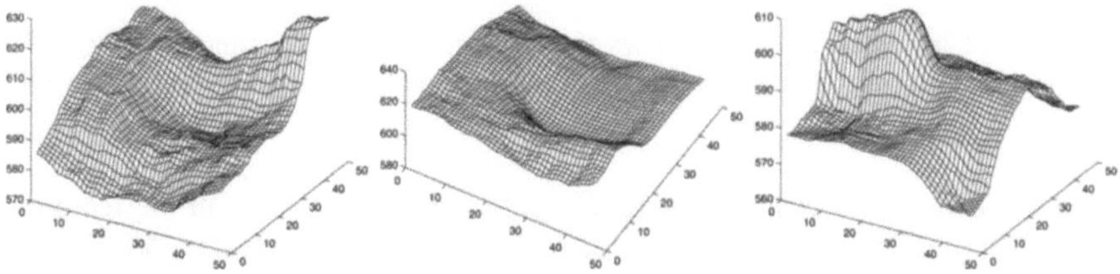

FIGURE 12.18. Basis depth images. A series of images with varying focus were obtained for each basis view and those images were used to estimate depth using the focus ranging algorithm. To display the depth variations, the orientations were not made to match those of the previous image.

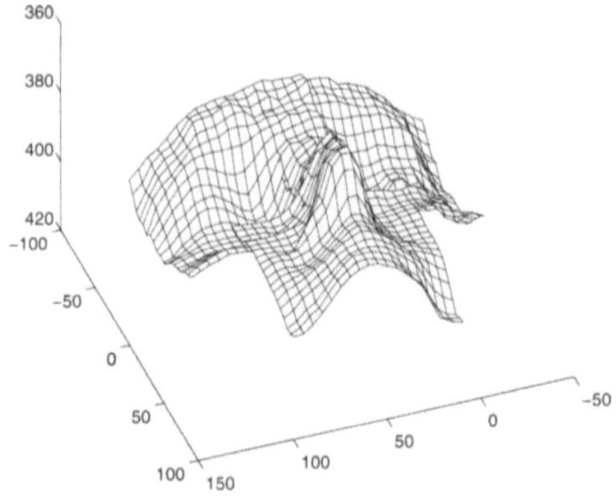

FIGURE 12.19. Registration of basis depth views. The separate depth maps were brought into registration and merged in the world coordinate system.

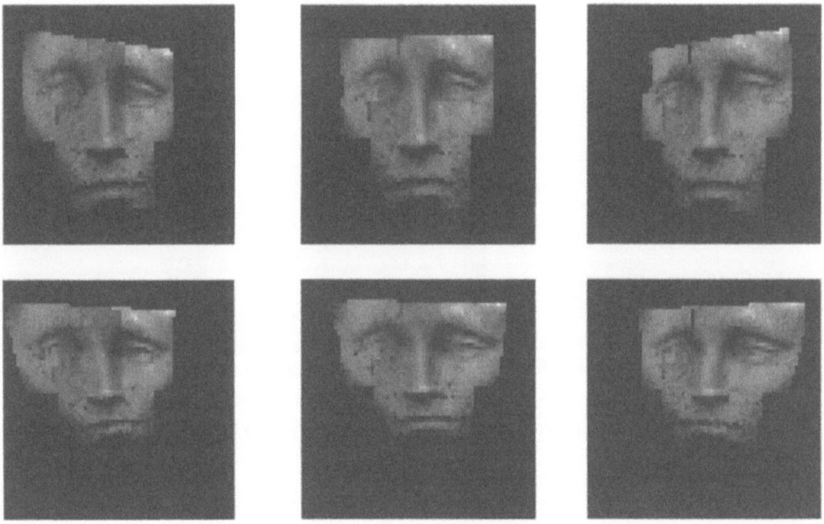

FIGURE 12.20. Virtual views. Six views of the mannequin's face are presented from different points of view. The views vary from direct frontal to 20° below and 27° to the side.

to the Z-axis. The virtual views display both the subtle change in the features as well as the larger effects (such as foreshortening in the lower views) that are associated with viewing a three dimensional object. The views also exhibit the not so subtle degradations from the current implementation of our merging techniques. There was also no attempt made to correct the photometric errors.

12.4 Discussion

The rationale behind the system we propose is that for many virtual reality applications, specifically for those involving telepresence in real environments, computing resources are better spent on acquiring an appropriate set of remote images that will allow the warping of surface reflectance into a desired point of view than on the extraction of parameters needed to generate a graphical reconstruction local to the viewer. On the other hand, we propose that it is not necessary to design a positioning system that will rapidly and accurately mimic the position of the participant's eyes in the remote environment. We base our belief partially on an examination of the recent efforts to establish telepresence. Although the intrusive latencies associated with tracking of the participant and with positioning the sensors can be reduced with the use of predictive filters, the compromises that remain are severe. Some attempts reduced the quality of views to wire-frame representations, while others that increased the quality of the rendering also required the availability of much specialized hardware. In addition, the increased quality of rendering required much off-line authoring (creation of models). The extraction of the parameters needed to animate graphical models is often difficult (e.g., facial expression) and the range of a given model is limited — novel events or objects cannot be accommodated. In the system we propose, the creation of models is data-driven. Photometric data are positioned in a world coordinate system through the use of associated depth information and then reprojected into a desired point of view. The computational cost of warping the information in world coordinates to the desired view is much less than rendering an image of equivalent detail. All that is required of such a system is to retain sufficient accuracy in the distribution of photometric data in the world coordinate system so that the information isn't distorted in an intrusive way. All the realistic grit of actual surfaces comes for free. For many environments and for many tasks, the proposed peripherally-positioned array of sensors can provide views that adequately approximate the data obtained from a remote pair of cameras that move in synchrony with the participant's gaze.

We also base our belief on an examination of what active vision offers in the acquisition of appropriate basis images. Apart from all of the optical limitations of focus and sensitivity, a fixed array of sensors is encumbered by the need to cover the environment so that each region is in the field of view of at least one camera. This coverage is difficult to achieve without either sacrificing the resolution needed for realism or providing a very dense array of sensors. In addition, unless the cameras can be configured in advance for a specific state of the environment, some surfaces will be viewed very obliquely, and thus sampled relatively poorly. The improvement of the acquired sampling of a surface through the adjustment of pan, tilt, focus and zoom can provide the equivalent coverage of a very large number of fixed cameras. Active vision, however, can do more than provide an efficient way to collect views of surfaces. It is not just the application of computer vision algorithms to disentangle highlights from surface properties in a desired view by the application of *shape-from* models of illumination and reflectance. Active vision is not optimizing what you have, it is getting what you need. Applied to telepresence, active vision is the movement of sensors about the remote scene to acquire the location and composition of the sources directly. It is acquisition of several views of a given surface patch from different angles to measure specular changes. It is the acquisition of depth gradients through the use of focus ranging and stereo to approximate surface normals as one of a multiplicity of sources of information that constrains the behavior of highlights. It is the interactive refinement of these estimates by successively removing contaminants and other obstacles to the performance or the perceptual experience of the participant. It is the provision and appropriate allocation of flexible resources (such as visual fixtures) to accomplish the goals of the participant. Active vision does not simply attempt to reconstruct the environment, it provides a mechanism to interact with the environment on both photometric and functional levels.

In the current system, we concentrated on the application of focus ranging techniques to construct the models in the world coordinate system because of the availability of motorized optics and positioning systems. Using focus ranging to determine the depth of local surface patches relative to the cameras affords us the benefit of the acquisition of moderately fine-grained depth information without having to solve the correspondence problem. We explored the use of focus ranging information, not because it was intended as a replacement for other sources of information, but because it was relatively unexplored (as compared to stereo). We fully expect a working system to incorporate any and all sources of information in the refinement of the distribution of photometric data in the world coordinate system.

Finally, we base our belief on an examination of the creation of intermediate views from simulated data. Using five basis views obtained by verging widely-spaced cameras on an object of interest (the sphere), we were able to generate arbitrary views that correspond to a range of observer points of view. These views not only vary over the part of the sensor plane bounded by the cameras, but also toward and away. The quality of the interpolated views remains high through the registration and combination of the multiple views even though the pose of the cameras is not optimal for all the surfaces of interest.

What else is needed to make our belief more believable? We need to improve the graphical rendering so that ideas may be tested with the comfort that rendering errors are negligible (currently this is not the case). It is important to have a source of basis images for which ground truth is known. Graphical rendering of scenes can fill this need if the consequences of the quantization error brought about by digital representation and bounds on available computation can be kept small. Because it was important for us to have knowledge and control of the errors involved in rendering, we created our own simple test-bed in Matlab (The MathWorks). Now that we have a better feeling for the trade-offs involved, we will take advantage of one of the many packages available (both commercial and freeware). Similarly, we need to improve our warping algorithm to more seamlessly combine the contributions of the various acquisition views into a virtual view. The current warping method results in unnecessarily large errors, especially in the presence of steep gradients.

Apart from the generation of views, there is a need to assess their quality and utility. A large body of psychophysical knowledge exists that should be applied to telepresence. With the new technology, reality is being distorted in a manner that is new to the participant. Photometric and geometric distortions that are subthreshold when viewed statically in a single image suddenly can become salient when viewed dynamically in sequence.

Traditional psychophysical measures of detection and discrimination alone are insufficient to measure the quality of virtual reality. New properties such as immersion need to be measured. Immersion is a measure of capture/involvement/reality of the presented views. It is a difficult measure to quantify. However, intrusion — the flip side of immersion — is somewhat easier to measure. An intrusive factor can be judged on how much it differs from reality or degrades performance. Intrusiveness is not limited to the quality of the photometric signal presented to the participant. Consequences of distortions other than appearance (such as stress to the oculomotor system) can limit the ultimate utility of this technology far more severely.

In summary, then, whether it is for an ophthamologist in an office in Bethesda, Maryland examining a patient in eastern Montana or a movie patron experiencing a car chase in Lethal Weapon VII, when both photometric and range information are available from a number of basis views, it is possible to construct a personalized viewpoint in real time that supports the perception of remote presence. In the former scenario, the clinician can be seen as tuning an instrument to obtain the necessary pose and resolution for an accurate diagnosis. In the latter application, a considerable amount of accuracy can be sacrificed to maintain the perceptually smooth continuity required for the maintained suspension of disbelief. In either case, the array of cameras on the remote clinic wall or in the stunt car can do more than simply provide multiple views.

- Image acquisition strategies based on active vision can be used effectively to obtain images for telepresence applications.

- A sparse array of active cameras (pan/tilt/zoom/focus) can be used to capture views of a remote environment equivalent to a much denser array of fixed sensors.

- The generation of views by warping digital images is computationally much less expensive than rendering.

- The combination of photometric and depth maps obtained in real time can eliminate the need to prepare models of the remote environment in advance and, at the same time, offer improved response to dynamic and novel events.

- Focus ranging provides a source of depth information that doesn't require solution of the correspondence problem.

- For many environments and applications, much of the appearance of remote views can be maintained with relatively coarse depth maps.

Acknowledgements

We would like to thank Norm Badler, Ruzena Bajcsy, Wendy Hunt, Visa Koivunen and Richard Paul for helpful comments and criticisms. The equipment used in this paper was supported by Navy Grant N00014-92-J-1647; Army/DAAL 03-89-C-0031PRI; NSF Grants CISECDA 88-22719, IRI 89-06770, ASC 91-08013, MSS-91-57156, CISECDA 90-2253, CDA91-21973 and GER93-55018; NATO Grant 0224/85; A. I. du Pont Institute, Barrett Technology Inc., duPont Corporation, General Motors, and The Preservation Hall Trust.

12.5 References

Adam, J. A. (1994). Medical electronics. *IEEE Spectrum, 31*, 70–73.

Adelson, E. H. & Bergen, J. R. (1991). The plenoptic function and the elements of early vision. In M. S. Landy & J. A. Movshon (Eds.), *Computational Models of Visual Processing* (pp. 3–20). Cambridge, Massachusetts: MIT Press.

Aloimonos, Y. (1990). Purposive and qualitative active vision. In *International Conference on Pattern Recognition* (pp. 346–360). Los Alamitos, California: IEEE Computer Society Press.

Aloimonos, Y., Weiss, I. & Bandopadhay, A. (1987). Active vision. In *DARPA Image Understanding Workshop* (pp. 552–573). San Mateo, California: Morgan Kaufmann Publishers.

Azarbayejani, A., Starner, T., Horowitz, B. & Pentland, A. P. (1993). Visually controlled graphics. *IEEE Transactions on Pattern Analysis and Machine Intelligence, 15*, 602–605.

Bajcsy, R. (1988). Active perception. *Proceedings of the IEEE, 76*, 996–1005.

Ballard, D. H. (1991). Animate vision. *Artificial Intelligence, 48*, 57–86.

Besl, P. J. & McKay, N. D. (1992). A method for registration of 3-D shapes. *IEEE Transactions on Pattern Analysis and Machine Intelligence, 15*, 239–256.

Caudell, T. P., Janin, A. L. & Johnson, S. K. (1993). Neural modeling of face animation for telecommuting in virtual reality. In *Proceedings of the First IEEE Virtual Reality Annual International Symposium* (pp. 478–485). Piscataway, New Jersey: IEEE Press.

Cruz-Neira, C., Sandin, D. J., DeFanti, T. A., Kenyon, R. & Hart, J. C. (1992). The CAVE – audio visual experience automatic virtual environment. *Communications of the ACM, 35*, 64–72.

Durlach, N. (1994). *Virtual reality: scientific and technological challenges*. Washington, DC: National Academy Press.

Foley, J. D., van Dam, A., Feiner, S. K. & Hughes, J. F. (1992). *Computer Graphics: Principles and Practice* (2nd Ed.). Reading, Massachusetts: Addison-Wesley.

Fuchs, H., Bishop, G., Arthur, K., McMillan, L., Bajcsy, R., Lee, S. W., Farid, H. & Kanade, T. (1994). Virtual space teleconferencing using a sea of cameras. In *First International Symposium on Medical Robotics and Computer Assisted Surgery* (pp. 161–167). Pittsburgh, PA.

Fuchs, H. & Neumann, U. (1993). A vision of telepresence for medical consultation and other applications. In *Proceedings of the 6th International Symposium on Robotics Research*. Cambridge, Massachusetts: MIT Press.

Hirose, M., Yokoyama, K. & Sato, S. (1993). Transmission of realistic sensation: development of a virtual dome. In *Proceedings of the IEEE Virtual Reality Annual International Symposium* (pp. 125–131). Piscataway, New Jersey: IEEE Press.

Hopkins, H. H. (1955). The frequency response of a defocused optical system. *Proceedings of the Royal Society London A, 231*, 91–103.

Klinker, G. F., Shafer, S. A. & Kanade, T. (1987). Using a color reflection model to separate highlights from object color. In *Proceedings of the International Conference on Computer Vision* (pp. 145–150). Los Alamitos, California: IEEE Computer Society Press.

Kollin, J. S. (1994). The virtual retinal display. Technical Report MS FJ-15, University of Washington, Human Interface Technology Laboratory.

Krishnan, A. & Ahuja, N. (1993). Range estimation from focus using a non-frontal imaging camera. In *Proceedings of the Eleventh National Conference on Artificial Intelligence* (pp. 830–835). Menlo Park, California: AAAI Press/The MIT Press.

Krotkov, E. P. (1987). *Exploratory visual sensing for determining spatial layout with an agile stereo camera system*. PhD thesis, The University of Pennsylvania.

Krotkov, E. P. (1989). *Active Computer Vision by Cooperative Focus and Stereo*. New York: Springer-Verlag.

Krueger, M. W. (1991). *Artificial reality II*. Reading, Massachusetts: Addison-Wesley.

Lee, S. W. & Bajcsy, R. (1992). Detection of specularity using colour and multiple views. *Image and Vision Computing, 10*, 643–653.

Madden, B. C. (1993). Extended intensity range imaging. Technical Report MS-CS-93-96, Department of Computer and Information Science, University of Pennsylvania.

Madden, B. C. & Cahn von Seelen, U. (1995). 3-dimensional redundant tracking. Technical report, Department of Computer and Information Science, University of Pennsylvania.

Minsky, M. (1979). Toward a remotely-manned energy and production economy. Technical Report 554, AI Laboratory, Massachusetts Institute of Technology.

Ohya, J., Kitamura, Y., Takemura, H., Kishino, F. & Terashima, N. (1993). Real-time reproduction of 3-D human images in virtual space teleconferencing. In *Proceedings of the IEEE Virtual Reality Annual International Symposium* (pp. 408–414). Piscataway, New Jersey: IEEE Press.

Pahlavan, K., Uhlin, T. & Eklundh, J. (1993). Active vision as a methodology. In Y. Aloimonos (Ed.), *Active perception* (pp. 19–46). Hillsdale, New Jersey: Lawrence Erlbaum and Associates.

Pentland, A. P. (1987). A new sense for depth of field. *IEEE Transactions on Pattern Analysis and Machine Intelligence, 9*, 523–531.

Rheingold, H. (1991). *Virtual Reality*. New York: Summit Books.

Sayers, C. & Paul, R. (1993). Synthetic fixturing. In *Advances in Robotics, Mechatronics and Haptic Interfaces, ASME Winter Annual Meeting, New Orleans, USA* (pp. 37–46).

Sayers, C. & Paul, R. (1995). An operator interface for teleprogramming employing synthetic fixtures. *Presence*. In press.

Sutherland, I. (1965). The ultimate display. In *Proceedings of the IFIP Congress* (pp. 506–508). Amsterdam: North-Holland.

Terzopoulos, D. & Waters, K. (1993). Analysis and synthesis of facial image sequences using physical and anatomical models. *IEEE Transactions on Pattern Analysis and Machine Intelligence, 15*, 569–579.

Torrance, K. E. & Sparrow, E. M. (1967). Theory for off-specular reflection from roughened surfaces. *Journal of the Optical Society, 57*, 1105–1114.

Tsai, R. Y. (1987). A versatile camera calibration technique for high-accuracy 3-D machine vision metrology using off-the-shelf TV cameras and lenses. *IEEE Journal of Robotics and Automation, RA-3*, 323–344.

Willson, R. G. (1994). *Modeling and calibration of automated zoom lenses*. PhD thesis, Carnegie Mellon University.

Wolberg, G. (1990). *Digital image warping*. Los Alamitos, California: IEEE Computer Society Press.

Xiong, Y. & Shafer, S. A. (1993). Depth from focusing and defocusing. In *DARPA Image Understanding Workshop* (pp. 967–976). Palo Alto, California: Morgan Kaufmann Publishers.

13

A Novel Environment for Situated Vision and Behavior

Trevor Darrell[1]
Pattie Maes[1]
Bruce Blumberg[1]
Alex P. Pentland[1]

ABSTRACT
We present a new environment for the development of active vision algorithms that are situated in a behavioral context. Our environment allows an unencumbered person to interact with a simulated graphical world which contains autonomous agents. Active vision techniques are used to observe the person and provide perceptual stimuli to the agents. An image of the person is composited together with the graphical world and projected onto a large video screen which the person can observe. The agents inhabiting the world are modeled as autonomous entities which have their own simulated sensors and goals and which can interpret the actions of the participant and react to them in real-time. The goals of these agents provide behavioral context for the execution of the low-level vision routines which observe the person. Several prototypical worlds have been implemented in this environment and tested on hundreds of users at a large-scale public installation.

13.1 Introduction

Active and dynamic processing in the visual system is a topic of importance, since the static application of visual routines to unconstrained scenes is impractical for all but the most basic tasks. Active models of visual processing provide controls for resource-limited sensing, based on both high-level behavioral goals (or prior knowledge) of an agent, and feedback based on observations of the scene. While there has been significant work on the modeling of low-level active vision, for example in the models of gaze control in biological vision and the development of active heads in computer vision, there are few, if any, established models of mechanisms for the integration of low-level routines with behavioral goals in an active setting.

[1]Media Laboratory, Massachusetts Institute of Technology

To this end, we have been exploring environments which integrate visual routines within a larger framework for the control of behavior and action. We focus on systems which are interactive, in the sense that they perceive human users. This focus allows a level of semantic interpretation that is not possible when interacting with a scene without people. In this chapter, we describe real-time active vision methods for tracking people and for recognizing their salient gestures and expressions. We illustrate their application in an interactive setting in which people can manipulate and communicate with quasi-animate "agents".

The system we describe, ALIVE (Artificial Life Interactive Video Environment), offers a world in which simulated agents interact with real people through a video screen and camera. In this environment the agents and a participant can "see" each other—the person can see the agents on the video screen, and the agents can see the person through a computer vision system. The participant's image appears on the video screen, effecting a type of "magic-mirror", in which people see themselves in a different world through the use of a simulated mirror (Fig. 13.1). Vision routines in this system are run on the image of the user seen by the video camera, and provide perceptual stimuli for the agents which live in the simulated world.

This interactive domain is well-suited for active vision research since it is situated, suitably constrained, and challenging. It is situated both in the sense that vision routines are mediated by the *behavioral* goals of the agent, and also in that the routines are run on real imagery of people and thus are perceptually grounded in actual sensory data. The domain is constrained since there are a limited space of forms and shapes that define the recognition and tracking tasks (e.g., we know we are looking at people, and they interact with the system in certain known ways) so it is possible to have real-time routines which perform successfully. Nonetheless, the domain is challenging and nontrivial as the precise forms that people project are dynamic, unpredictable, and very difficult to model explicitly.

The ALIVE system has been installed in public forums and had hundreds of users from a general technical audience. People of varied age and experience have enjoyed interacting with the agents in the world, and found the creatures "believable", in the sense that they react to actions and gestures in a somewhat natural way. The combination of agent-based programming and real-time active vision techniques offers the promise of interfaces to computer generated environments that are intuitive to the user since they exploit natural modes of interpersonal communication.

13.2 The "Looking at People" Domain

The "Looking at People" domain provides several challenges for computer vision relative to traditional application areas. Unlike static scenes or scenes with simple object motion, scenes with people are dynamic and have complicated articulated body kinematics, nonrigid motion, as well as gestures and other semantically-laden

FIGURE 13.1. The magic-mirror metaphor: a participant sees his/her mirror image surrounded by autonomous agents.

forms of communication. All of these properties prove difficult for traditional vision algorithms.

First, the shape and motion of human bodies are complicated and can be hard to characterize with precision. Fig. 13.2 shows a set of silhouettes of users of our system. A complete motion model of these forms would require a model of articulated limb dynamics and a model of nonrigid skin and clothing dynamics. Since these are unlikely to be computable in real-time with conventional methods, the people-watching domain requires methods which work robustly in the absence of a strict model.

Second, the fact that people are not simply objects, and have intentions and communicate via meaningful signs, strongly argues for an "active" or "purposive" approach to vision (Aloimonos, 1993; Bajcsy, 1988; Ballard, 1991). The traditional stated goal of a computer vision system has been to assume the world is in a particular state and to attempt to recover as complete and accurate a description of that state as possible. But, communication requires context, negating the utility of an "absolute state" of the world in this domain. A purely descriptive approach is thus inappropriate with regard to building interactive vision systems. The goal of vision routines for a people-watching system should not be to estimate perfectly and to represent the three-dimensional shape of the body, face and hands, but rather to recover and provide a signal which is meaningful in the context of the

FIGURE 13.2. Binary silhouettes of users after figure-ground processing. Task-dependent vision routines to find hands and pose information use this representation as input.

current interaction. What exactly is meaningful will change over time, so no static description would suffice.

Finally, the presence of people in an interactive system provides a strong pressure to achieve real-time performance. Quite simply, if the system does not react in real-time, or perhaps what is more appropriately called "interactive-time", the user will get bored and leave. Unlike many static domains in which the agent can stop momentarily if necessary to make a decision, the visual routines and agent models used in an interactive man-machine system must be both robust and fast.

13.3 Attention and Intention

A key issue for practical real-time vision systems is the question of what to look for and where/when to look for it. Vision algorithms which are too general (e.g. look for everything everywhere), will usually suffer from an explosion in search complexity and fail to offer adequate performance. As many authors have noted, the solution to this dilemma lies in the use of an attentional mechanism (Tsotsos, 1988).

Attentional mechanisms require some state to exist in the perceiving agent. One can say "Attention requires Intention," in that without meaningful states and goals, a vision system has no principled way to prioritize what to look for next. To address this, we rely on the action-selection system in the simulated agents; the same mechanisms that govern their locomotion, feeding, and searching behaviors can provide perceptual goals as well. In this case, vision routines are "situated" in that they are based on behavioral goals of the perceiving creature. (They are also situated in that they are implemented on real video input of the user in the scene, not a simulated input scene.)

To this end, an important aspect of our system is the use of a behavior-based agent model to drive the simulated creatures and agents in the computer graphics world. A behavior-based approach to modeling agents provides a simulated world that is populated with autonomous, unpredictable creatures, whose action selection model is ethologically plausible. The use of agents whose action pattern mimics the behavior of animals in the real world allows for a "natural" and intuitive interface between people and those agents: A user can use his/her preexisting knowledge of how to interact with a creature.

13.4 Action Selection with Time-Varying Goals

To achieve this level of believable interaction and provide the context for both interpretation and attentional mechanisms in the agents, a model of goals and intentional behavior is needed. Since multiple goals can exist and conflict with one another in an agent at any given time, the model should be able to mediate between heterogeneous goals in real-time.

Recent results with reactive systems (Kaelbling, 1987), routines (Agre & Chapman, 1987) and subsumption architectures(Brooks, 1986) have demonstrated remarkably reliable and successful performance in performing real-time action selection in autonomous agents. In these architectures, the emphasis is put on direct coupling of perception to action, distributedness and decentralization, dynamic interaction with the environment and intrinsic mechanisms to cope with resource limitations and incomplete knowledge.

Unfortunately, with many of these models it is impossible for the agent to have time-varying goals that affect the behavior. As a result, agents built this way only seem to engage in very predictable, reflex-oriented behavior. In our system, we employ a behavior model combining some of the best of both worlds. The model produces fast and robust activity in a tight interaction loop with the environment, while at the same time allowing for some goal-dependent planning to take place. One of the distinguishing characteristics of this model is that it is closely based on models of animal behavior. In particular, it borrows heavily from the work of classical ethologists such as Baerends (1976), Lorenz (1973), Ludlow (1980), McFarland and Sibley (1975) and Tinbergen (1950).

Specific ideas from ethology that are incorporated in our model include the following.

- *Structured behavior repertoire*: Behaviors are organized as a loose hierarchy with the top of the hierarchy representing more general behaviors and the leaves representing more specific behaviors.

- *Real-time dynamic planning*: All behaviors compete at every time step for control of the creature.

- *Exclusivity*: A model of mutual inhibition among competing behaviors is used to insure that the agent engages in a single behavior at a time and does not dither between multiple behaviors.

- *Hysteresis*: Behavior-specific fatigue is modeled to insure that the temporal pattern of behaviors is believable and so that a creature doesn't mindlessly pursue one behavior indefinitely to the detriment of other needs.

The main difference between our approach and current situated behavior systems is that we neither hard-wire nor precompile the action selection. Arbitration among actions is a run-time process which differs according to the goals of the system and the situation in which it finds itself.

Full details of the behavior model and a discussion of typical locomotion, foraging and exploration behaviors are described by Blumberg (1994). In the next section we discuss the vision routines we have implemented for perceiving a person in an interactive setting.

13.5 Routines for Looking at People

We have developed a set of vision routines for perceiving body actions and gestures performed by a human participant in an interactive system. Our routines solve subsets of the perception task that are computationally tractable and still useful: tracking a user's location and posture and performing simple, context-dependent gesture recognition.

The "magic-mirror" paradigm is attractive because it provides a set of domain constraints that is restrictive enough to allow simple vision routines to succeed, but is sufficiently unencumbered that it can be used by real people without training or special apparatus. In this paradigm, a person faces a large screen on which is presented both an image of the person and an image of the virtual world. Vision routines analyze the image of the person and allow interaction with the virtual world.

13.5.1 Domain constraints

The "Looking at People" domain provides constraints on the recognition and tracking problems that need to be tackled. We take advantage of these constraints in constructing our vision system. The most basic constraint is that we are looking at people and can exploit specific domain knowledge about human anatomy. Arms are connected to a torso which is usually below a head and above two feet, etc. A system can explicitly or implicitly use this knowledge to guide the recognition and/or tracking process.

Another important set of constraints derive from the fact that we can arrange the imaging geometry of the camera such that the user is almost always in a frontal pose. In our system, the same camera used for vision processing is also used for acquiring the image of the user which is composited into the graphics display. For the "magic mirror" effect to work, the image of the user used for the display must come from a camera position which is located approximately at the position of the screen. Since in this paradigm the user will be watching the screen almost continuously, we can assume with some degree of confidence that they will face the screen and thus their body will be oriented parallel to the screen most of the

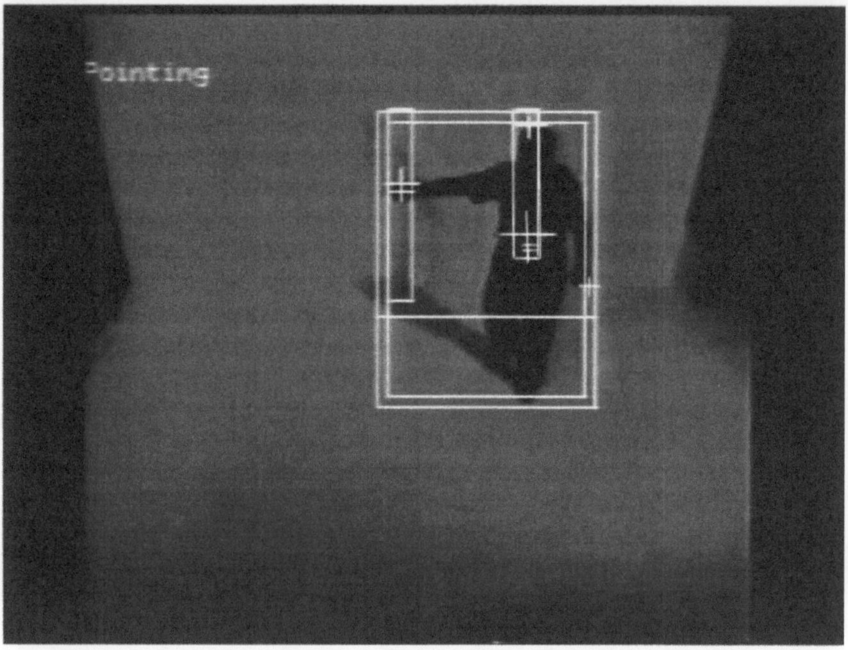

FIGURE 13.3. The vision system works from a silhouette of the user provided by the figure/ground segmentation routine. It computes a range of features including the bounding box, outlined in the image, the centroid of mass, marked by a large cross, and the position of the user's head and hands, marked by smaller crosses in the image. A horizontal line across the lower third of the box indicates that the user is upright (and not bending) and in this case is a lower bound on expected hand location. Search rectangles on the side of the box indicate where an edge corresponding to a hand is expected to be found. A similar search region extending up from the centroid indicates the expected position of the top edge of the user's head. Other marks on the crosses are indicators that that feature is moving or stationary, and are used for detecting gestures. In this frame, the user is found to have been performing a pointing gesture, hence the label in the upper-left-hand corner of the image.

time. With this assumption, and the assumptions about the human figure, we can make relatively strong inferences about the position of the user's head and hands.

Below, we will describe the algorithms we use which exploit these domain constraints and recover information about the user's position and pose. While we believe the constraints will be valid most of the time, we cannot always guarantee their validity. People may attempt to use the system in unforeseen ways, their dimensions may be far from the norm, and/or they may have therapeutic or prosthetic devices such as crutches or wheelchairs. When someone who does not fit the "prior model" implicit in the vision system attempts to use the system, it is important that the system not fail catastrophically. (In the cases described above, it is typically possible to continue to estimate basic position information about the user, but limb position information becomes unreliable.)

13.5.2 Figure-ground processing

Certain basic tasks and 3-D information required to place the user in the graphical world are computed continuously (descriptively) and without regard to the state of agents in the world. For example, simple figure-ground segmentation is continuously performed by our system, as is the computation of bounding box information and the 3-D position of the user in the world. Other tasks, such as localizing the head or hands of the user, or interpreting body state (such as whether the user is bending over or pointing), are performed conditional on agent state.

Before other processing can occur, the vision system must isolate the figure of the user from the background (and from other users, if present). Our approach is to use low-level image processing techniques to detect differences in the scene, and use connected-components analysis routines to extract objects.

The simplest method of detecting differences is to constrain the background to be a known color, and detect instances of that hue in the input images. Several commercially available video image processing boards are available which can automatically isolate figure/ground and perform video compositing based on this approach. This approach essentially performs clustering in color space to determine pixel membership in figure/ground classes. The advantage of this analog hardware approach is that segmentation and compositing are performed at frame rate, however a disadvantage is the inability to have the graphical objects occlude the user in the scene.

Our first implementation relied on this type of analog hardware device for figure-ground segmentation. Later implementations adopt a more sophisticated approach, allowing the background to be an arbitrary, but static, pattern. Mean and variance information about the background pattern are computed, and these statistics are used to determine space-variant criteria for pixel class membership. If the scene conditions are such that static background subtraction is inappropriate, for example if there are multiple moving objects which cover large portions of the scene, then even more sophisticated clustering methods will be needed. In these cases motion-based grouping methods could be applied to find regions which are moving with coherent motions (Darrell & Pentland, 1991).

Once a difference signal has been computed, we apply connected components analysis to find a foreground region. We binarize the difference image, find connected regions, and compute the image bounding box and first-order moments of the largest connected region. (For a review of connected components and binary image processing methods, see Ballard and Brown (1982) and Horn (1991).)

13.5.3 Scene projection and calibration

Once the figure of the user has been isolated from the background, we compute an estimate of its 3-D location in the world. If we assume the user is indeed sitting or standing on the ground plane, and we know the calibration of the camera, then we can compute the location of the bounding box in 3-D.

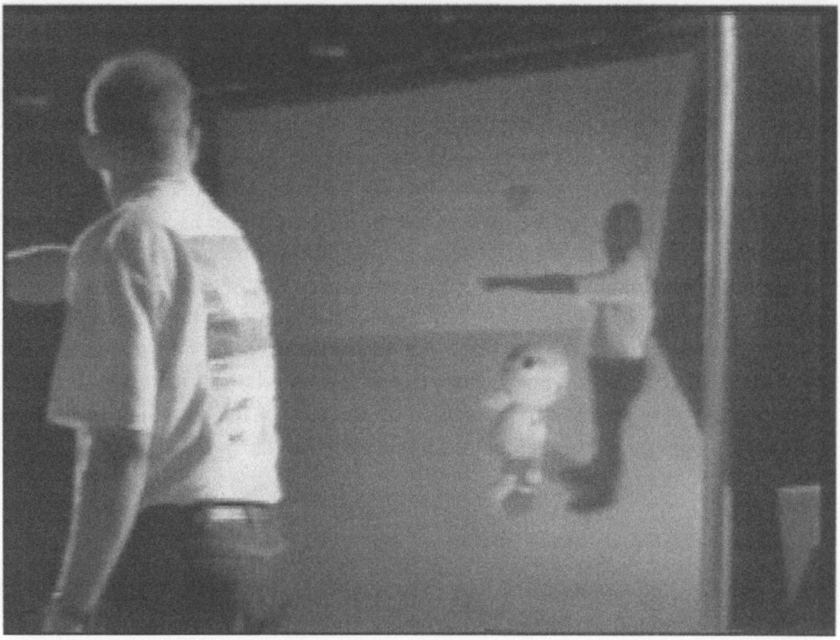

FIGURE 13.4. Gestures are interpreted by the agents based on their context. Here, the Puppet walks away in the direction the user is pointing.

Establishing the calibration of a camera is a well-studied problem, and several classical techniques are available to solve it in certain broad cases (Ballard & Brown, 1982; Horn, 1991). Typically, these methods model the camera optics as a pinhole perspective optical system, and establish its parameters by matching known 3-D points with their 2-D projection.

Knowledge of the camera geometry allows us to project a ray from the camera through the 2-D projection of the bottom of the bounding box of the user. Since the user is on the ground plane, the intersection of the projected ray and the ground plane will establish the 3-D location of the user's feet. The 2-D dimensions of the user's bounding box and its base location in 3-D constitute the low-level information about the user that is continuously computed and made available to all agents in the computer graphics world.

13.5.4 Hand tracking

One of the most salient cues used by the agents in our world is the location of the user's hands. We have implemented a hand search algorithm that uses spatial search patterns to localize hands in the input images. We make heavy use of the domain constraints outlined above, for example, that people are (mostly) oriented in a fronto-parallel plane with respect to the camera.

The hand-tracking algorithm we have developed is comprised of several different context-dependent search heuristics. In general, a normalized correlation search is done along the sides of the upper-torso bounding box to find a strong horizontal edge. The upper-torso bounding box is defined in such a way that it discounts to some degree the effect of shadows and feet in the horizontal dimension; it takes the vertical dimensions from the real bounding box and computes horizontal dimensions from the top 66% of the user's image. Fig. 13.3 illustrates hand recognition processing on the silhouette output from the figure-ground method.

Depending on the current context, different search windows and search patterns are used. The main contextual cue is the size of the bounding box which provides a rough estimate of overall pose. If the box is narrow, we infer that the user's hands are at their side, and we do not attempt to find them in the silhouette. In this case, we return the hand position to be located on the side of the bounding box, at the same relative height along the bounding box as it was last reliably seen.

13.5.5 Gesture interpretation

In the implementation reported here, the agents in the world respond both to the absolute position of hands and whether they are performing characteristic gesture patterns. We use simple low-level recognition strategies to detect these characteristic patterns.

Our model of gestures is highly reduced and comprises two possible spatiotemporal hand patterns: pointing and waving. Each gesture is defined in terms of the 2-D motion of the location of the hand in the image plane. Pointing requires a particular relative hand location (an extended arm), and a steady position over time. Waving requires a predominantly side-to-side motion of the hand while the user is otherwise stationary. These are special cases of our more general work on gesture recognition, which builds space- and time-separable template patterns for recognition (Darrell & Pentland, 1993). However, this work assumed a high-resolution image of the object performing the gesture. Future versions of the ALIVE system will have an additional high-resolution camera to provide a foveated image of the hands (or face) of the user, but at present we only model gesture as change in position (or lack thereof) over time.

Each gesture may be interpreted by different agents in the world depending on their context. For example, in the implementation described below, the waving gesture elicits a response from the Puppet agent, but not from the Hamsters. And the response given by the Puppet is state dependent; it will return when waved at only when it has been sent away or otherwise ignored.

13.6 An Example Implementation: ALIVE

We have combined these ideas in a system designed to allow a user to interact with an immersive visual environment without any physical apparatus. Our system,

ALIVE ("Artificial Life Interactive Video Environment"), uses the agent and vision modeling techniques described above. With few exceptions (e.g. Krueger, 1990), to experience these environments previously required the use of gloves, goggles, and/or a helmet, and most likely a wired tether to a computer graphics workstation (Rheingold, 1991).

We implemented the magic-mirror model in ALIVE using a single CCD camera to obtain a color image of the scene. The image of the user was separated from the background, composited into the 3-D graphical world, and projected onto a large screen which faced the user. (The polarity of projection was reversed so that the image indeed appeared as a mirror.)

This first implementation of ALIVE contained two virtual worlds. The user could switch between them by pressing a virtual button. (To activate the button, the users hand had to be in the correct 3-D location, not just the same image position.) One world was inhabited by a Puppet and the other world by a Hamster and a Predator. The Puppet had behaviors to follow the user around, try to hold the user's hand, and imitate some of the actions of the user (sitting down, jumping, etc). It would be sent away when the user pointed away and come back when the user waved. The Puppet employed facial expressions to convey some of its internal state. For example, it would pout when the user sent it away and smile when the user motioned it to come back. It giggled when the user touched its belly.

Similarly the Hamster had behaviors to avoid objects, follow the user, and to beg for food. The user was able to feed the Hamster by picking up food from a virtual table and putting it on the floor. The user could open an adjoining cage and release a Predator, which would then chase the Hamster (but avoid the user).

The user could interact with the agent using certain hand gestures, which were interpreted in the context of the particular situation. For example, when the user points away (Fig. 13.4) and thereby sends the Puppet away, the Puppet will go to a different place depending on where the user is standing. If the user waves or comes towards the Puppet after it has been sent away, this gesture is interpreted to mean that the user no longer wants the Puppet to go away, and so the Puppet will smile and return to the user. In this manner, the gestures employed by the user can have rich meaning which varies depending on the previous history, the agent's internal needs and the current situation.

ALIVE was demonstrated for 5 days at SIGGRAPH-93 as part of the Tomorrow's Realities show (Maes, 1993), and was used by over 500 people. In this installation the user moved around in a real-world space of approximately 16 by 16 feet. A video camera at the front of the space captured the user's image. (This version of the ALIVE system employed hardware for chroma-key background subtraction, but later versions have relied on more general background subtraction methods.) The composited image was displayed on a large screen (10 by 10 feet) which faced the user. The machine used for the graphics and behavior modeling was an SGI Indigo Elan. We implemented the visual search routines on a dedicated image processor built by Cognex, Inc. which was connected to the Indigo via serial line and ethernet.

In terms of performance, as might be expected, rendering and sensing were the two bottlenecks. The agents update their state 6 times per animation frame. An individual update (including numerical integration) takes approximately 5 milliseconds per agent of which 60% is taken up by the simulated sensing. Rendering of the agents and the world takes another 75 milliseconds. The combined frame rate was roughly 10 frames per second. The vision sensing had a minimum update rate of 6 hertz, but can be quite a bit faster depending upon the complexity of interpretation requested by the behavior routines.

13.7 Conclusion

The ALIVE environment provides an interesting new domain to test behavior-based vision routines. The domain is dynamic, situated, and difficult to predict using conventional models. The "magic-mirror" metaphor allows a user to interact and navigate in a virtual world, using familiar and intuitive means. Unlike in viewer-centered virtual reality systems, the user did not get disoriented. Users knew at all times where they were in the artificial world and could observe the actions of the other agents as well as their own.

Simple real-time vision routines can be successfully used in this context to provide an interaction between people and virtual agents or creatures. The gross 3-D position of the user, the location of hands and head, and coarse information about overall pose (bending over, arms outstretched, etc.) can be recovered using classical image processing techniques: figure-ground extraction, connected components, and context-based correlation search.

With these routines, users can directly manipulate objects and agents in the world. The agents populating the world have varying internal needs and motivations which determine what aspects of the user's state they are interested in, what visual search processing should occur, and how the results will be interpreted. This provides the necessary "situatedness" for a successful active vision system.

Acknowledgments: This work was sponsored by BT (British Telecom) and the Office of Naval Research.

13.8 References

Agre, P. & Chapman, D. (1987). Pengi: An implementation of a theory of activity. In *Proceedings of the AAAI-87* (pp. 268–272). Los Altos, California: Morgan Kaufmann.

Aloimonos, Y. (1993). *Active Perception.* Hillsdale, New Jersey: Lawrence Erlbaum Associates.

Baerends, G. (1976). On drive, conflict and instinct, and the functional organization of behavior. In M. A. Corner & D. F. Swaab (Eds.), *Perspectives in Brain Resarch,* Volume 45 (pp. 427–447). New York, New York: Elsevier.

Bajcsy, R. (1988). Active perception. *Proceedings of the IEEE, 76,* 996–1005.

Ballard, D. H. (1991). Animate vision. *Artificial Intelligence, 48*, 57-86.

Ballard, D. H. & Brown, C. M. (1982). *Computer Vision*. Englewood, New Jersey: Prentice-Hall.

Blumberg, B. (1994). Action-selection in hamsterdam: Lessons from ethology. In Cliff, D., Husbands, P., Meyer, J. & Wilson, S. (Eds.), *Proceedings of the 3rd International Conference on Simulation of Adaptive Behavior* (pp. 108-117). Cambridge, Massachusetts: MIT Press.

Brooks, R. A. (1986). A robust layered control system from a mobile robot. *IEEE Transaction on Robotics and Automation, RA-2*, 14-23.

Darrell, T. & Pentland, A. P. (1991). Robust estimation of a multi-layer motion representation. In *Proceedings of the IEEE Motion Workshop* (pp. 173-178). Los Alamitos, California: IEEE Computer Society Press.

Darrell, T. & Pentland, A. P. (1993). Space-time gestures. In *Proceedings of the IEEE Conference on Computer Vision and Pattern Recognition* (pp. 335-340). Los Alamitos, California: IEEE Computer Society Press.

Horn, B. K. P. (1991). *Robot Vision*. Cambridge, Massachusetts: MIT Press.

Kaelbling, L. P. (1987). An architecture for intelligent reactive systems. In *Proceedings of the 1986 Workshop on Actions and Plans* (pp. 395-410). Los Altos, California: Morgan Kaufmann.

Krueger, M. W. (1990). *Artificial Reality II*. Reading, Massachusetts: Addison Wesley.

Lorenz, K. (1973). *Foundations of Ethology*. New York: Springer-Verlag.

Ludlow, A. (1980). The evolution and simulation of a decision maker. In F. Toates & T. Halliday (Eds.), *Analysis of Motivational Processes* (pp. 273-296). London, England: Academic Press.

Maes, P. (1993). Alive: An artificial life video environment. In *SIGGRAPH Visual Proceedings* (p. 189). New York, New York: ACM.

McFarland, D. & Sibley, R. (1975). The behavioral final common path. *Philosophical Transactions of the Royal Society B, 270*, 265-293.

Rheingold, H. (1991). *Virtual Reality*. New York, New York: Summit Books.

Tinbergen, N. (1950). *The Study of Instinct*. Oxford, England: Clarendon Press.

Tsotsos, J. (1988). A 'complexity level' analysis of immediate vision. *International Journal of Computer Vision, 1(3)*, 303-320.

Author Index

Subject Index

MIX
Papier aus verantwortungsvollen Quellen
Paper from responsible sources
FSC® C105338

If you have any concerns about our products,
you can contact us on
ProductSafety@springernature.com

In case Publisher is established outside the EU,
the EU authorized representative is:
Springer Nature Customer Service Center GmbH
Europaplatz 3, 69115 Heidelberg, Germany

Printed by Libri Plureos GmbH
in Hamburg, Germany